T0296466

Vasco da Gama (?1469–1524) is well known as one of a generation of discoverers, along with Magellan, Cabral and Columbus. Yet little is known about his life, or about the context within which he 'discovered' the all-sea route to India in 1497–9.

This book, based on a mass of published and unpublished sources in Portuguese and other languages, delineates Gama's career and social context, focusing on the delicate balance between 'career' and 'legend'. It argues that Gama's political position in Portugal makes him an unlikely candidate for the role of nationalist hero. However, by exploiting creatively the symbolic capital of a legend that existed in his own lifetime, Gama, his descendants and panegyrists (notably Camões) transformed an obscure nobleman from the Alentejo into the Great Argonaut. Thus the book addresses larger questions of myth-building and nationalism, while never losing sight of Gama himself.

THE CAREER AND LEGEND OF VASCO DA GAMA

THE CAREER AND LEGEND OF VASCO DA GAMA

SANJAY SUBRAHMANYAM
Ecole des hautes études en sciences sociales, Paris
and
Delhi School of Economics, Delhi

CAMBRIDGE
UNIVERSITY PRESS

CAMBRIDGE UNIVERSITY PRESS
Cambridge, New York, Melbourne, Madrid, Cape Town,
Singapore, São Paulo, Delhi, Mexico City

Cambridge University Press
The Edinburgh Building, Cambridge CB2 8RU, UK

Published in the United States of America by
Cambridge University Press, New York

www.cambridge.org
Information on this title: www.cambridge.org/9780521646291

First published 1997
Reprinted 1997

A catalogue record for this publication is available from the British Library

Library of Congress Cataloguing in Publication data
Subrahmanyam, Sanjay.
The career and legend of Vasco da Gama / Sanjay Subrahmanyam.
p. cm.
Includes bibliographical references and index.
ISBN 0 521 47072 2
1. Gama, Vasco da, 1469–1524. 2. Explorers – Portugal – Biography.
I. Title.
G286.G2S83 1997
910.92–dc20
[B]
96–25106 CIP

ISBN 978-0-521-47072-8 Hardback
ISBN 978-0-521-64629-1 Paperback

For Ashin Das Gupta
from a prodigal son

The battle, the fortress,
diplomacy and robbery,
doctoring, serving the kings,
trading in ships on the ocean
and powerful spells – if they work for someone,
then the fruit will be great.
But if control is lost
and the wrong things happen,
all that wealth will disappear
and the man's life will hang in the balance,
O God of Kalahasti.

Dhurjati, *Kâḷahastîśvara Śatakamu* (16th c.), tr. Heifetz and Narayana Rao (1987), 114.

Contents

ILLUSTRATIONS

PLATES

MAPS

Preface

This book has had a curious and tortuous gestation. In the spring of 1992, I was approached by a Portuguese publisher to write a brief, heavily illustrated and popular biography of Vasco da Gama. I was reluctant at first, since it seemed to me that nothing of any great interest could be said on the subject. But correspondence with some colleagues soon convinced me otherwise. However, in the summer of 1992, when I contacted the publisher and the editor of the series which had solicited the volume, they had already lost interest. Then, in October 1992, a seminar was organised at the India International Centre, New Delhi, by the Centre for Spanish Studies, Jawaharlal Nehru University, to 'celebrate' the fifth centenary of Columbus's trans-Atlantic voyage. Using that occasion, I was able to present a first rough outline of a re-interpretation of Vasco da Gama's career in the form of a paper.

It soon became obvious to me, however, that the appropriate format for what I wished to say was a book, not an essay, and this book was thus definitively launched in the summer of 1993, while visiting Cambridge. Since then, debts have rapidly and ominously accumulated. The Fundação Calouste Gulbenkian supported me for a summer of archival work in Lisbon in 1994; I am particularly grateful to Dr Maria Clara Farinha and Dr José Blanco of the Serviço Internacional of that organisation, not least of all for opening the closed doors of some collections in the Torre do Tombo. José Alberto Tavim and Manuel Lobato were extremely helpful in various ways through that stay (and even later), and I should like to acknowledge that debt. Luís Filipe Thomaz accompanied me on a splendid trip to the Alentejo, and especially Vidigueira, and commented on a draft first outline of the work. I am also grateful to him and Arlindo Fagundes for taking trouble over photographing various sites and objects in India, though only some of these sites appear in this book.

A good deal of the early part of the book was written at Minneapolis in Fall 1994, while visiting the Center for Early Modern History, University of Minnesota. I am grateful to Lucy Simler and Jim Tracy for their help and support during that visit, and especially to Stuart Schwartz, who was a marvellous next-door neighbour. Carol Urness at the James Ford Bell Library helped with the illustrations, and organised a public lecture on Vasco da Gama for me, which helped clarify my thoughts, and once more reminded me of the sensitivity of the subject matter of this book. David Lelyveld at Columbia University and Claude Markovits in Paris also helped organise two seminars on the Great Argonaut, where the ideas in this book took shape; perhaps the most enjoyable of the 'Vasco seminars' was one arranged by David Shulman at the Israeli Academy of Sciences in Jerusalem in March 1995. I am also grateful to Frank Conlon in Seattle who patiently found, watched, and partly slept through a video-recording of *L'Africaine* with me in late 1994. Other debts are to Muzaffar Alam, Kunal Chakrabarti, António Coimbra Martins, Anthony Disney, Dick Eaton, Maria Augusta Lima Cruz and Geoffrey Parker, who have all encouraged and helped in one or the other way. Nalini Delvoye closely followed the progress of this book with amusement and irritation, and kept a sharp lookout for Vasco-related references. I am particularly grateful to Jorge Manuel Flores for a careful reading of the text, and for providing useful comments and additional references, as well as aid with the illustrations.

My greatest intellectual debt is, however, to that master of the Luso-Asian Conjunction, *Ṣâḥib-Qirân* Jean Aubin, who accompanied the evolution of the book from the start, offered critical advice, crucial bibliographical aid, photocopies and transcriptions of archival documents, and in general took precious time from his own project on D. Manuel to help this book ripen. In particular, I have had occasion time and again to value his own astringent vision of Portuguese expansion, so far from both the Portuguese and Asian nationalist historiographies, which are still promoted today by 'official' historians under interested patronage. Not least of all, he also corrected some embarrassing errors in the penultimate draft.

The book itself is dedicated to one of my favourite maritime historians (and a major influence on anyone who works seriously on early modern Indian Ocean trade), Ashin Das Gupta, whose work continues to be an inspiration in spite of (or perhaps because of) our many disagreements. I recall with amusement that in the course of my thesis defence at the Delhi School of Economics in April 1987, he laid bare my own ignorance on Vasco da Gama, thus planting the unconscious seeds perhaps of this book. Needless to add that it makes no pretence of meeting his own exacting standards.

ABBREVIATIONS

AN/TT	Arquivos Nacionais/ Torre do Tombo, Lisbon
CC	Corpo Cronológico
CEHU	Centro de Estudos Históricos Ultramarinos
CVR	Cartas dos Vice-Reis da Índia
BNL	Biblioteca Nacional de Lisboa
BPADE	Biblioteca Pública e Arquivo Distrital, Évora
Gavetas	*As Gavetas da Torre do Tombo*, ed. António da Silva Rego, 12 vols., 1960–77
IICT	Instituto de Investigação Científica Tropical

Note on Transliteration

Portuguese place and personal names follow modern usage in Portugal. However, in quotations, the original orthography of place and personal names has usually been retained, with modern equivalents in square brackets when necessary. For Indian and Arabo-Persian words, a simplified version of the standard transliteration system (Steingass-Platts) used by historians of South Asia has been followed. Long vowels and certain special consonants are marked when words appear in italics, but not otherwise. For persons' names in the bibliography, for Portuguese authors, listing is usually by the last name: thus Godinho, rather than Magalhães Godinho, and Thomaz rather than Reis Thomaz. For some Indian writers, e.g. Krishna Ayyar, an exception has been made.

CHAPTER ONE

OVERTURE: BONES OF CONTENTION

This Da Gama, whose fortune it was to initiate direct European contact with the East, was a man of iron physique and surly disposition. Unlettered, brutal, and violent, he was nevertheless loyal and fearless. For some assignments he would have been useless, but for this one he was made to order. The work lying ahead could not be accomplished by a gentle leader.

Charles E. Nowell, *The Great Discoveries* (1954)[1]

VASCO AND ZULAIKHA

THERE ARE FEW MORE STRIKING WAYS OF TRANSFORMING history into legend than through the medium of opera. In this dramatico-musical genre, quintessentially a form of the nineteenth century (even if its origins lie earlier), characters and situations are automatically transfigured, and happily dissolve into the most melodramatic colours. The storm and thunder of emotions played out through virtuoso voices slotted into well-defined registers and categories, and nearly always working on a simple palette of theatrical possibilities, dominates over reason – even dramatic reason. This is clearly the genre of Heroes then, in consonance with the particular romantic imagination of the Great Age of Nationalism, and hence suited above all in its sensibility to the exigencies of a nationalistically tinted history. Thus, so much of what is essential to nineteenth-century German nationalism filters through the works of Richard Wagner, even as the Italian *Risorgimento* finds voice through the operas of Giuseppe Verdi.

We could do worse than to approach our subject thus, to listen to the tenor voice of Vasco da Gama (1469–1524) not as it appears

[1] Charles E. Nowell, *The Great Discoveries and the First Colonial Empires* (Ithaca: Cornell University Press, 1954), p. 36.

in his very few writings or those of his contemporaries, but through the lips of a nineteenth-century operatic *alter ego*. And what does he say to us in this form?

> Ah! Have pity on my memory,
> O, you to whom I pray!
> Take no more than my days,
> and leave me with my glory!
> Ah! The torments that the furies bring together,
> are less cruel to me;
> For 'tis to die twice over,
> To lose at once both life and immortality!

So sings the explorer Vasco da Gama in the fourth act of the once-celebrated but now largely forgotten opera of Giacomo Meyerbeer and Eugène Scribe, *L'Africaine* (1865).[2] We may rapidly rehearse the twists and turns of the improbable plot of this work, in five acts and lasting four or more hours in performance, a work that is *not* – we should stress – the product of Portuguese authorship.[3] The first Act opens in the Hall of the Admiralty Council in Lisbon, where Vasco da Gama is awaited after a long absence by his beloved Inès, the daughter of Don Diégo, the Admiral of Portugal. She sings, in soprano, to her companion Anna, of Vasco's last adieu to her before departing the banks of the Tagus ('Adieu, mon beau rivage'), and, in at least one version, adds:

> It is for me that Vasco, seeking glory,
> shares the travails of Diaz, a sailor grand,
> Battling both the winds and the waves,
> he sails with him in search of new-found lands.
> My hand will be the prize of his victory,
> By love protected, Vasco triumphant shall stand.

But having accompanied his master Bernard Diaz – a composite figure, made up in equal parts of the navigator Bartolomeu Dias

[2] G. Meyerbeer–E. Scribe, *L'Africaine: Opéra en cinq actes* (Paris, 1865), Act V, Morceau 15, 'Grand Air de Vasco'. All translations cited here are mine, from the French libretto. For a good summary description of the opera, see Steven Huebner, 'Africaine, L' ("The African Maid")', in Stanley Sadie (ed.), *The New Grove Dictionary of Opera*, Vol. I (London: Macmillan, 1992), pp. 31–3. For a detailed discussion, John Howell Roberts, 'The Genesis of Meyerbeer's "L'Africaine"', PhD dissertation, Department of Music (Berkeley, University of California, 1977).

[3] My account is based on a video recording of the performance of the San Francisco Opera, conducted by Maurizio Arena, and directed by Lotfi Mansouri. Vasco da Gama was played by Placido Domingo, Sélika by Shirley Verrett, Inès by Ruth Ann Swenson, and Nélusko by Justino Diaz. The recording was broadcast by KQED TV, San Francisco, and produced by RM Arts (1989).

who turned the Cape of Good Hope in 1487, and the warrior-chronicler Bernal Díaz del Castillo who was with Cortés in Mexico – we learn soon after that Vasco da Gama has suffered shipwreck on a distant desert island. Some in Lisbon are gleeful, thinking him dead, and Inès quite naturally is despondent. But three-fourths of the way to the end of the act, matters take a dramatic turn for the better, as Vasco da Gama returns, bringing with him the Indo-African Queen of the opera's title, Sélika, and her servant Nélusko in tow, having purchased them both at a slave mart, after they had been captured on the high seas. Vasco meets, however, with a disappointing reception, largely on account of the machinations of the villainous Don Pédro, who himself lusts after Inès, whose hand he has been promised by King Emmanuel. Instead of being given ships and men to lead another expedition, Vasco is judged wanting by the ferociously intolerant Grand Inquisitor as well as the Council of State, headed by none other than Don Pédro himself. The sternly ridiculous figure of the Grand Inquisitor with his choir of robed priests reproduces clichés about the priest-ridden character of Iberia, while at the same time confirming the mid-nineteenth-century prejudices of the French bourgeoisie in respect of the Catholic clergy.

In Act II, we hence see Vasco da Gama thrown into an Inquisition prison along with Sélika and Nélusko; and the Queen sings him an Indian lullaby ('Sur mes genoux, fils du soleil') to rest his downcast spirit. Sélika, it has emerged already from broad hints and *sotto voce* asides in the first Act, has fallen in love with Vasco, while her servant Nélusko hates him not only because of *his* own love for the Queen, but because, significantly, Vasco is Christian. Nélusko's restless pagan mind leads him to contemplate killing Vasco, but he eventually is persuaded to desist; while he frets and fumes, Don Pédro swaggers in with Inès, and announces his own appointment to lead an expedition to continue the work of Bernard Diaz, with the aid of the two native informants whom Vasco has brought back, namely Sélika and Nélusko. Worse still, he has persuaded Inès to marry him, in exchange for arranging the release of Vasco from prison.

By Act III, we see the voyage of exploration in full progress; Don Pédro is on board ship in mid-ocean, accompanied by Inès, and guided by an obsequious Nélusko, who however has his own nefarious designs. Indeed, he broadly hints that he intends to wreck the vessel, through his worship of the terrible Adamastor,

Plate 1 Drawing of a set for *L'Africaine* (Act IV) by Charles-Antoine Cambon and Joseph Thierry (1865), showing the great temple in the land discovered by Vasco da Gama.

God of the Tempests, to whom he dedicates the song: 'Adamastor, roi des vagues profondes'. Another nobleman, Don Alvar, who accompanies Don Pédro, has his eye though on Nélusko, and his worst suspicions are confirmed when two of the three vessels in the fleet are lost on account of Nélusko's ill advice. A storm gathers and, at this stage, another vessel (whose sail they have caught sight of from time to time, and which mysteriously and significantly *precedes* them) sends out a boat to Don Pédro's ship. On it arrives Vasco da Gama, who has somehow managed to obtain a ship to command. He informs Don Pédro that he has come to save him, in spite of their mutual hatred, because of his own love for the noble Inès. Don Pédro, ironically sneering, refuses his aid, and they prepare to fight a duel with swords when Adamastor intervenes. There is a terrible storm, and the ship is wrecked on a reef, where natives swarm on board to capture all survivors; in the ensuing confusion Nélusko stabs and kills Don Pédro. We thus find ourselves by Act IV in the exotic land of which Sélika is Queen.

This act, Act IV, then, is the real centrepiece of the opera, where its exoticism reaches full bloom. The setting is a temple with Indian architecture: to the right there is a palace, and the background is of sumptuous monuments. Priestesses enter, followed by Brahmins, amazons, jugglers, warriors and finally the Queen, Sélika. The Great Brahmin, High Priest of Brahma, sings to her:

> We swear by Brama, by Wischnou, by Shiva,
> The Gods whose power Indoustan reveres,
> We swear obedience to the daughter of our Kings.

Sélika is thus back among her own people in an African Hindustan (confusingly also called a great island, and at times identified through the stage directions in Act II with Madagascar), and the tables are definitely turned on the Portuguese. The malevolent Nélusko meanwhile plots to have the Brahmins make a human sacrifice of the Portuguese women survivors and of Vasco da Gama, who is the sole male Portuguese survivor of the storm. The latter, all unsuspecting, enters the scene in his role of Great Discoverer, singing his Grand Air ('O paradis sorti de l'onde') for which the opera is perhaps best known:

> Marvellous land, fortunate garden!
> Radiant temple, greetings!
> O paradise emergent from the waves,
> A sky so blue, a sky so pure,
> that my eyes are ravished,
> You belong to me!
> O New World that I would gift to my country!
> This vermilion countryside ours,
> this rediscovered Eden ours!
> O charming treasures, o marvels, greetings!
> New World you belong to me
> Be mine, mine, so be mine, o lovely land!
> New World, you belong to me, so be mine, so
> be mine, mine, mine, mine.

The tenor's crescendo ends what is not the most inspired verse, perhaps, but assertive enough in its own unsubtle way of the spirit of the explorer and conquistador that Eugène Scribe wished to depict. Still, Vasco is soon disabused of his happy illusions, and assaulted by murderous Brahmin priests clamouring for his blood ('Du sang! du sang!'); and it is now that he launches his plea to save his claim to immortality, which is, naturally enough, more precious to him than life itself. But Sélika intervenes to save him, by falsely claiming to the assembly they are already betrothed; and their marriage is celebrated in style by the Great Brahmin of the temple of Brahma. Indeed, Vasco even discovers, aided by a magical potion that he is given by the Great Brahmin, that he is just a little bit in love with Sélika after all, and refuses her offer to escape. Indians dance, the festivities are on, and it seems that a happy ending of sorts has been achieved.

This is of course illusory, and it remains to play out the tragic finale. East is East, and West after all is West. In Act V, Vasco discovers that Inès has survived the massacre of the Portuguese women, and meets her secretly in the palace garden. Sélika discovers them in conversation, and is initially furious and vengeful. But realising that she cannot intervene in the inevitable course of their romance, she eventually lets them go to their ships, accompanied on the way by a happily relieved Nélusko. She herself now mounts a promontory, with a view over the sea, in the second part of Act V. Breathing the fumes of a poisonous manchineel tree conveniently perched atop the hill, she swoons, singing a farewell to Vasco. At this juncture, Nélusko returns, to find her half-dead, and offers to die with her too. He sings:

> O heavens, her hand is cold and icy,
> It is death, it is death, it is death!

She replies:

> No, it is happiness!
> Here is the rest of eternal love!
> Here is the rest of a pure love!

Vasco da Gama has meanwhile sailed away to Portugal, where glory and immortality no doubt await him. Don Pédro has died in the shipwreck, and Vasco's romance with the widowed Inès can be brought to its happy conclusion. Tragic though the end is for the African Queen (*L'Africaine*) Sélika, the optimistic celebration of Europeans' exploration of a world that had been to them, hitherto, unknown, remains essentially intact. Hardly in the same literary league as *The Tempest*, the logic of *L'Africaine* entirely lies in its use of stock operatic characters and exoticist tropes, with the Inquisition-ridden Portuguese as much as the Indo-Africans being used to provide theatrical colour. Above all, it is laden with the suggestion that despite all romantic yearnings to the contrary, East and West are destined never to make a happy marriage; Vasco must find his true love with Inès.

The Vasco da Gama of Scribe-Meyerbeer may appear at first sight to have little to do with the Vasco da Gama whose career we shall explore in the remainder of this work. The spectator of the opera sees him as a patriot, an intrepid explorer, a romantic and chivalrous hero, whom women the world over pine for. Contrasted to this dashing (and rather egotistical) masculine spirit is Sélika (a distortion of Zulaikha, from the classical Persian romance-tradition), who in a somewhat curiously tragic mode represents the lands that Vasco is destined to discover. Now, significantly, Scribe's remaining cast of characters draws in good measure on a particular source: the sixteenth-century work of the Portuguese national poet, Luís Vaz de Camões (1524–80), titled *Os Lusíadas*.[4] This makes perfect sense, for Scribe had earlier written the libretto for Gaetano Donizetti's *Dom Sébastien, roi de Portugal* (1843), where one of the major characters besides the ill-fated Portuguese monarch was Camões himself. For *L'Africaine*, Scribe used his source loosely, with a certain abandon even. Thus, the villainous Don Pédro's name is almost certainly derived from that of the Portuguese king Dom Pedro (1357–67), celebrated for his affair with Inês de Castro (herself surely the source for the name of Vasco's beloved), a romance that is set out in detail by Camões; and above all the figure of Adamastor is clearly and very directly borrowed from Camões, who invented this pagan god to be the great opponent of Vasco da Gama in his epic text. The opera thus built not on conventional historical writings, but on a classicised mythification of the Portuguese discoveries that had already been inherited from the late sixteenth century, in which Vasco da Gama appeared as a figure that was at once larger than life and somewhat two-dimensional.

In retrospect, the success of *L'Africaine* (called *Vasco de Gama, ou le Cap de tempêtes*, in its first performances) may seem a little astonishing. The composer, Giacomo Meyerbeer (born Jakob Liebmann Beer in Berlin in 1791), had died in Paris in 1864 before it could ever be staged; the librettist Scribe had already preceded him from this world in February 1861. The performed version was eventually based on a substantial reworking of Meyerbeer's manuscripts by the celebrated Belgian musicologist François-

[4] Luís de Camões, *Os Lusíadas*, introduction and notes by Maria Letícia Dionísio (Lisbon, 1985). For useful interpretative works, see António Cirurgião, 'A divinização do Gama de *Os Lusíadas*', *Arquivos do Centro Cultural Português* 26 (1989), 513–38; also Sílvia Maria Azevedo, 'O Gama da História e o Gama d'*Os Lusíadas*', *Revista Camoniana* (N.S.) 1 (1978), 105–44.

Joseph Fétis, who was appointed to the task by Meyerbeer's widow. It was Fétis, incidentally, who changed the opera's name back from *Vasco de Gama* to *L'Africaine*, insisting that the former was too obscure for the French public of the day! Now, Meyerbeer had signed a contract with the Opéra de Paris as early as May 1837 for *L'Africaine*, and a first version of the libretto had been written in the same year by Scribe, who was himself something of a favourite amongst the mid-nineteenth-century French bourgeoisie. This version contains the kernel of the later plot structure, but in fact has no mention of Vasco da Gama. Instead, the setting is Seville in the epoch of Philip III, the hero is Fernand, a young naval officer, and the main love triangle involves him, Inès – daughter of the viceroy of Seville – and Sélika (or Sélica), queen of a kingdom in darkest Africa, near the source of the river Niger. This setting survives in the second, shorter, version, which dates to 1843, and which Meyerbeer actually set to music. By 1849, the composer and the librettist had begun to veer towards India as a possible setting and, in 1850, Meyerbeer finally urged Scribe to take on Vasco da Gama as the central character. In a somewhat revealing letter of 27 October 1851, he wrote to Scribe:

> I do not know if I am mistaken, but it seems to me that it would be a good and clear exposition, and at the same time a magnificent musical introduction, to have a solemn meeting of the Portuguese Admiralty, before which Vasco appears to present his plans for the discovery of India, and to ask for vessels and troops. This Council would be presided over by the rival of Vasco (later the husband of Inès). As in all meetings, there would be a division of opinion among the members, each one defends his opinion, but in the end the cabal which is against Vasco triumphs, his projects are declared impracticable, chimerical, the Grand Inquisitor who supervises all the Councils of State, declares that Vasco's opinions on the position of the earth are heretical. Vasco assaults and threatens him, but he is stripped of his posts, and banished; he leaves feeling furious, and declares that he will devote his small private fortune to equip a vessel, and to prove that he is not mistaken.[5]

This is to say that Meyerbeer wished to transform Vasco da Gama into a Columbus-like figure, driven by personal convictions which he presented before an uncomprehending State, hoping for

[5] Bibliothèque Nationale, Paris, Département des manuscrits, n.a.fr. 22840, fl. 328r, cited in Roberts, 'Genesis of Meyerbeer's "L'Africaine"', pp. 103–4. It is not clear whether Meyerbeer had read some version of João de Barros's *Da Ásia*, as one of his letters to Scribe suggests. Barros could be the source for the composer's notion of a factional struggle in the Portuguese Admiralty on the issue of explorations in Asia.

its support. This is a theme that is somewhat subdued in the final version (in which Columbus does find explicit mention, though) but of a curious and ironic significance, as we shall see at length below.

After its long gestation period of over a quarter-century, in which it was the object of numerous rumours, scandals, and even witticisms in Paris, *L'Africaine* was eventually brought to the stage, and performed in the Opéra de Paris on 28 April 1865.[6] Its commercial success in the short term was astounding: it was performed 100 times in a year, and had seen 485 performances (an average of nearly seventeen a year) by 1893. Its first American performance at the New York Academy of Music, led by Max Maretzek, was a resounding success. Only in Lisbon was it met with a mixed reaction. The Marquês de Niza, descendant of Vasco da Gama, objected to the frivolous portrayal of his ancestor, and at his insistence the version that played in Lisbon somewhat mysteriously substituted the name of Vasco da Gama with that of Guido d'Arezzo (ca. 990–1050), a Benedictine monk and medieval musicologist of renown.[7] Since some of the rhymes in the text depend on the word 'Gama', we can only speculate how these might have been resolved.

ON THE BANKS OF THE TAGUS

The operatic text eventually had other resonances in these years. In the troubled political climate of the year 1880, fifteen years after the first production of *L'Africaine*, five years after the foundation of the Portuguese Sociedade de Geografia, and four years after the creation of the Portuguese Republican Party, a great national celebration was held in Lisbon, as also in the other large towns of Portugal. Its ostensible purpose was to commemorate the third centenary of the death of none other than Camões, whose death had coincided with the takeover of Portugal by the Spanish Habsburgs in 1580. Legend had it that the poet, who in fact was a victim of the plague, had bid farewell to the world because he could not bear the thought of his *Pátria* under foreign rule. The challenges of the 1880s gave redoubled weight to this

[6] A first approach to the opera's evolution may be found in Georges Servières, 'Les transformations et tribulations de *L'Africaine*', *Rivista Musicale Italiana* 34 (1927), 80–99; this has been incorporated though in Roberts, 'Genesis of Meyerbeer's "L'Africaine"'.

[7] D. Maria Telles da Gama, *Le Comte-Amiral D. Vasco da Gama* (Paris: A. Roger and F. Chernoviz, 1902), p. 54.

legend: for even if Portugal was not directly under external threat in these years, her African empire certainly was. Indeed, the Sociedade de Geografia was formed in 1875 – among other reasons – to justify Portugal's historic rights to Angola and Mozambique, and even to the intervening territories in South-Central Africa that separated the two, in the face of challenges from the British, Germans, Belgians and other European powers.[8]

The great ideologue of the celebration of 1880 was the prolific academician, intellectual and journalist Joaquim Teófilo Fernandes Braga (1843–1924), later to serve two brief terms as President of the Portuguese Republic in 1910 and 1915.[9] Fiercely anti-clerical and also vaguely socialist in his leanings, Teófilo Braga was above all a convinced Portuguese nationalist, who believed that the need of the hour was a regeneration of the Portuguese spirit, to which end the national celebrations could contribute in a significant way. Opposed to him were other republicans of a rather more internationalist colouring, such as Antero de Quental and J.P. Oliveira Martins (the latter himself romantic enough in his own view of history), who saw in the projected celebration 'patriotism reduced to a theatrical sentiment and national life to an opera' (perhaps *L'Africaine*?).[10] Teófilo Braga nevertheless had his way. The central design of the national spectacle that he planned was to move an urn with some of the remains of Camões to the famous Jeronimite monastery on the banks of the Tagus in Lisbon (founded by the king Dom Manuel, in the early sixteenth century), to become an object of national veneration. This would be followed by a public procession on 10 June (the actual anniversary of the death of Camões) in which the earlier glory of Portugal would be contrasted to its present decadence. This would presumably shame the Portuguese nation into action, to defend itself and above all its colonial possessions.

The transformation of Camões into a key symbol for progressive republicans of the late nineteenth century may be curious,

[8] Cf. António Ferrão, *A Sociedade de Geografia: As suas origens e a sua obra de 50 anos (1875–1925)* (Lisbon: n.d.), a work that I consulted as unbound proofs in the library of the late Alexandre Lobato in Lisbon. I am grateful to Manuel Lobato for permitting access to this work.

[9] Cf. Teófilo Braga, 'O Centenário de Camões em 1880', *Revista de Philosophia e Positivismo* 2 (1880), 1–9, and for a larger ideological justification, Braga, *Os Centenários como Síntese Afectiva nas Sociedades Modernas* (Oporto: A.T. da Silva Teixeira, 1884). For a useful discussion of the political context of the 1880s celebrations, see Maria Isabel João, 'A festa cívica: O tricentenário de Camões nos Açores (10 de junho de 1880)', *Revista de História Económica e Social* 20 (1987), 87–111.

[10] Oliveira Martins, cited in José-Augusto França, *Le Romantisme au Portugal: Etude de faits socio-culturels* (Paris: Editions Klincksieck, 1975), p. 744.

but it is not without precedent. From the seventeenth century, each generation of Portuguese writers had refashioned Camões in the light of their own needs, and the generation of 1870 to which Teófilo Braga belonged was no exception. To some other Portuguese intellectuals of the epoch, the celebrations represented a window of opportunity of another sort. These were men like Augusto Carlos Teixeira de Aragão (1823–1903), a medical doctor and numismatist, who saw the occasion as one where the importance not only of the poet Camões but of the hero of the *Lusíadas*, Vasco da Gama, could be reaffirmed. Teixeira de Aragão had already helped organise the Portuguese participation in the *Exposition Universelle* at Paris in 1867, and so was an experienced hand at the orchestration of public display and celebration; he had thereafter served in Portuguese India in 1871–2 in a variety of posts, including that of interim Secretary-General. This Goan sojourn gave him a taste for Indo-Portuguese numismatics, but also helped transform him into a convinced worshipper of the legend of Vasco da Gama.

Some necrology may be in order here. After his death at Cochin on 24 December 1524, Vasco da Gama had been buried with honours in that town in the chapel of the monastery of Santo António (later renamed the church of São Francisco). However, his remains were subsequently exhumed and taken back to his properties at Vidigueira in the Alentejo in 1538 by one of his sons, Dom Pedro da Silva Gama. There they remained, for two and a half centuries, in a resting place (the *Jazigo dos Gama*) constructed in the 1590s, with the epitaph (probably from the late seventeenth century): *Here lies the Great Argonaut Dom Vasco da Gama, First Count of Vidigueira, Admiral of the East Indies, and their Famous Discoverer*. But in 1840, the church where he lay buried was vandalised by unknown persons, in the course of the so-called Septemberist disturbances of the epoch. Large parts of its interior had to be reconstructed hastily in 1841, and a few years later a local cleric launched a plea to move the remains to a safer place, namely the Manueline monastery of the Jeronimites at Belém. The matter was not pursued though.

In the early 1870s, a more concerted campaign was launched to the same end, spearheaded by Teixeira de Aragão, by now far and away the leading Gamaphile in Portugal. The still-powerful Marquês de Sá da Bandeira was mobilised, and a committee was formed to look into the matter. The government of the Marquês d'Ávila e Bolama actually agreed to transfer the remains ceremoniously, and the Marquês de Niza (direct descendant of the Gamas) gave his assent too. But once more, a change of government frustrated Teixeira de Aragão's intentions. The 1880 Camões campaign thus appeared to him to be the ideal occasion finally to bring his plans to fruition. In April that year, the Academia Real das Ciências gave its support to the move, and later in the same month, royal permission was obtained. Besides Texeira de Aragão, the ultra-romantic journalist and orator Manuel Pinheiro Chagas (1842–95) was closely associated with the management of the ceremonies.

On 5 June 1880, a committee left for the sleepy Alentejan town of Vidigueira, with the necessary paperwork apparently having been done. The tombstone was raised up and the work of transfer began the next day, but was interrupted by an urgent telegram from the Count of Vidigueira (another descendant of Gama), who demanded that the committee stop its work until he was present. On his arrival, the work was completed in some haste, and the ceremonial cortège, accompanied by cavalry and infantry, set out on the morning of 7 June from the town to Lisbon after the usual speeches and orations had been gone through. The railway station at Cuba was the first stop, and from there the party set out on the morning of 8 June by train, making stops at Alvito and Barreiro, for further speeches and ceremonies. In the latter place, the remains were transferred on board an official river armada, which made its way down the *Mar de Palha* (the Sea of Straw, as the backwaters of the Tagus upriver from Lisbon are called), to a spot where another committee awaited it with the bones of Camões. The two sets of remains were then carried down by river to Belém, where they were met in the late afternoon by a reception committee with various municipal and other dignitaries, members of the Academia Real das Ciências and the Sociedade de Geografia. Martial airs were played by massed military bands, and the cortège then proceeded by stages to the monastery and church of Santa Maria. Here, they were awaited by the King of Portugal, Dom Luís, the Queen and others, and solemn religious ceremonies were conducted. Two days later, the civic procession organised by Teófilo Braga and his associates crowned the affair.[11]

However, matters soon took on a mildly farcical turn. Teixeira de Aragão, ever the keen researcher, had by 1884 discovered two

[11] For a contemporary account, see Manuel Pinheiro Chagas, 'A trasladação dos ossos de Vasco da Gama em 1880', *A Ilustração Portuguesa*, Vol. II, nos. 49–51 (1885–6) (in three parts).

Plate 2 The tomb of Vasco da Gama, Mosteiro dos Jerónimos, Lisbon, late nineteenth century.

documents – one from the 1640s and the other from the 1750s – which cast rather embarrassing doubt on the identity of the bones that had been transferred. Both these testimonies pointed to the fact that in the *Jazigo dos Gama*, the remains of Vasco da Gama had been on Gospel side of the nave, while the bones that were transferred had in fact been on the other side. Indeed, nothing could be clearer than the first of these texts, written in 1646 by Frei Álvaro da Fonseca, and dedicated to the first Marquês de Niza:

> In the church of Our Lady of the Relics of the town of Vidigueira, in the main chapel on the side of the Gospel, close to the chief altar, this convent has kept the bones of the famous great Dom Vasco da Gama, first discoverer of the East Indies, Royal Admiral thereof, and first Count of Vidigueira; there is no epitaph on his tomb. His bones are respected, and the tomb always kept covered with a silken cloth.[12]

Worse still, it was very clearly noted that opposite this tomb, on the Epistle side, there was another tomb covered in black velvet

[12] AN/TT, Miscelâneas Manuscritas Mss. 1104, No. 41, Fundação do Convento de Nossa Senhora das Rellíquias da Ordem do Carmo, pp. 259–61.

with the family arms; in it were the bones not of Vasco da Gama, but of his great-grandson, Dom Francisco da Gama (1565–1632), ironically enough twice viceroy of India at a time when Portugal was under Spanish Habsburg rule. Dom Francisco was patently no Portuguese nationalist hero, and it appeared that the gravestones of the two had been exchanged, probably in the course of the repairs and restoration of 1841. Further, Luciano Cordeiro, also involved in the commemorations, noted that the remains were transferred to Lisbon in a casket containing at least eight femurs and two crania, and speculated that besides Dom Francisco da Gama, the remains might have been those of Dom Miguel da Gama, Dona Guiomar de Vilhena, and Dona Leonor de Távora. In contrast, the other tomb, eventually opened up by Teixeira de Aragão on 11 July 1884, on his own initiative, contained only a single skeleton.[13]

Teixeira de Aragão, Cordeiro and others at once set to work to lobby the government to rectify the error. The government for its part, as governments are wont to do, refused to admit that there had been an error at all. The affair dragged on through the 1890s, as an interminable correspondence built up. Finally, in 1898, on the occasion of the fourth centenary celebrations of Gama's arrival at Calicut, orchestrated by the Sociedade de Geografia, the matter was rectified. Once more, a committee comprising Teixeira de Aragão, Luciano Cordeiro and various others (including the Minister for Public Works) set out for Vidigueira. Two members of the Gama family, D. Manuel and D. Eugénia Telles da Gama, were also present, as the second set of bones was taken out of the chapel, then moved by railway wagon to Barreiro, where it was eventually transferred on to the steamship *Dona Amélia*. At 10 p.m. on 8 May 1898, the vessel arrived at the Terreiro do Paço, where it was met by the Minister of the Marine and other officials; the next morning, it was moved on land, and taken to the Jeronimite monastery. Only a few persons attended the ceremony, which had none of the grandiose character of that of 1880. This was more than compensated for, however, by the elaborate ceremonies that marked the fourth centenary of Gama's arrival at Calicut, on 17, 18, 19 and 20 May 1898. Amongst the events of note was not only the creation of a grand aquarium, but the *Vasco da Gama Exposition*, in which there was to be a 'colonial section,

[13] See Luciano Cordeiro, 'Os Restos de Vasco da Gama', *Boletim da Sociedade de Geografia de Lisboa* (15ª Série) 4 (1896), 195–200.

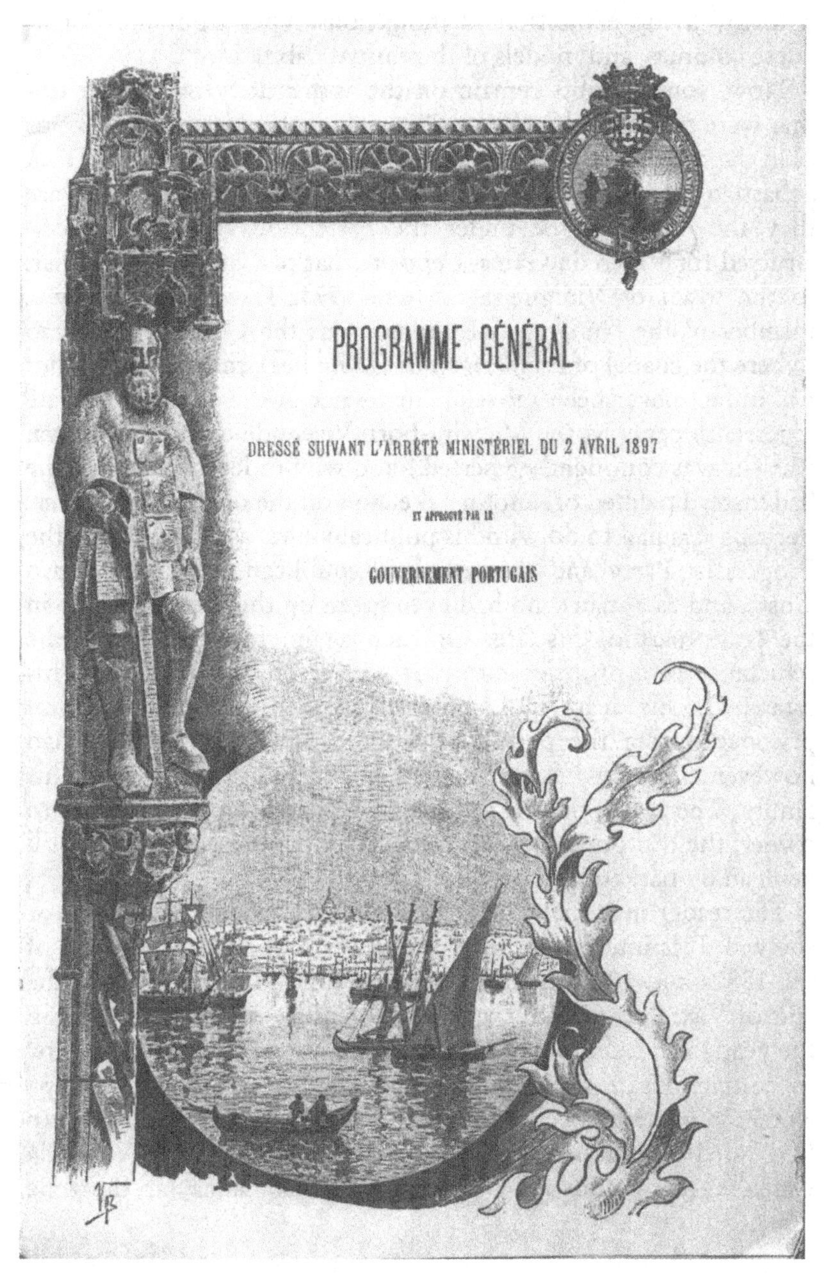

Plate 3 Cover of the programme in French for the fourth centenary celebrations of Vasco da Gama's voyage to India (1898).

with the living exhibition of indigenous types from the Portuguese colonies, and models of their usual habitations'.[14]

Now, some doubts remain on the authenticity of the remains that were transferred in May 1898, even if they are not as serious as in the case of the messianic figure of the Portuguese King Dom Sebastião (1554–78). Were the real bones in fact moved, and are they the ones that lie under the pretentious tombstone constructed for Vasco da Gama, opposite that of Camões? On a visit to the town of Vidigueira in June 1994, I was assured by a member of the family which today owns the Quinta do Carmo (where the chapel of the *Jazigo dos Gama* lies), that the real bones had in fact never been moved. The owner of the estate in the late nineteenth century, the Madeira-born Visconde de Ribeira Brava, was – it was confidently asserted – too wily to let the relics go; he had instead palmed off another skeleton on the official committee. Perhaps this had to do with his political views, as a member of the Progressist Party and then of the Republican Party of Afonso Costa, and as a man who had even given up the use of his title in the latter part of his life. Or then again, perhaps it was the reluctance of a proprietor to part with a valuable section of his estate! On his death in Lisbon in October 1918, as a political prisoner during the period of Sidónio Pais, the Visconde had however allegedly left the 'secret' of the bones' location to his family. The secret then is said to have passed on from owner to owner; the real bones, so it is claimed by the present owners, still lie in an unmarked spot on the estate!

The reader may make of these bones of contention what he or she will. It is interesting, for our part, to note that the episode of the 1880s and 1890s had not been the first attempt to build the cult of Vasco da Gama. For, a mere century after his voyage, in the year 1597, a series of ceremonies had already been held in Goa to commemorate the voyage of Vasco da Gama on the Cape Route. On that occasion, the chronicler Diogo do Couto (1542–1616), an adherent of the Gama clan, and client of Vasco da Gama's great-grandson, Dom Francisco da Gama (at the time

[14] Cf. *Programme général dressé suivant l'arrêté ministériel du 2 avril 1897 et approuvé par le gouvernement portugais* (Lisbon: Imprensa Nacional, 1897), pp. 6–7. For a violent critique of these celebrations, from an ultra-nationalist Portuguese, who felt they were excessively 'universal' and 'international', and not sufficiently Portuguese, see Thomaz Ribeiro, *Senhor, Não! Memorial de Recurso à Coroa* (Lisbon: Typografia da Companhia Nacional Editora, 1897). This also reproduces the Portuguese version of the general programme cited above.

viceroy at Goa) had gone so far as to declare in a public speech before the Goa Municipal Council:

> If that Américo Vespusio who discovered those western Indies, which the Geographers deem to be a fourth part of the world, became so famous in the doing that all that land is called America after him, preserving in itself the name of its discoverer, with how much more reason could this part of Asia which this valiant captain of ours discovered be called *Gama* in order to preserve by such an illustrious name the memory of the greatest feat that there has been since God created the world until now.[15]

The speech had an overt political meaning in the epoch. Two great legends already dominated the history of early Portuguese expansion in Asia by the time of Couto's speech: those of Vasco da Gama and Afonso de Albuquerque. Both men had left behind not merely legends though; they had also left behind dynasties of clansmen, who zealously guarded and propagated the legends of their heroic forbears, already by the late sixteenth century. Albuquerque had his admirers in the chroniclers Fernão Lopes de Castanheda and Gaspar Correia; Gama had his in the later sixteenth-century writers Luís de Camões and Diogo do Couto.[16] Couto, ever indiscreet, had noted, albeit *sotto voce*, that the two legends vied for space not merely metaphorically but also literally: when Dom Francisco da Gama had his great-grandfather's portrait placed in a position of honour in the Municipal Council-House at Goa in 1597, it was resented by many. Couto writes:

> This portrait of the Count Dom Vasco da Gama, which was thus put in that place with such rejoicing in the City, was later shifted I do not know by whose order; for the relatives of Afonso de Albuquerque claimed that the first place in that City belonged to

[15] Extract from a speech made by Couto at the Goa Municipal Council before the viceroy Dom Francisco da Gama, Conde da Vidigueira, and great-grandson of Vasco da Gama; text from the Biblioteca Pública e Arquivo Distrital, Évora (henceforth BPADE), Codex CXV/2–8, no. 1, p. 268. A slightly different version may be found in Diogo do Couto, *Da Ásia, Década XII*, facsimile of the Régia Oficina Tipográfica edition, 1788 (Lisbon, 1974), pp. 116–17. For Couto, his life and loyalties, see Charles R. Boxer, 'Diogo do Couto (1543–1616), Controversial Chronicler of Portuguese Asia', in R.O.W. Goertz (ed.), *Iberia – Literary and Historical Issues: Studies in Honour of Harold V. Livermore* (Calgary, 1985), and Maria Augusta Lima Cruz, *Diogo do Couto e a Década 8ª da Ásia*, Vol. II (Lisbon: Imprensa Nacional, 1995). Also see António Coimbra Martins, 'Diogo do Couto et la famille Da Gama: Un traité inédit', *Revue des Littératures Comparées* (1979), pp. 279–92.

[16] Cf. the important and neglected writings on this subject of António Coimbra Martins, such as 'Camões et Couto', in *Les cultures ibériques en devenir: Essais publiés en hommage à la mémoire de Marcel Bataillon* (Paris, 1979), pp. 691–705, and 'Sobre as Décadas que Diogo do Couto deixou inéditas', *Arquivos do Centro Cultural Português* 3 (1971), 272–355.

> them as he was the Conqueror of that City; and in order not to offend anyone, both these Captains were moved to the porch of the House, that of Afonso de Albuquerque on the right where the Councillors sit, and that of the Count Admiral on the left . . .[17]

Not only this: a statue of Vasco da Gama, which his great-grandson had erected in the Viceroy's Arch in Goa, was mysteriously pulled down one night by miscreants, who it was suspected enjoyed the protection of Francisco da Gama's successor as viceroy, Aires de Saldanha, as well as of a number of other officials and noblemen. Pieces of the statue were strewn all over the city, in a process that looks remarkably like a metaphor for the desecration of a grave and the strewing about of bones. This affair, which became a *cause célèbre* in the early seventeenth century, was never quite resolved to anyone's satisfaction, despite a long report by a commission of enquiry, before which most of Goa's leading citizens – including Diogo do Couto himself – deposed.[18] Eventually, a new statue of Gama was put in place of the one that had been destroyed, and remains there even to date.

FOR A BIOGRAPHY

There may be a lesson, indeed more than one, that unites these disparate materials, which range from the late sixteenth to the late nineteenth century. Paradoxically, it would seem, few figures in world history are at once so well known and so obscure as Vasco da Gama, Portuguese commander of a fleet of three vessels, which opened the all-sea route between Europe and Asia in the years 1497–9. It appears likely but by no means certain that he was born at Sines, a port in southern Portugal facing the Atlantic, and we do know that he was the child of Estêvão da Gama and Isabel Sodré, as well as that he was a younger son – thus establishing an early precedent for the export of such 'surplus' progeny overseas to make a fortune. We are aware too that besides his first expedition of 1497–9 to India, he returned there twice more, once in 1502–3 as Admiral of the Seas of Arabia, Persia and India, and then again in 1524 as viceroy, admiral and titled nobleman, the Count of Vidigueira. So much the reader of

[17] Couto, *Da Ásia*, Década XII, p. 120.

[18] 'Devassa (treslado da) que tirou o Licenciado Silvarte Caeiro de Grã ouvidor geral do crime, a respeito do motim que se fizera pera quebrar a estátua de D. Vasco da Gama', in A. da Silva Rego (ed.), *As Gavetas da Torre do Tombo* (henceforth *Gavetas*), Vol. VI (Lisbon, 1967), pp. 370–98.

Plate 4 Restored statue of Vasco da Gama atop the Arco dos Vice-Reis (Viceroys' Arch), Old Goa.

any dictionary of biography or encyclopaedia can gather with the greatest of ease.[19]

However, even Columbus, his contemporary, who has been portrayed as everything from the bastard son of the Portuguese prince Dom Fernando (and hence the half-brother of Dom Manuel I), to a crypto-Jew, is easier to obtain a grasp on than the Portuguese discoverer in many respects.[20] The parallel with Columbus is a more curious one than has often been suspected: for just as Vasco da Gama's remains were much disputed, there are at least three sites (two in the New World, and one in the Old) which claim to house the remains of the Genoese explorer.[21] Indeed, one of the few things we do know with certainty from the historiography is that despite an obscene Indian children's doggerel to the effect (Vasco da Gama/ went to Panama/ took off his *pajâma*/showed his banana/and paid a *jurmâna* (i.e. a fine) . . . and so on), the Portuguese admiral neither went to the New World nor was fined there for indecent conduct! Nor indeed did Columbus ever attain the true Indies, which makes it all the more picquant that the Venetian diarist, Girolamo Priuli, should write the following words in August 1499 with reference to the first Portuguese expedition there:

> Letters of June arrived from Alexandria, which wrote that through letters from Cairo written by men who had come from India, it was understood that at Calicut and Aden in India, principal cities, there were arrived three caravels of the King of Portugal, which had been sent to enquire after the Spice Islands, and of which the commander was Columbus (*di quelle hera patron il Colombo*).[22]

[19] See Eila M.J. Campbell, 'Gama, Vasco da, 1er Conde (1st Count) da Vidigueira (b. c. 1460, Sines, Port. – d. Dec. 24, 1524, Cochin, India)', *The New Encyclopaedia Britannica: Micropaedia, 15th edn* (Chicago, 1989), Vol. V, pp. 100–1, a slightly revised version of an article already published in the 14th edition (1964). For a brief, popular, biography in Portuguese on the occasion of the celebrations marking the fifth centenary of Gama's birth, see J. Estêvão Pinto and Maria Alice Reis, *Vasco da Gama* (Lisbon, 1969). The best summary biography of Gama to date is that of Jean Aubin, 'Préface', in *Voyages de Vasco de Gama: Relations des expéditions de 1497–1499 & 1502–1503*, tr. Paul Teyssier and Paul Valentin (Paris: Editions Chandeigne, 1995), pp. 7–28. The reader may also consult, at his or her own peril, the rambling, self-indulgent and romanticised 'essay' by René Virgile Duchac, *Vasco de Gama: L'orgeuil et la blessure* (Paris: L'Harmattan, 1995).

[20] On Columbus, see P.E. Taviani, *Cristoforo Colombo: La Genesi della Grande Scoperta*, 2 vols. (Novara, 1974); and more recently in English Carla Rahn Phillips and William D. Phillips jr, *The Worlds of Christopher Columbus* (Cambridge, 1992), and Felipe Fernández-Armesto, *Columbus* (Oxford, 1991).

[21] For the late nineteenth-century dispute between Havana and Santo Domingo for this honour, see Henry Harrisse, *Los restos de Don Cristoval Colón* (Seville: F. Alvarez y Cª, 1878).

[22] Arturo Segre (ed.), *I Diarii di Girolamo Priuli* [AA. 1494–1512], Vol. I (Città di

A creative confusion, and one that has persisted in another form as we have seen, being reflected in the conceptual framework used by Meyerbeer, wherein the motivations and circumstances of Columbus are confounded with those of Vasco da Gama. But this at least we know to be false. What more can we gather, however, from the researches of recent years? Since the writings of Brito Rebelo, Teixeira de Aragão and Luciano Cordeiro late in the nineteenth century, relatively few new documents have been found with a direct bearing on Vasco da Gama's career.[23] Some of those which have been published, such as a few Italian letters from Florence, with reports of his first expedition, do provide new details but do not answer key questions.[24] Why was this obscure petty nobleman chosen by the ruler Dom Manuel to captain the expedition? Why was the expedition itself so small, even in comparison to the one that followed it, captained by Pedro Álvares Cabral (who ostensibly 'discovered' Brazil en route to India)? Why was there such a long interval between Bartolomeu Dias's arrival at the Cape of Good Hope (1487), and the expedition to the Indian Ocean?

The only hypothesis of relatively recent vintage by an historian that has sought to provide solutions to these problems is that of Armando Cortesão. This author, a well-known authority on cartography and Portuguese navigation, followed up and extended the interpretation offered by his brother, the celebrated historian Jaime Cortesão, and thus argued that Portuguese maritime activity under Dom João II (r. 1481–95) was characterised by much secrecy, particularly in view of the rivalry with the Catholic rulers of Castile and Aragon, Ferdinand and Isabel.[25]

Castello, 1921), p. 153. The fact that Columbus was absent on a voyage at this time may have helped to fuel this rumour.

[23] Cf. especially A.C. Teixeira de Aragão, *Vasco da Gama e a Vidigueira: Estudo histórico* (Lisbon, 1898). In English, see the mediocre account in K.G. Jayne, *Vasco da Gama and His Successors, 1460–1580* (London: Methuen and Co., 1910), and the materials published in Henry E.J. Stanley, *The Three Voyages of Vasco da Gama and His Viceroyalty* (London: The Hakluyt Society, 1869).

[24] Cf. Carmen M. Radulet, *Vasco da Gama: La prima circumnavigazione dell'Africa, 1497–1499* (Reggio Emilia: Edizioni Diabasis, 1994), pp. 167–98; for a more general discussion, see Radulet, Maria Emília Madeira Santos and Luís Filipe Thomaz, 'Fontes italianas para a história das viagens portuguesas à Ásia' (paper presented to the Fourth International Seminar on Indo-Portuguese History, Lisbon, November 1985).

[25] Jaime Cortesão, *A política de Sigilo nos Descobrimentos* (Lisbon, 1960); Armando Cortesão, 'The mystery of Columbus', *The Contemporary Review* 151 (1937), 322–30. Jaime Cortesão had already developed the basic hypotheses of the 'secrecy policy' by the mid-1920s, for example in his 'Do Sigilo Nacional sobre os Descobrimentos: Crónicas desaparecidas, mutiladas e falseadas – Alguns dos feitos que se calaram', *Lusitânia* 1 (1924), 45–81.

Extrapolating from odd and obscure phrases in chronicles and letters, and making arguments from silence (i.e. the 'significant' absence of documents) Armando Cortesão suggested that the 'mystery of Vasco da Gama' could be solved if one posited that between 1487 and the death of Dom João II in October 1495, there had been some Portuguese expeditions to the east coast of Africa, round the Cape of Good Hope.[26] Further, he proposed that one of these, in 1494–5, was led by Vasco da Gama himself, but that by the time of its return, Dom João II was already dead. Dom João's successor Dom Manuel, not wishing to give the dead king credit for the entry into the Indian Ocean, suppressed information of the expedition's success, and himself then mounted another one (commanded by the same Vasco da Gama) two years later, in 1497.

Cortesão's hypothesis has met with a rather cool reception from historians, who have been largely unconvinced by the fragments of evidence that he presented.[27] It shall be argued at some length below, moreover, that some of his evidence can be interpreted in an entirely different manner than that proposed by him. However, in order to do that, we need to re-examine the contexts within which Vasco da Gama's expedition occurred, namely Portugal at the close of the fifteenth century, and the dawn of the sixteenth century, and the Indian Ocean in the same period. This then is the task which this book sets itself, and it is thus as much about the environment of Vasco da Gama as it is about the man himself.

As to the particular balance chosen here between 'life' and 'times', it is a function of factors both fortuitous and deliberate. Vasco da Gama has left us very few writings of his own, and these are mostly official receipts or formula letters, which are not particularly useful in furthering an analysis of the sort we shall be engaged in. Further, the extensive debate on the relationship between biography and history – from Carlyle to Lenin, to E.H. Carr, and more recently to Arnaldo Momigliano, Pierre Bourdieu and Jacques le Goff – that has raged now for several centuries has

[26] Armando Cortesão, *The Mystery of Vasco da Gama* (Coimbra–Lisbon: Junta de Investigações do Ultramar, 1973). The book was also published in a Portuguese version, *O mistério de Vasco da Gama*, in the same year.

[27] For a balanced discussion of the evidence and hypotheses, see Francisco Contente Domingues, 'Colombo e a política de sigilo na historiografia portuguesa', *Mare Liberum* 1 (1990), 105–16; also Luís de Albuquerque, *Navegadores, Viajantes e Aventureiros Portugueses: Séculos XV e XVI*, Vol. I (Lisbon, 1987), a work highly critical of Armando Cortesão's hypothesis on Gama, by one of his former collaborators.

brought us no closer to a definitive answer on this question of balance.[28] If some writers, taking an extreme 'sociological' position, have argued that biography is largely an illusion, which has nothing whatsoever to do with the respectable business of social history, others like Momigliano continue to see in biography a perfectly legitimate fashion of giving history a human face, and opening up a variety of questions at the centre of social and political history. The debate is not one, it seems to me, that can be resolved *a priori*, from purely theoretical positions. The remainder of this book will attempt, in practice, to render this debate largely irrelevant. We shall oscillate freely from the man to his context, and from history to the historical formation of legend.

[28] A cautious survey of positions, and some suggestions, may be found in the recent article by Giovanni Levi, 'Les usages de la biographie', *Annales ESC* 44, 6 (1989), 1325–36. For a sense of the evolution of the terms of the debate in the past half-century, compare Levi with Sidney Hook, *The Hero in History: A Study in Limitation and Possibility* (Boston: Beacon Press, 1955). For an ambitious claim to 'total biography', see Jacques Le Goff, *Saint Louis* (Paris: Gallimard, 1996).

CHAPTER TWO

THE HERITAGE OF SANTIAGO

> Santiago is . . . one of the sanctuaries most frequented, not only by the Christians of al-Andalus, but by the inhabitants of the neighbouring continent, who regard its church with veneration equal to that which the Muslims entertain for the Ka^{c}aba at Mecca.
>
> Abu Marwan Hayyan ibn Khalaf (987/8–1076)[1]

MEDIEVAL ROOTS

THERE IS, IN THE BEGINNING, A RATHER SIMPLE QUESTION of identity. Who, really, was Vasco da Gama, besides being the hero of *L'Africaine*, the body whose remains were disputed, the object of a perhaps cynical veneration on the part of Diogo do Couto, and the 'Great Discoverer', the leader of the expedition that made the first all-sea voyage from western Europe to Asia, who thus linked – as it were – the worlds of Iberia and India, or, metaphorically speaking, the pilgrimage centres of Santiago de Compostela and the Hijaz? As we have noted briefly, it might be answered that Vasco da Gama was, in the first place, a native of the south of Portugal; his origins lie after all in the Atlantic port of Sines, south of Lisbon, where he was born in about 1469, a 'fact' that we may, incidentally, only infer from indirect evidence in the absence of the relevant parish records. To understand the part played by his origins in his later career, we might then choose to pose him as local historians and patriots would, which is to say primarily as a native of the lower Alentejo region (though by some sixteenth-century definitions, Sines fell in the Algarve).[2]

[1] As cited in Al-Maqqari, *The History of the Mohammedan Dynasties in Spain*, tr. Pascual de Gayangos, 2 vols. (London: W.H. Allen and Co., 1840–3), Vol. II, p. 193. The translation is slightly modified here.

[2] For history in this style, see José A. Palma Caetano, *Vidigueira e o seu concelho: Ensaio monográfico* (Vidigueira: Edição da Câmara Municipal, 1986), especially pp. 250–83.

This would require us to engage however in a potentially rather embarrassing exercise, claiming the existence of very long-term regional and even local continuities and essences of identity, which are almost impossible for the historian to demonstrate. Once an Alentejano, always an Alentejano? Who could defend such a stereotype! An alternative approach, a rather more sensible one, would be to combine this regional and local origin with a notion of social stratum, and here we are already on rather firmer ground. For Vasco da Gama was, himself, and in terms of his immediate family tradition, a member of the lower nobility resident in the Alentejo, a social group which crystallised in a gradual process, consequent on the emergence of Portugal as a kingdom.[3] He was also, like his father Estêvão da Gama, a member of a military order, the Order of Santiago.

It is conventional to begin any account of Portuguese overseas expansion with a mention, however token, of the Reconquest, the process by which the first rulers of the kingdom of Portugal seized the central and southern parts of that kingdom (including the Alentejo) from the Muslim dynasties that ruled it in the early centuries of this millenium. Arguably, Reconquest (*reconquista*) is an unhappy term, suggesting that the powers that emerged in Portugal in the course of the twelfth and early thirteenth centuries did no more than take back what had been theirs in an earlier epoch, prior to the establishment of Islam in Iberia. This was not the case in fact, and the consolidation of a state by first Dom Afonso Henriques (1106–85), and then his successors through to Dom Afonso III (r. 1246–79) brought altogether new elements to the fore, rather than merely restoring the power of groups, families and institutions that had earlier dominated. Amongst these new elements, the most significant gainers from the Reconquest in the south of Portugal were undoubtedly the Military Orders, especially those of Christ, Santiago (or Santiago da Espada) and Avis. This was in contrast to the centre and north of Portugal, where the creation of an independent kingdom favoured a territorial aristocracy on the one hand, and the *concelhos* on the other.

Founded in the Middle East, between the First and the Second Crusades, with the patronage of the King of Jerusalem, Baldwin II (r. 1118–31), the idea of Christian military orders as such is

[3] On the origins of Portugal as a kingdom and state, and the role therein of social stratification, see José Mattoso, *Identificação de um país: Ensaio sobre as origens de Portugal, 1096–1325*, 2 vols. (Lisbon: Editorial Estampa, 1985).

usually attributed to a knight from Champagne, Hugh of Payens. Initially granted rooms in the so-called Temple of Solomon in Jerusalem by Baldwin, the order formed by the companions of this knight was hence termed that of the Templars; by the late 1120s, they had attracted a good deal of positive attention in church circles, and garnered much praise and aid from St Bernard, Abbot of Clairvaux, especially once they had received their rule at the Council of Troyes, in 1128. In his work, *Praise of the New Knighthood*, St Bernard contrasted these knights, pure and just in their celibacy, uncaring about worldly wealth, with the venal run-of-the-mill knighthood.[4]

Within a few years, the irony of making such a contrast was to be evident. The Templars' rise to political, financial and religious power was rapid, and they soon came to accumulate substantial amounts of land in different parts of western Europe to support them in their role as the cutting edge of the Christian forces in the Crusades. The Grand Master of the order became a formidable figure, seen by several Christian states as a dangerous and destabilising element. A second order, that of the Hospitallers (later also termed the Knights of Malta), in fact predated the Templars in a technical sense; but it was substantially reorganised in the course of the twelfth century so as to become a military order, whereas its character earlier had been largely charitable.[5] In the late twelfth and early thirteenth centuries, these two orders vied for pride of place in Palestine, but also came to acquire a situation of importance in almost all the Christian states that participated in the Crusades.

With the fall of St John of Acre in 1291, the two orders initially moved the centre of their activities to Cyprus; the Hospitallers continued to concentrate largely on the Mediterranean, whereas the Templars were soon to be found playing an increasingly active part in France, England and the Iberian Peninsula. In the last of these areas, they helped play a role in the Reconquest, and in the resettlement of frontier regions with Islamic states, and thus continued to sustain a predominantly military profile. Elsewhere, such opportunities for surrogate crusading did not present themselves. Their legendary wealth, and penchant for financial activity

[4] Georges Duby, *The Three Orders: Feudal Society Imagined*, tr. Arthur Goldhammer (Chicago and London: The University of Chicago Press, 1980), pp. 226–7. For a recent and comprehensive study of the Templars, see Malcolm Barber, *The new knighthood: A history of the order of the Temple* (Cambridge: Cambridge University Press, 1994).

[5] Hans Eberhard Mayer, *The Crusades*, tr. John Gillingham (New York and Oxford: Oxford University Press, 1972), pp. 82–3.

(especially in France and England), soon led the Templars to be attacked by political opponents in these last countries; a particularly virulent anti-Templar campaign was orchestrated by the King of France Philip IV, the Fair (r. 1285–1314), in the late thirteenth and early fourteenth centuries. Accused of crimes from heresy and *lèse-majesté*, to usury and depravation, the Order of Templars was eventually suppressed under heavy French royal pressure in 1312 by the papal bull *Vox in excelso* of Clement V; in 1314, the Grand Master of the order, Jacques de Molay, and another leading member Geoffrey de Charnay were burnt at the stake after confessions had been extracted from them under torture.[6] In the Iberian Peninsula, where political opposition to the Templars had been the weakest, and where the order had functioned under a single Master (overseeing Portugal, Castile and León), it was not the state that resumed the bulk of their extensive properties and effects. Rather, in Portugal, unlike in France, these properties were largely reassigned to an order formed by a papal bull of John XXII (*Ad ea exquibus*) in 1319, namely the Order of Christ. Initially based at Castro Marim, on the frontier with the territories under Muslim control, the order later transferred its headquarters in 1357 to Tomar.

TWO ORDERS, OR PORTUGUESE EXPANSION DISMANTLED

By the time of the formation of the Order of Christ, other military orders, often limited in their activities to the Iberian kingdoms, had arisen. The Order of Avis, for example, appears to have originated in the mid-twelfth century under another name (the Friars of Évora), and was closely associated, albeit probably in a subordinate role, until the late fourteeenth century with the Castilian order of Calatrava, which was itself created in the twelfth century (receiving papal approval from Pope Alexander III in 1164). The Portuguese succession crisis of 1383–5, which saw the accession to the Portuguese throne of Dom João I, Master of Avis, distanced the two orders definitively and forced a certain 'national' consciousness on the orders' members; the Order of Avis as it now existed controlled middling properties in Portugal, in the Estremadura, Ribatejo and Upper Alentejo.

[6] Elizabeth M. Hallam, *Capetian France, 987–1328* (London and New York: Longman, 1980), pp. 317–20.

The Hospitallers, meanwhile, continued their relatively minor activities, and gradually the mastership of the order in Portugal came to be associated with the centre of Crato, so much so that the order's head was known by the fifteenth century as the Prior of Crato. In its dimensions, and in its holdings of land and other wealth, the Portuguese branch of the order never could match up to the Order of Christ. Such a place in the sun was available, if at all to any order, then to that of Santiago, which is to say St James.

It is now generally agreed that the Order of Santiago was not founded, as once was thought, before 1030, but rather that its origins lie in around 1170, in a conjuncture wherein the most important element was the emergence of the Almohads, and the decline in the power of the Almoravids, who had been a force to reckon with in Iberian politics from about 1075. The Almoravids (or *al-murâbitûn*) found themselves challenged three-quarters of a century after their emergence, in the mid-twelfth century, by a number of internal and external foes, most notably the Almohads (*al-muwâhhidûn*), who for their part drew inspiration from the Sufic teachings of al-Ghazzali, and made their first inroads into Spain in about 1146. The death of the last Almoravid, Ishaq ibn ᶜAli, in 1147, and the very rapid conquest by the Almohads in the years 1146–7 of Tarifa, Algeciras, Mértola, Jerez, Badajoz and finally Seville, nevertheless left room for a number of independent Islamic rulers in other centres further east.[7] In this situation, a number of joint campaigns were conducted by Dom Afonso Henriques (Afonso I) of Portugal, and Don Fernando II of León against the power of Ibn Mardanish (1147–72), king of Murcía and Valencia. This ruler, known in the Spanish historiography as *el rey Lobo de Murcía*, fought wars simultaneously on several fronts, and eventually fell victim to strong pressure from his rivals, the Almohads, whom he had dominated early in his career. The Almohad Caliph Abu Ya'qub Yusuf, who ascended the caliphal seat in 1163, launched a successful campaign against the king of Murcía, the momentum of which took the Almohads as far as the Tagus; in 1170, the Almohads were at the gates of Toledo, and – one historian writes – had 'put in peril not only the existence of Castile, but that of León and Portugal as well'.[8]

Yet, it was not *solely* to combat the southern threat that the

[7] Joseph F. O'Callaghan, *A History of Medieval Spain* (Ithaca and London: Cornell University Press, 1975), pp. 227–9.

[8] José Luis Martín, *Origenes de la Orden Militar de Santiago (1170–1195)* (Barcelona: Consejo Superior de Investigaciones Científicas, 1974), pp. 6–7.

Order of Santiago seems to have been founded. Rather its immediate origins appear to lie at least partly in the desire of the king of León, Fernando II, to resist the pressure of the Castilians and the Portuguese, who at this time were still his principal adversaries. The consideration may have been a short-term one, but is nevertheless worth noting. The grand theme of the Reconquest thus conceals the fact of a shifting set of alliances, where León might as often be allied to the Almohads as to the Portuguese, and where the Portuguese *caudilho* Geraldo Sempavor could switch sides when the occasion demanded, going over to Almohad service. But the order soon developed a mind and a political will of its own, in playing off León against Castile and Portugal, as well as against the prestige of the Archbishop of Compostela. Even if the order's first possessions were granted to its founder-master Pedro Fernández in about 1170 by Fernando II of León, it then went on to acquire other territories elsewhere – Oreja from Alfonso VIII of Castile, Monsanto from Afonso Henriques of Portugal. The last grant, made at Coimbra in September 1172, is a particularly interesting one, specifying that the castellan at Monsanto should not only be a member of the Order of Santiago but a subject of Dom Afonso Henriques, and further that the garrison should be available to aid the King of Portugal as much against his Christian enemies as against the Muslims (*tam christianorum quam sarracenorum*).[9] Thus, the early years of the Order of Santiago involved a political balancing-act of some complexity.

In the mid-1170s, there was a brief respite in the rivalry between the Christian kingdoms. Castile and León were allied now against the ascendant power of the Almohads, and the role of the Order of Santiago could emerge somewhat more clearly as ideologically an anti-Islamic organisation, in keeping with the medieval hagiography of Santiago de Compostela, where the saint himself was conventionally portrayed on horseback, and treading over the bodies of Muslim foes, in his form of *matamoros* ('Moor-Killer'). The Papal Bull of Alexander III dated 5 July 1175, which laid down the first regulations of the Order, confirmed this notion of an organisation that embraced a number of Christian Iberian states, and was devoted to the defence of the Christian faith against its enemies.

The legend of Santiago de Compostela itself is of course of

[9] Martín, *Origenes de la Orden de Santiago*, Doc. 56, pp. 230–1.

Plate 5 Statue of Santiago (St James). See Cathedral, Tuy, Galicia, Spain.

thoroughly doubtful historicity. The 'discovery', probably in the early ninth century, of the site of the tomb of the Apostle St James in Galicia, acted as a catalyst to spark off a powerful myth, in which Iberia was said to have been converted to Christianity centuries before by the actions of the Apostle. A church was built under the patronage of Bishop Theodomir of Iria Flavia, and the site came to be known as Compostela, from *Campus stellarum*, or field of stars, a reference to a miraculous vision that a hermit is supposed to have seen on the site. By the late tenth century, the site had such prestige that the *de facto* ruler of Córdoba, Muhammad ibn Abi Amir al-Mansur, launched a major campaign against it; in August 997, al-Mansur sacked the town, razed its church, and carried off its massive doors and bells. Yet by the reign at León of Alfonso VI (1065–1109), Compostela was once more a flourishing site, drawing pilgrims even from beyond the Pyrenees, and drawing in equal measure the scorn of the historian Ibn Hayyan (cited briefly above) for this falsification of the real history of Ya'qub (that is, St James). When, in 1120, Pope Calixtus II raised Compostela to the seat of an archbishopric, the prestige of the site and the cult of Santiago were assured once and for all, as we see from the compilation shortly thereafter of the *Historia Compostelana*, at the orders of the first Archbishop, Diego Gelmirez.

The Order of Santiago is thus a curious mixture. In its origins and name, its early associations with León and its early holdings of castles, it presents a profile that is northern Iberian, but in the logic of the first century of its history, it is inevitably drawn southwards. As the Reconquest of Iberia proceeded, the Order of Santiago gained substantial territories and castles, especially south of the Tagus. The process accelerated with the conquest in 1217 of Alcácer do Sal, which opened up what has been termed 'a vast uncultivated plain [which] turned itself into a zone of *latifundios*, belonging above all to the Order of Santiago, in the western Alentejo'.[10] The ruler Sancho II gave the order possession of forts in centres like Palmela (which became its headquarters), Almada, Alcácer itself, Setúbal, Sesimbra, Santiago and Aljustrel, to which they subsequently added Odemira, Ourique, Almodôvar and Castro Verde. By 1249, when the process of Reconquest in Portugal had virtually been completed under D. Afonso III, with the taking of Faro, Silves and other centres of the Algarve, the

[10] Mattoso, *Identificação de um país*, Vol. I, p. 98.

possessions of the Order in this western Iberian kingdom represented a formidable economic weight, proportionally speaking. In the late thirteenth century, the order controlled a total of 47 *vilas* and 150 *comendas*, as well as 75 *padroados* of churches, worth some 12,000 *cruzados* in all by 1500.[11] Further, D. Afonso III, mindful of the civil strife that had riven Portugal during the reign of his predecessor Sancho II, and which had eventually led to the latter's deposition, treated the military orders with suitable caution, and was anxious not to infringe overmuch on the powers of the *comendadores-mores*. Yet, once the limits of Portuguese territory had been more or less defined, it was inevitable that such institutions as the Order of Santiago would have to acquire a more 'national' character, wherein the Portuguese chapter would be separated from the other four, of León, Castile, Aragon and Gascony. The beginnings of this process may be seen during the four decades (1235–75) when Dom Paio Peres Correia dominated the Portuguese section of the Order, and the tendency is one that gains considerable momentum with the accession to the Portuguese throne of Dom Dinis (r. 1279–1325). In 1288, under this ruler, the Portuguese branch of the order came to separate itself to a large extent from the others by virtue of a papal bull from Nicholas VI; it henceforth claimed an autonomy in its functioning, as exemplified by the emergence of a separate Master, even if the process of separation was formally completed only as late as 1440, with a further bull from Pope Eugene IV.[12] In Castile, by the mid-fourteenth century, rulers took to naming their bastard sons as masters of the order, as a means of integrating its resources more closely with those of the Crown; an early example is Fadrique, son of Alfonso XI of Castile, who was named Master of Santiago there. Portugal was eventually to follow suit, but in the reign of Dom Dinis and his immediate successors, the Portuguese Order of Santiago continued to elect its masters. It took the Revolution of 1383–5, and the accession to the Portuguese throne of the Master of Avis, Dom João I, to change these rules once and for all.

The last elected Master of Santiago in Portugal was a certain Rui Freire, whose father, Nuno Gonçalves Freire, was Master of

[11] For the early history of the Order of Santiago in Portugal, see Mario Raul de Sousa Cunha, 'A Ordem Militar de Santiago (Das Origens a 1327)', Master's dissertation in Medieval History (University of Oporto, 1991), pp. 19–68. For a discussion of its properties, *ibid.*, pp. 214–43.

[12] Sousa Cunha, 'Ordem Militar de Santiago', pp. 129–85.

the Order of Christ. Dom João I, anxious to assert his control over the orders, set aside this election (in 1386), and instead imposed on the Order his own candidate, Mem Rodrigues de Vasconcelos. Once into the fifteenth century, the Castilian precedent cited above was followed; in 1418, the Infante Dom João (1400–42) was named Master of Santiago by a Papal Bull of Martin V, followed in 1444 by the Infante Dom Fernando (son of the king Dom Duarte). Thus, in 1472, when the Infante Dom João (the future king, Dom João II) was named Master of Santiago, it was as if a well-established precedent were being followed.

Once the Portuguese reconquest had been completed, that is as early as 1250, the dominant logic and justification for such Orders as Santiago and Avis could no longer be seen as fighting the Moor. True, in 1341, in the aftermath of the joint Castilian-Portuguese victory in the Battle of Salado (30 October 1340) against the ruler of Granada, Yusuf I, and his Moroccan Marinid allies, Dom Afonso IV had asked for and received a papal bull of Crusade (*Gaudemus et exultamus*), which apparently was designed to permit the Portuguese to open up a front against Islam in North Africa at any time. The fact that this bull was renewed in 1345, 1355, 1375 and 1377 by Clement VI, Innocent VI and Gregory XI, on Portuguese request, suggests that the revival of the offensive was always a political possibility, most likely in North Africa, but also perhaps in respect of the residual Islamic presence in Granada.[13] In Spain, where the Nasrids at Granada could be mobilised far more effectively as a demonological device than in Portugal, the identity problem of the military orders may not have been so acute. But there too, discomfort with their role was manifested by rulers in times of relative external peace, and cases of ambitious nobles who sought to further their political careers through gaining the Masterships of these orders are legion. One need only think of the celebrated and tumultous career of Don Álvaro de Luna (executed in 1453) in Castile, and recall that he was Master of Santiago.[14]

The crusading role of the military orders was however definitively revived in the fifteenth century, once the kings of Portugal launched their North African campaigns with the taking of Ceuta

[13] Cf. the discussion in Luís Filipe Thomaz, 'Le Portugal et l'Afrique au XVe siècle: Les débuts de l'expansion', in *Arquivos do Centro Cultural Português* 26 (1989), 164–7.

[14] For details of his career, see Nicholas Grenville Round, *The Greatest Man Uncrowned: A study of the fall of don Álvaro de Luna* (London: Tamesis Books, 1986).

in 1415; these campaigns were retrospectively given a genealogy that took them back in a seamless flow to the first Crusade. In an interim phase, the principal role of the military orders came to be that of corporate bodies controlling landed and fiscal resources in Portugal, and thus a counterweight to the territorial aristocracy, church and monarchy. This division is naturally complicated by the fact that no clear separation existed between these categories in terms of membership. Thus aristocrats could be members of the military orders, even as they could enter the church. The military orders themselves worked at times in close conjunction with the church. And, as we have seen, the monarchy sought with increasing success to penetrate the functioning of the military orders though the late fourteenth and fifteenth centuries.

Research on the internal organisation of the Order of Santiago is to date not sufficiently advanced for us to detail how precisely its membership changed over these centuries.[15] The internal records of the order list a large number of castles and other holdings, stretching from the Setúbal peninsula, down the Sado valley, the Atlantic front of the Alentejo, and parts of the Algarve, and also point to the fact that rights to these were held on behalf of the order by its *comendadores* and *cavaleiros*. There is a suggestion, moreover, that membership in these Orders was loosely hereditary during the fourteenth and fifteenth centuries. Access was not truly open, though members of the petty aristocracy, even if of illegitimate birth, had the notional possibility of staking a claim to membership. Thirty to forty members were counted among those who held substantial *comendas*, and from amongst these thirteen attended the meetings of the council in a privileged position, in an obvious metaphorical reference to the Apostles. At the same time, perhaps the most crucial aspect of the Orders (certainly of Santiago, and to a lesser extent of Christ) was that they did *not* reproduce other social hierarchies exactly. Within even the aristocracy (defined to include not merely *fidalgos*, but others of a lesser stripe), the *comendas* did not necessarily follow from rank. This was to change in the transition from the fifteenth to the sixteenth century, but a form of

[15] But see, in the interim, Isabel Maria Gomes Fernandes de Carvalho Lago Barbosa, 'A Ordem de Santiago em Portugal na Baixa Idade Média', Master's dissertation in Medieval History (University of Oporto, 1989), Part II, pp. 127–53. For a recent set of essays that touch on this question, see *As Ordens Militares em Portugal: Actas do I.º Encontro sobre Ordens Militares* (Palmela: Câmara Municipal de Palmela, 1991).

independent hierarchisation can be observed as late as the 1480s inside the military orders.

The gradual penetration of royal power into the functioning of Orders such as those of Santiago and Christ took place, as we have noted above, through the naming of royal offspring (whether legitimate or not) to the Masterships of these Orders. It is significant, though, that the king himself did not attempt to take over the Orders, by seeking papal sanction to name himself to such a capacity. Rather, the first major royal thrust which took place, and which involved the sons of Dom João I, saw the division of the Orders, as the Infantes were separately given powers over them, as if to create a system of pseudo-appanaging and to check the power of whichever one of them would succeed to the throne. We have already noted the nomination of the Infante Dom João to the Mastership of Santiago in 1418; Dom Henrique was named administrator of the Order of Christ by a Papal Bull of 1420 on the death of the Master, Dom Lopo Dias de Sousa; a third brother, Dom Fernando (the eighth and youngest child of the king, Dom João I), was named Master of Avis in 1434; but the royal succession went to Dom Duarte. There is thus a paradox inherent in the relationship between the Crown and the military orders; if, on the one hand, there was centralisation in the sense of gradually taking the Masterships away from the aristocracy and retaining them within the royal family, on the other hand, it was not the king's own personal power that was necessarily enhanced. This explains the situation of the middle years of the fifteenth century, when the powerful Infante Dom Henrique, manipulating his position at the head of the Order of Christ, created for himself a substantial commercial and political enterprise stretching overseas. Thus, rather than the kings Dom Duarte (1433–8) or Dom Afonso V (1438–1481), or the controversial Infante Dom Pedro, it is Dom Henrique who emerges as the first major architect of Portuguese overseas expansion by virtue of his relationship with the Order of Christ.

It is a commonplace that fifteenth-century Portuguese expansion brings into relief the role of the cavaliers of the Order of Christ, both as warriors and commanders in the wars of attrition that were fought in North Africa, and in more commercial ventures in the Atlantic, where the Order came to acquire properties and fiscal rights as well. As early as 1426, Gonçalo Velho, a *comendador* of the Order, and member of the household of Dom Henrique, had sailed to an unidentified spot on the African coast,

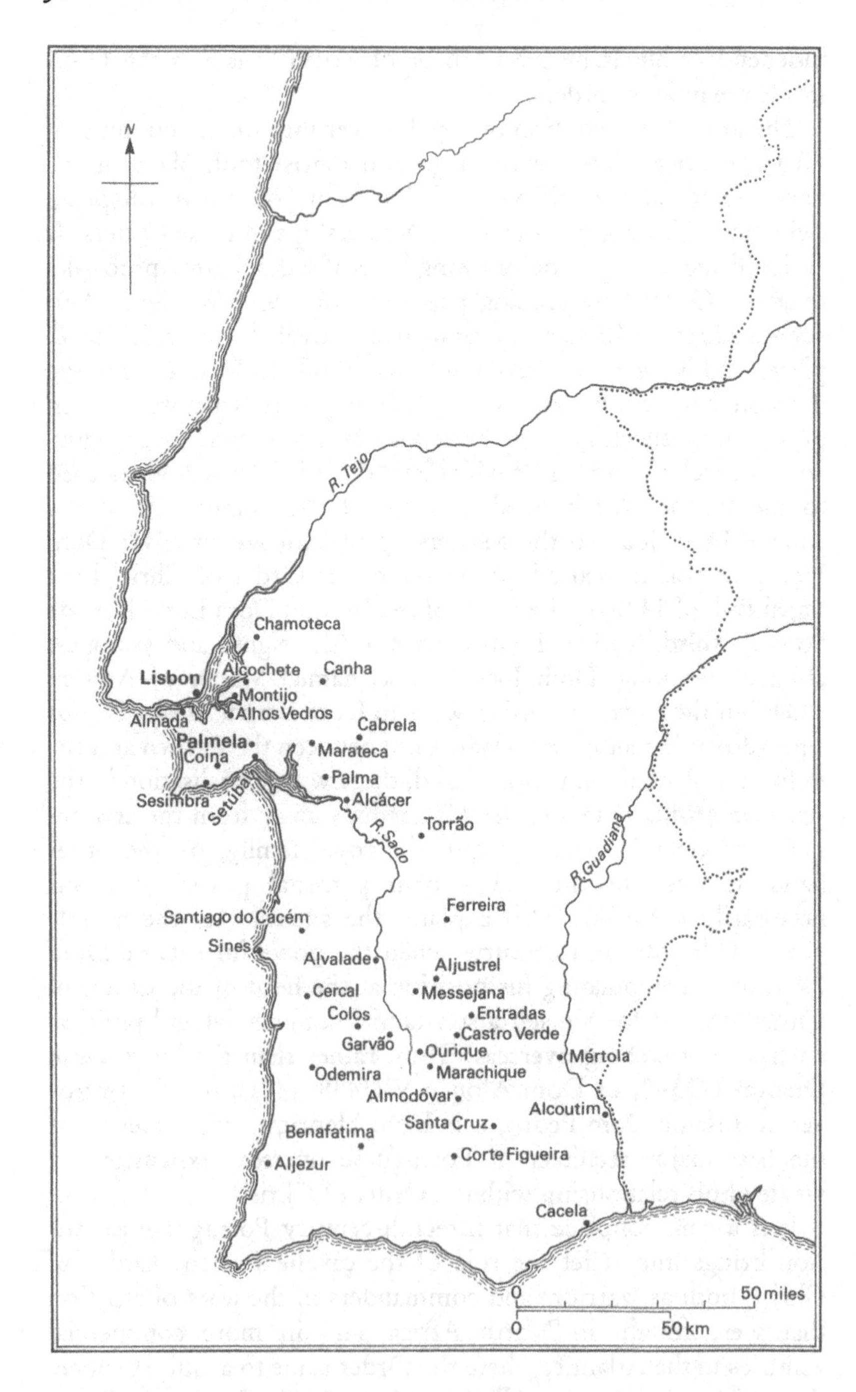

Map I Main centres of activity of the Order of Santiago

not far from Cape Bojador. Again in 1447, when an expedition to Guinea was mounted, a prominent role therein was given to a certain Fernão de Afonso of the same Order. Other examples can be found, and the reason is not far to seek. As early as 1433, the King Dom Duarte had granted the Order of Christ and its administrator and governor Dom Henrique extensive powers over Madeira, Porto Santo and other islands, a role that was confirmed by a series of somewhat generous and open-ended Papal Bulls. Not merely administration and settlement, but even exploration thus came to be associated with the Order, even if other captains and navigators were never totally excluded. In 1455, the Papal Bull *Romanus Pontifex* of Nicholas V seemingly strengthened the stranglehold of the Order on overseas expansion, since together with royal orders of 1454, and a later bull *Inter Caetera* of 1456, from Calixtus III, it definitively gave the Order 'all power, dominion and spiritual jurisiction', over the regions of Portuguese expansion, while at the same time defending Portuguese rights from Castilian claims.

A question that has sometimes been posed, notably by the Portuguese historian Joaquim Veríssimo Serrão, is that of why the Order of Christ emerged in this position of superiority with respect to the Order of Santiago, where overseas expansion was concerned.[16] On the face of it, the Order of Santiago had a more maritime profile in the late fourteenth century; its possessions were in general closer to the Atlantic coast, and more oriented to the south than those of the Order of Christ which, in the early fifteenth century, still had a distinctly rural and interior bias in its activities. The Order of Santiago had furthermore received extensive fiscal rights over fishing and anchorages, as well as tithes over ships, from Dom Dinis and Dom Afonso IV, by the middle of the fourteenth century; besides, the crucial place of the order in the economy of the Sado valley, from Alcácer do Sal down to Setúbal, gave it a direct interest in the flourishing salt-export trade.[17] Indeed, the Order of Santiago and its Master, the Infante Dom João, appear to have taken an early interest in overseas

[16] Joaquim Veríssimo Serrão, *História de Portugal*, Vol. II: *Formação do Estado Moderno (1415–1495)* (Lisbon: Editorial Verbo, 1978), pp. 133–5. For a brief discussion on the Order of Santiago and its role in overseas expansion, also see João Ramalho Cosme and Maria de Deus Manso, 'A Ordem de Santiago e a Expansão Portuguesa no Século XV', in *As Ordens Militares em Portugal*, pp. 43–56.

[17] This is well documented in Virgínia Rau, *A exploração e o comércio do sal de Setúbal* (Lisbon: Instituto para a Alta Cultura, 1951), who cites extensively from the records of the Order of Santiago in the Arquivo Nacional da Torre do Tombo, Lisbon, to support her case.

expansion in the second and third decades of the fifteenth century; the presence of a *fidalgo* of Dom João's household, Bartolomeu Perestrelo, in the second official expedition to Madeira in 1420, is often seen as significant, as is the fact that he was sent back from there post-haste. Later, Perestrelo, who was of Italian origin, shifted allegiance, became a client of the Infante Dom Henrique and was then made captain-donatory of Porto Santo in the Madeiras; he is best known to posterity as the father-in-law of Christopher Columbus.

A possible explanation proposed by Serrão for the exclusion of the Order of Santiago is a nationalist one: he argues that the Order of Christ was seen as exclusively Portuguese in character, whereas the Santiaguistas still had the taint of an overly close association with Castile. The fact that the Master of Santiago, Dom João, was extremely closely allied with his brother, the celebrated Infante Dom Pedro (and is even seen by some historians as the latter's grey eminence) may not be irrelevant either; it is possible that a real struggle with political dimensions took place for control over the early Atlantic explorations between these Orders. It is also worth noting that the Order of Christ inherited certain duties and traditions from the Templars, and it was the Templars who in the twelfth and thirteenth centuries had been mainly concerned with corsair engagements against Muslim and other rival shipping in the Mediterranean. Whatever be the case, the appointment of the Infantes to the Masterships of the military orders did not resolve the issue of the rivalry either between the Orders or between the Infantes, and if anything may have accentuated certain elements thereof.

FROM DOM HENRIQUE TO DOM JOÃO

At the centre of the political web of overseas expansion from the mid-1430s, as is very well known, was the Infante Dom Henrique, who was active as a slave-trader, and managed sugar plantations in addition to his other varied ventures; it is not surprising that he is perhaps the most mythologised figure in the history of Portuguese expansion. Transformed by the pen of a number of foreign historians (and above all by the nineteenth-century Englishman R.H. Major) into Prince Henry 'the Navigator', it has been imagined that the Infante was a man of science and Renaissance culture, supporting a school of astronomers and navigators as he

Plate 6 The Infante D. Henrique and other figures from the Portuguese discoveries on a celebrated Salazarist monument, Belém (c. 1960).

paced in lonely splendour on a promontory at Sagres.[18] Major may have had his own reasons for trying to develop the already incipient personality-cult around Dom Henrique, who was after all of English descent through his mother, Philippa, herself of the house of Lancaster, and sister of Henry IV of England. Contem-

[18] Richard Henry Major, *The Life of Prince Henry of Portugal, Surnamed the Navigator, and its results: comprising the discovery, within one century, of half the world* (London: A. Asher, 1868). We may equally cite such writings as those of Charles Raymond Beazley, *Prince Henry the Navigator: The hero of Portugal and of modern discovery, 1394–1460 A.D.* (New York: G.P. Putnam, 1895), and the thoroughly romantic Elaine Sanceau, *Henry the Navigator* (New York: W.W. Norton, 1947). The first major work by a Portuguese author to use the term 'Navigator' is the posthumously published book of Joaquim Pedro Oliveira Martins, *The golden age of Prince Henry the Navigator*, tr. J.J. Abraham and W.E. Reynolds (London: Chapman and Hall, 1914).

porary records, alas, do not provide any evidence of such scientific activities as the eulogists of Dom Henrique would have us believe, or indeed even of the existence of a School of Sagres.[19] They suggest instead that Dom Henrique combined personal piety and celibacy (though he was not an ordained priest) with astute political calculation (evident in the civil wars of the 1440s), and that he was capable as well of ruthless decision-making – as we see from his refusal to give up Ceuta to ransom his brother Dom Fernando, Master of the Order of Avis, who was held as a hostage in Fez from 1437 and eventually died there. It is scarcely a coincidence that Dom Henrique was chosen as a role-model by the twentieth-century Portuguese dictator, António de Oliveira Salazar.

The three decades before the death of Dom Henrique in 1460 are principally associated with the exploration of the Atlantic Islands, and of the African west coast. This period culminates with the first indistinct Portuguese sightings of the Cape Verde Islands, and with the explorations by Pedro de Sintra in Sierra Leone, which are usually dated to 1460–1. The seven western islands of the Cape Verde archipelago were eventually discovered in 1462, and were assigned to Dom Henrique's designated heir and adopted son, the Infante Dom Fernando, brother of Dom Afonso V. Indeed, Dom Fernando came to control very substantial rights, and the key to this was the fact that he was named to succeed his uncle (and adopted father) in the administration of the Order of Christ. It is interesting to note that as early as March 1436, when aged forty-one, Dom Henrique had already named Dom Fernando (then aged three) as his heir, thus making it clear on the one hand that he had no ambitions of creating a lineage ('And as I neither have a son nor hope to have one, I take as my son and heir the Infante Dom Fernando, my nephew and godson'), while at the same time ensuring that the balance of power between older and younger sons (here, Dom Afonso and Dom Fernando) in the dispensation was held stable.[20]

[19] The discussion by Bailey W. Diffie of the Infante D. Henrique, in Bailey W. Diffie and George D. Winius, *Foundations of the Portuguese Empire 1415–1580* (Minneapolis: University of Minnesota Press, 1977), pp. 113–22, is a reasonable one, and follows in this respect the Portuguese historian Duarte Leite, *Descobrimentos portugueses*, 2 vols., ed. V.M. Godinho (Lisbon: Cosmos, 1958–60), Vol. I, *passim*. Also see Luís de Albuquerque, *Introdução à história dos descobrimentos portugueses*, 3rd edn (Lisbon: Publicações Europa-América, n.d.), who in turns refers to the earlier work of Luciano Pereira da Silva.

[20] António Joaquim Dias Dinis (ed.), *Monumenta Henricina*, Vol. V (Coimbra: CMIH, 1963), doc. 102, pp. 205–7.

Still, on the death of Dom Henrique in late 1460, a departure from the scheme that we have set out does appear briefly to have been meditated; an initial Papal Bull, *Dum Tua*, of January 1461, thus names the king Dom Afonso V himself as Master of the Order of Christ. However, this position was then renounced by the king, and Pope Pius II hence reassigned the administration of the Mastership for life to Dom Fernando on 11 July 1461, despite the fact that he was already the Master of Santiago, and had been so since 1444, on the basis of a concession by Pope Eugene IV. This earlier grant of the Mastership of Santiago had been made – interestingly enough – at the instance of the Infante Dom Pedro, often portrayed by historians as an unremitting centraliser. Obviously, there was a perceived difference between centralisation on behalf of royal power in *opposition* to the territorial nobility, and centralisation *within* the royal house itself. As the later case of the third Duke of Bragança demonstrated, this was partly an invidious and dangerous distinction. At any rate, by late 1461, Dom Fernando signed his grant-letters as follows: 'I, the Infante Dom Fernando, administrator and governor (*regedor e governador*) of the Orders of the Knighthood of the Masterships of Our Lord Jesus Christ and of Santiago, Duke of Viseu and of Beja, Seigneur of Covilhã and Moura, and Constable . . .'[21]

On the death of Dom Fernando in 1470, these titles were first inherited *in toto* by his oldest son, the infirm Dom João, but, after the latter's death in 1472, they came to be divided up. The Infante Dom Fernando's second son, Dom Diogo, thus emerged from 1472 as Duke of Viseu and Beja, and administrator of the Order of Christ, initially under the tutorship of his mother, Dona Beatriz; but the Mastership of Santiago was given, in a highly significant move, and already in 1472, to the Infante Dom João, the heir-apparent to the throne, who had married Dona Leonor, daughter of the deceased Dom Fernando, at Setúbal in 1471. The titles and privileges held by Dom Diogo eventually passed on after his death in September 1484 to his brother Dom Manuel, who thus came to accumulate extensive powers which followed logically on the tradition of the Infante Dom Henrique, his adoptive grandfather, and the Infante Dom Fernando, his father. By the late 1480s, Dom Manuel was Duke of Beja, administrator

[21] António Joaquim Dias Dinis (ed.), *Monumenta Henricina*, Vol. XIV, (Coimbra: CMIH, 1973), doc. 65, pp. 186–7. For the Papal Bull of 11 July 1461, see *ibid.*, doc. 57, pp. 158–62.

of the Order of Christ, and had extensive powers and revenues in the Atlantic Islands.

We should not fall into the error of assuming that the transition in 1460–1 from Dom Henrique to Dom Fernando had left the latter's powers over overseas expansion intact. The extant documents suggest at any rate that Dom Fernando's main projects were to consolidate (especially in the Atlantic Islands) rather than to push ahead. Besides, even if Dom Fernando controlled the prestigious Masterships, and enjoyed extensive prerogatives in the Azores, and the Madeira and Cape Verde archipelagos, the situation with respect to west African exploration remained distinctly problematic. In 1469, the five-year contract for trade and exploration on the Guinea coast signed between the Crown and Fernão Gomes of Lisbon does not appear to have involved Dom Fernando or the Order of Christ at all; then, in 1474, the heir-apparent, the Infante Dom João (by now Master of Santiago, and also of Avis) was given over the residual Crown rights on the trade of Arguim and Guiné, introducing a further complication into the picture. Further, this almost immediately inaugurated a period of war with the Spanish monarchs of Castile and Aragon, Isabel and Ferdinand, both on land and sea, which was partly a result of a succession dispute stemming from the marriage of Dom Afonso V with a Castilian princess, Juana (derisively called *la Beltraneja*, on account of doubts cast on her birth). At the end of this bitterly contested period, inaugurated with border raids in 1475, the two parties signed the treaty of Alcáçovas-Toledo (1479–80), which gave the Portuguese rights to the Azores, Madeiras and Cape Verdes, and to the future discoveries on the west African coast, while leaving Castile with the Canary Islands and some other ill-defined islands beyond the Canaries. These last clauses were the thin edge of the legal wedge, used by Castile to justify its claims in the next decade over the Caribbean, and then over the American mainland. In turn, the prolonged if uneven period of war is likely to have strengthened the Portuguese Crown's claims to direct overseas expansion, rather than to leave matters to the Order of Christ, or such para-state organisations as had earlier dominated.

On his accession to the throne in August 1481, Dom João II, *O príncipe perfeito* ('The Perfect Prince') to the chroniclers, had thus some strong arguments in favour of further centralising the overseas enterprise with the Crown. However, as expeditions proceeded rapidly down the African west coast, internal opposition to Dom João equally gathered momentum, and broke in wave after wave. The immediate causes for this cannot be found in the process of overseas expansion, even if some surprising connections between internal opposition and external exploration existed. On the face of it, the problem stemmed from the purely internal confrontation of Crown and nobility, on the issue of oaths of fealty, and on the right of the Crown to review earlier fiscal alienations, which had been made extensively during the reigns of Dom João I, Dom Duarte and above all Dom Afonso V. As has been argued at some length in recent years, the aftermath of the battle of Alfarrobeira (1448), and the death there of the Infante Dom Pedro, saw the emergence of an ever more powerful nobility in Portugal, whose support the king sought in order to prosecute his North African ventures. In the *cortes* of 1455, the king was overtly criticised for allowing a too-great influx into the nobility; and in the *cortes* of 1472, it was noted that a large proportion of royal powers had fallen into the hands of this nobility, and most especially the House of Bragança. Thus, writes the historian, Humberto Baquero Moreno, 'The reign of Dom. Afonso V is characterised by an extremely high point in the dominance of the nobility, and it is thus that one comprehends the true meaning of Alfarrobeira.'[22]

Even before his accession to the throne, Dom João II had shown that his conception of royal power was rather more forcefully centralised than that of his father. The most dramatic example occurs in 1478, towards the close of the Luso-Castilian war, when Lopo Vaz de Castelo-Branco, *alcaide* of the castle of Moura, and a powerful political figure, decided to go over to Castile. Dom Afonso V opened negotiations with him, and eventually a compromise was struck, but this was not to Dom João's liking. He hence had a group of cavaliers from Évora assassinate Lopo Vaz at his own stronghold of Moura, and also acted to confiscate the properties of another conspirator. In this sense, the confrontationist note was struck even before 1481.[23]

Opposition to Dom João II gathered very quickly around Dom Fernando, the third Duke of Bragança, descended from Dom

[22] Humberto Baquero Moreno, 'La noblesse portugaise pendant la règne d'Alphonse V', *Arquivos do Centro Cultural Português* 26 (1989), 399–415. These and related issues are discussed at greater length by the same author in an useful collection of essays, *Exilados, Marginais e Contestatários da sociedade portuguesa medieval: Estudos de história* (Lisbon: Editorial Presença, 1990).

[23] Humberto Baquero Moreno, 'La lutte de la noblesse portugaise contre la royauté à la fin du Moyen Age', *Arquivos do Centro Cultural Português* 26 (1989), 49–65.

Afonso, illegitimate son of Dom João I. As massive landowners, and possessors of a number of privileges, including the right to arm knights of the Order of Christ from their own resources and have them legitimised by the Papacy without reference to the Portuguese kings, the Dukes represented a voice in late fifteenth-century politics that was distinctly pro-Castilian in character, as well as diffident about overseas expansion. Nevertheless, the harshness of the conflict between the Duke and Dom João II, who was his brother-in-law, shook the Portuguese élite. Accusing the Duke of *lèse-majesté*, and then of invidious dealings with the Catholic Monarchs of Castile and Aragon, Dom João had him seized in May 1483, while his family – including his brother Dom Álvaro de Portugal, as well as his young son Dom Jaime – fled to Castile. After a trial, the Duke was publicly beheaded at Évora on 20 June 1484; it is reported by chroniclers that on hearing the sentence, the King wept but decided that the interests of the state had to prevail over his personal affections. The execution itself is described by Garcia de Resende, in his *Crónica del-Rei D. João II*, as having been carried out by an anonymous executioner, who cut off the Duke's head with a single stroke, while a public crier announced that such was the 'Justice that was ordered by the King Our Lord, and he orders beheaded Dom Fernando, Duke of Bragança, for having committed and contemplated treason, and loss, to his Kingdoms and to his royal Person'. The same chronicler writes that the king had ordered a church bell rung when the act was done; on hearing it, he is reported to have risen from his chair, and gone down on his knees and said, with his eyes full of tears: 'Let us pray for the soul of the Duke, which has now ceased to suffer.'[24]

These tears may have fostered some illusions about the firmness of the political will of the monarch, but they were soon dispelled. The next major political coup was the death of Dom Diogo, the Duke of Viseu, cousin of the king and also his brother-in-law, who was suspected of planning to avenge the Duke of Bragança; in this case, Dom João did not go through with a trial but instead quite simply stabbed Dom Diogo with his own hand in August 1484. Other important noblemen implicated in the conspiracy, such as Dom Guterre Coutinho, Dom Pedro de Ataíde and Dom Fernando de Meneses, were beheaded, while still others continued

[24] Garcia de Resende, *Crónica de Dom João II e miscelânea*, ed. Joaquim Veríssimo Serrão (Lisbon, 1973), Ch. xlvi, pp. 69–70.

to receive sentences into the next year. The Bishop of Évora, Dom Garcia de Meneses, was shut up in a cistern in the castle of Palmela, headquarters of the Order of Santiago, where he died of the rigours. Others, like Dom Álvaro de Portugal, preferred exile in Castile, and did not return to their native land in the king's lifetime.[25]

These drastic, even Machiavellian, measures, probably were intended to prepare the way for a wide-ranging set of institutional reforms on a number of fronts. A major preoccupation by the mid-1480s, when Diogo Cão's expeditions had advanced far down the African west coast, was to prepare the ground to enter the Indian Ocean, or rather to enter the trade to the vaguely defined Spice Islands. In this sense, the first and the second half of Dom João II's reign are a remarkable study in contrast. In the former, we see him master of the circumstances, imposing his centralising will with a stunning force and remarkable success; in the latter years, circumstances conspire to render many of his schemes invalid. The midway mark is defined by the expedition of Bartolomeu Dias in 1487–8 to the Cape of Good Hope.

As is well known, the voyage of Bartolomeu Dias forms part of a larger strategy, and indeed a larger conception on the part of Dom João II. In its beginnings, when the overseas enterprise was dominated by Infante Dom Henrique, the logic of the push down the African west coast was to judge the limits of the southward penetration of Islam and to see if allies could be found in Africa to attack the 'Moors' from the rear, as it were. This became inextricably mixed with a revived version of the late medieval legend of Prester John, who had been the object of widespread rumours from the 1140s. In 1145, the German historian Bishop Otto of Friesing wrote that he had heard of the Prester from the Bishop of Gabul, in Syria:

> He said that a few years ago a certain John, king and priest of the people living beyond the Persians and the Armenians in the extreme Orient, professing Christianity, though of the Nestorian persuasion, marched against the two Samiard brothers, kings of the Medes and the Persians, and conquered their capital, Ectabana

[25] For a discussion of Castilian–Portuguese relations in the period, see Luis Suárez Fernández, *Política Internacional de Isabel la Católica: Estudio y Documentos*, Vol. II (1482–88) (Valladolid: Universidad de Valladolid, 1966), pp. 61–9. The author argues it is unlikely that the Catholic Monarchs had to do with the conspiracies in Portugal, but notes the widespread condemnation of Dom João II's actions by Castilian court-chroniclers.

> . . . Victorious, the said John moved forward in order to come to the aid of the Holy Church.

It was noted by the chronicler that the freezing up of the Tigris prevented Prester John from carrying out his intentions; nevertheless, his intentions were pure and he was moreover 'descended from the Magi'.[26]

Geographically displaced from Central Asia, where the earlier texts place him, to Africa, Prester John was at least one of the objects of Dom João II's multi-pronged strategy of the 1480s. This displacement of the Prester should not be credited though to Dom João's time; rather it was an idea that preceded it, and must have been linked to rumours of a Christian kingdom which did in fact exist, of the Coptic persuasion, in Ethiopia, from which two embassies – in 1402 and 1408 – are reported in Venice. Indeed, in the 1450s, the Genoese merchant and explorer Antoniotto Usodimare, who was active on behalf of the Portuguese off the African west coast, speculated that he was no more than 300 leagues from the kingdom of Prester John.[27]

In view of this rather confused set of rumours, legends and speculations, Dom João II adopted the sensible course and hedged his bets. In early August 1487, Bartolomeu Dias, who was in all probability a professional mariner rather than a member of the nobility, was sent out to explore beyond what had been found thus far by Diogo Cão. The expedition was a low-key one, and little is known about it from contemporary evidence; it involved three vessels, *São Cristóvão*, commanded by Dias himself, *São Pantaleão* commanded by João Infante, and a third ship, the name of which is unknown. What is known for certain is that by early December 1487, Dias had passed the limits of Cão's explorations, and that either late that month or early the next month, he rounded the tip of the African continent, to put in at what is usually identified as Mossel Bay, in south-eastern Africa. Dias pushed on still further from this point, touching Cape Recife, and going as far as the *Rio do Infante* (Great Fish River). In mid-May, on his return, he passed and sighted the Cape of Good Hope, and then took a good seven months to sail home, arriving at Lisbon only in December 1488, where his discreet entry into the port was witnessed by, among others, Christopher Columbus.

[26] Cited in L.N. Gumilev, *Searches for an Imaginary Kingdom: The Legend of the Kingdom of Prester John*, tr. R.E.F. Smith (Cambridge: Cambridge University Press, 1987), pp. 4–5.

[27] Diffie and Winius, *Foundations*, pp. 36, 103.

Three months before Dias's departure, already in early May 1487, the other complementary part of the conception was put into place by Dom João II. This was to verify by the traditional, overland, route what the eastward ocean beyond Africa really held in store, and the men chosen to implement this were both junior members of the royal household, the picturesque Pêro da Covilhã, and the more obscure Afonso de Paiva. The two, both Arabic speakers to some degree, appear to have followed a common route via Spain and Italy to Alexandria, a major point of arrival for Asian spices into the Mediterranean. It was as if the spices were being traced to their source, as the two spies moved on from Alexandria to Cairo, and then made their way down the Red Sea to Aden. Here they parted ways in mid-1488, just as Bartolomeu Dias was readying himself to sail back from the Cape. Paiva reached Gujarat, but died on his way back from Hurmuz; Covilhã, on a more elaborate mission, made his way as far as Cannanur and Calicut, and may even have touched on Mozambique, before making his way back to Cairo from where he sent word of what he had found. Finally arriving in Ethiopia, possibly in about 1494, he was held back there and died considerably later in the fabled lands of the Prester.[28] It is reported by some sixteenth-century Portuguese chroniclers that Pêro da Covilhã sent back letters to the Portuguese court from Cairo through a certain José de Lamego, a Jewish shoemaker and voyager, though it is unclear whether these letters actually arrived in Portugal. At any rate, direct information of commercial conditions in the western Indian Ocean remained scarce in Portugal in the early 1490s, and a certain dependence on earlier accounts (notably that of the mid-fifteenth century traveller Niccolò de' Conti, as reworked by Poggio Bracciolini) may still be discerned. This may explain, in part at least, the absence of further expeditions into the Indian Ocean during the remaining years of Dom João II's reign.

THE POLITICAL CONTEST OF THE 1490S

But we need to advance further than this if we are to understand the precise context of Vasco da Gama's first expedition. In this venture, we are helped by the fact that in recent years there has been a quiet revolution in the interpretation of the Iberian roots of early Portuguese expansion in Asia. It has now become

[28] Conde de Ficalho, *Viagens de Pedro da Covilhan* (Lisbon: Imprensa Nacional, 1898).

increasingly clear that the early phase of the Portuguese presence in Asia cannot be understood save by seeing the country's élite as highly divided on the issue of overseas expansion by the late fifteenth century, in terms of both ideology and practice.[29] Following the researches of Jean Aubin, we can thus trace the origins of Vasco da Gama's selection to the last five years of Dom João II's reign, in particular the period from July 1491, when the monarch's son and heir-apparent Dom Afonso died as a result of a fall from a horse. Dom João was hence left with neither legitimate sons to inherit, nor daughters to marry, and the closest legitimate blood-relation left to him was his cousin, Dom Manuel, the Duke of Beja, also his brother-in-law.[30] Dom Manuel had managed to survive the purges of the 1480s, when – as we have noted above – his brother-in-law, the powerful Duke of Bragança had been executed, and Dom Diogo, the Duke of Viseu and Dom Manuel's own brother, killed by Dom João himself. Nevertheless, the ruler did not repose great confidence in Dom Manuel, nor indeed does he appear to have had a high opinion of his abilities. Dom João II moved instead to promote his illegitimate son, Dom Jorge, who had been born in August 1481 at Abrantes to the king's mistress, Dona Ana de Mendonça.

We may follow the account of the chronicler Rui de Pina in this respect, who writes:

> As soon as the Prince [Dom Afonso] died, the King supplicated Pope Innocent for the government and perpetual administration of the Masterships of Avis and Santiago for Senhor Dom Jorge his son; when the King was in Lisbon, the bulls for this arrived. And the *comendadores* and knights of the said orders pledged obedience to him in the Monastery of St Dominic on the 12th of April 1492, where a solemn Mass (*Missa d'Estado*) was heard that day. The King gave him as tutor (*ayo*) and governor of his household Dom Diego d'Almeyda, who a few days later, on the death of the Prior Dom Vasco d'Ataíde, was at once named Prior of Crato.[31]

[29] Jean Aubin, 'L'ambassade du Prêtre Jean à D. Manuel', *Mare Luso-Indicum* 3 (1976), 1–56, preceded by the same author's 'Duarte Galvão', *Arquivos do Centro Cultural Português* 9 (1975), 43–85. The theme was then taken up and developed over a decade later in Luís Filipe F.R. Thomaz, 'L'idée impériale manueline', in Jean Aubin (ed.), *La découverte, le Portugal et l'Europe* (Paris, 1990), pp. 35–103, and in L.F.F.R. Thomaz, 'Factions, Interests and Messianism: The Politics of Portuguese Expansion in the East, 1500–1521', *The Indian Economic and Social History Review* 28, 1 (1991).

[30] Jean Aubin, 'D. João II devant sa succession', *Arquivos do Centro Cultural Português* 27 (1991), 101–40.

[31] Rui de Pina, *Crónica d'El Rei Dom João II*, in M. Lopes de Almeida (ed.), *Crónicas de Rui de Pina* (Oporto: Lello e Irmão, 1977), pp. 991–2.

The hint that Dom Jorge was the new favourite was clear enough, and from that time on, the partisans of Dom Manuel (including his sister, Dona Leonor, the Queen) began to fear that the king would manipulate the succession. The nobility thus divided itself, loosely speaking, into groups on the basis of the succession – with Dom Jorge being supported above all by the Almeida clan. The Almeidas' fortune was centred in the initial phase around Dom Lopo de Almeida, the first Count of Abrantes, a close ally of Dom João II, in whose household Dom Jorge had been brought up. As Jean Aubin has shown through painstaking reconstruction, Dom Lopo's sons numbered at least six, of whom one, Dom Diogo Fernandes de Almeida, was a particularly close companion of Dom João II. In 1490, when Dom Jorge was first brought to court, it was as a protégé of the second Count of Abrantes, Dom João de Almeida; two years later, in 1492, when Dom Jorge was made Master of Santiago, his tutor was – we have seen – none other than Diogo Fernandes de Almeida. Again, in 1494, when Dom João II sent emissaries to the Papacy to ask that Dom Jorge be legitimised, those whom he sent were two of the sons of the first Count of Abrantes, namely Dom Pedro da Silva (who had taken his mother's surname) and Dom Fernando de Almeida. Still another son, Dom Jorge de Almeida was in the mid-1490s Bishop of Coimbra, in what appears to be a typical strategy of mixing warriors, diplomats and churchmen within a noble family. But it was the youngest son, Dom Francisco de Almeida, who had the most chequered and interesting career: he was implicated in the conspiracy against Dom João II in 1484 (the reason for the execution of the Duke of Bragança), but was let off lightly; later, he was associated with the equally ill-fated Duke of Viseu, and had to flee to Castile for safety. He participated there in the successful Crusading campaign against the Nasrids at Granada, but had returned to Portugal by 1492. Here, thanks to his family connections, he regained favour, and was in fact named in 1493 to lead an abortive Portuguese expedition to see what Columbus had actually discovered. A decade later, in 1505, he was to be named first viceroy of Portuguese Asia.

Now the Almeidas, and their extended clan (which included their nephews by marriage, Dom Álvaro de Castro and, still more crucially, Dom Diogo Lobo, the Baron of Alvito) were not merely resolute opponents of Dom Manuel before his succession; they remained so afterwards as well. As Luís Filipe Thomaz has argued, the Baron of Alvito was in the second decade of the sixteenth century the main adversary in Portugal of the designs of Dom Manuel and Afonso de Albuquerque.[32] The opposition was defined not merely in factional terms; it also had an ideological dimension as we shall see below.

Having come as far as the Cape of Good Hope, we can see in the last years of Dom João's reign a marked hesitation in respect of Asia. Some of this hesitation centred around information, and this explains the sending of agents overland to ascertain the political, religious and commercial situation on the shores of the Indian Ocean. Information aside, there was also a genuine diffidence about making use of the Cape route, partly because of the ambiguity concerning its economic advantages compared to the traditional overland routes via the Levant (with the perilousness and uncertainty of navigating the all-water route being a standard theme in the period), and partly in terms of the defence of the route when faced with Castilian competition. The explosive potential of the maritime rivalry between Portugal and Castile had already been shown in respect of the Guinea trade, and once the news of Columbus's discovery was brought back in April 1493 by one of his companions, Martín Alonso Pinzón, tensions between the Catholic monarchs and Dom João II mounted. Later, in the 1520s, the expectations of those who had foreseen difficulties in defending Asia from Castile were fulfilled, after Magellan's expedition caused Charles V to lay claims to the Moluccas. In this rivalry, too, Vasco da Gama was to play a minor role, as a resolute opponent to Castilian claims.

It should be stressed that building a consensus in Portuguese society on the issue of overseas expansion was no easy matter, as the frequent shifts in geographical emphasis over the course of the fifteenth century show.[33] The beginnings of expansion, with the capture of Ceuta in North Africa (1415), had a logic that was less commercial than military and religious. However, soon enough, the expansion into the Atlantic islands brought the commercial aspect to the fore; no 'Moors' (Muslims) were to be found in Madeira and the Azores, and the military nobility could find little to interest them there. To the nobility, with its largely land-based set of values, Castile represented a pole of attraction, and the Holy War even against the feeble remnants of Islam in Iberia was

[32] Thomaz, 'Factions, Interests and Messianism'.

[33] For an overview of the ideological and political problems of the epoch, see the fundamental work of Luís Filipe F.R. Thomaz, *De Ceuta a Timor* (Lisbon: Difel, 1994), bringing together many of the author's scattered essays on the question.

often more appealing than the waters of the Atlantic.[34] As long as overseas expansion to the west was under the charge of the Infante Dom Henrique, matters were sufficiently centralised and on a sufficiently small scale to contain discontent. In the second half of the fifteenth century, and especially after the death of Dom Henrique, a compromise of sorts had to be reached. When Dom Fernando was Master of the Order of Christ, Lisbon merchants, whether Portuguese Christians, Jews, or resident Italians, were given the main charge by the Crown of the Atlantic explorations, which soon extended down the west coast of Africa. A section of the nobility interested itself in these affairs, largely to the extent that they coincided with possibilities of corsair raiding (*guerra do corso*). For the most part, however, the nobility remained burdened with a mentality that was resolutely Iberian in its horizons, with North Africa entering into the ken because it was seen as a frontier of Iberian military expansion.

The structure had a simplicity which the realities of Asia did not permit; Asia thus was simply too complex to be accommodated within the existing regime of compromises. Many elements in the Portuguese nobility saw sense in the North African campaigns, which after all brought glory and at times fortunes. The Atlantic too seemed to make sense as a relatively low-cost, high-return affair, underwritten by Lisbon's cosmopolitan mercantile class. But Asia seemed at once too vast, too distant, and too risky a venture. This opposition, taken together with Dom João II's other preoccupations – with regard to Castile and the rest of his European policy, to his own succession, and with regard to Columbus's discovery – goes a long way towards explaining why, between 1487 and 1495, no further expeditions were sent out to the Cape, let alone beyond it.[35]

We see this all too well in the retrospective discussion by the greatest Portuguese ideologue of overseas expansion in the sixteenth century, João de Barros, of the decision to send an

[34] Luís Filipe Thomaz, 'Expansão portuguesa e expansão européia: Reflexões em torno da génese dos descobrimentos', *Studia* 47 (1989); Sanjay Subrahmanyam, *The Portuguese Empire in Asia, 1500–1700: A Political and Economic History* (London: Longman, 1993), pp. 30–7.

[35] For a longer view of the hostility of the territorial nobility to the Crown, see Baquero Moreno, 'La lutte de la noblesse portugaise'. For a view of the constraints faced by Dom João II, and his weakness with respect to Castile, see besides Aubin, 'D. João II devant sa succession', two articles: Jean Aubin, 'D. João II et Henry VII', and Luís Filipe Thomaz, 'O projecto imperial joanino: Tentativa de interpretação global da política ultramarina de D. João II', both published in *Congresso Internacional, Bartolomeu Dias e a sua época – Actas*, Vol. I (Oporto, 1989).

expedition to Asia. I quote the discussion in his first *Década* (published in 1552) at some length, since it throws invaluable light on the affair.

> The King Dom João having died without a legitimate son, who might succeed to the kingdom, there was raised as king (in accordance with what he had left in his testament) the Duke of Beja, Dom Manuel, his cousin, like a brother, son of the Prince Dom Fernando brother of the King Dom Afonso, to whom the royal inheritance was due by legitimate succession. Which he took possession of by its sceptre, given to him in Alcácer do Sal on 27th of October of the Year of Our Redemption of 1495, being twenty-six years, four months, and twenty-five days of age.

Barros now continues, getting to the crux of the matter:

> And since, with these Kingdoms and Seignories, he also inherited the furtherance of so high an enterprise as his predecessors had undertaken, which was the discovery of the Orient by this our Ocean-Sea, which had cost such industry, such work and expense over seventy-five years; he wished at once in the first year of his reign to show how much he wanted to add to the Crown of this kingdom new titles in addition to the Seignory of Guinea which had been assumed by his cousin the King Dom João on account of its discovery, and in the hope of other larger states that were to be discovered by that route. For which, in the following year of [14]96, being at Montemor-o-Nôvo, he held some general councils, in which there were many and different votes, and the majority was that India should not be discovered (*e os mais foram que a Índia não se devia descubrir*) . . .

Nothing could state the opposition more clearly, though Barros is at some pains – naturally enough – to state the roots of the opposition in somewhat guarded language.

> [B]esides bringing with it many obligations as it [India] was a state that was rather remote for it to be conquered and kept, it would so weaken the forces in Portugal (*o Reyno*) that it would lack the ones necessary for its conservation. Even more so, for once it was discovered, this Kingdom would face new competitors, which had already been experienced in what transpired between the King Dom João and the King Dom Fernando of Castile over the discovery of the Antilles: it reaching such a point that they came to divide the World into two equal portions for it to be discovered and conquered.

The reference is to the Treaty of Tordesillas (signed in June

Plate 7 Distant view of the castle of Alcácer-do-Sal, in the Alentejo, of the Order of Santiago, where D. Manuel was crowned (1495).

1494); and the implication is that if nations came to such a pass over something so minor, how much more grave the matter would be if the prize were a truly rich one – like the India trade! But Barros now disingenuously skirts the edge of matter in arriving at his conclusion.

> However, as against these reasons, there were other contrary ones, which as they were in conformity with the desire of the King, they were more acceptable. And the principal ones that moved him were that he had inherited this obligation with the inheritance of the kingdom, and the Prince Dom Fernando his father had worked towards this discovery when by his orders the islands of Cape Verde were discovered; and even more so for the singular affection he had for the memory of his uncle the Prince Dom Henrique, who was the author of the new title of the Seignory of Guinea which this kingdom had, it being a most profitable acquisition made without the cost of arms, and other expenses which even smaller states had cost. And in the end, giving as reason, to those who pointed out the inconveniences if

> India were to be discovered, that God, in whose hands he put the affair, would provide the means that were needed for the welfare of this Kingdom [Portugal], finally the King decided to proceed with this discovery, and later while at Estremoz, named Vasco da Gama, *fidalgo* of his household, as Captain-Major of the sails that he was to send there . . .[36]

The implications are clear: Dom Manuel encountered great opposition to the idea of the expedition to Asia (an opposition which clearly already existed in the last years of Dom João II's reign), which he himself had somewhat autocratically to suppress. What motivated him, among other things, was a factor that Barros does not mention: namely Dom Manuel's particular Middle Eastern strategy. Dom Manuel, we now know, had over a period of time developed a rather strong Messianic streak, which made the capture of Jerusalem a particularly important objective in his policy decisions. This was at one level a particular manifestation of the religious dimension of Portuguese expansion, the residual momentum left by the uneasy cohabitation with Islam in Iberia, and the *reconquista*, a dimension about which robustly materialist historians such as Vitorino Magalhães Godinho have expressed considerable scepticism. But such a round dismissal does scant justice to the sources of the period, and support for a partly religious (and 'idealist') interpretation can be found in the writings of even João de Barros, who though asked by the court to write 'of the deeds that the Portuguese did in the discovery and conquest of the seas and lands of the Orient', thought fit to begin his great chronicle *Da Ásia* with the rise 'in the land of Arabia [of] that great anti-Christ Muhammad'.[37]

For Barros and for a number of his contemporaries and predecessors like Duarte Galvão, the birth and spread of Islam provided the logical point of departure for an understanding of how the Portuguese came to be in Asia; despite the fact that historically Christianity preceded Islam, the Iberian version of Christianity which they understood was reactive, even as the medieval Iberian version of the Apostle St James could be portrayed killing Moors. It is notable that the first *Década da*

[36] João de Barros, *Da Ásia*. Décadas I–IV (Lisbon: Livraria Sam Carlos, 1973), facsimile of the 1777–8 Régia Oficina Tipográfica edition, Década I/1, Book IV, Ch. 1; also see the comments in João Paulo Costa and Vítor Gaspar Rodrigues, *El Proyecto Indiano del Rey Don Juan* (Madrid: Mapfre, 1992).

[37] Barros, *Da Ásia*, Década I/1, pp. 1–2.

Ásia commences with a discussion of the Muslim conquest of Iberia, the Christian reconquest, and only then moves on to the Atlantic explorations and the charting of the west coast of Africa, eventually arriving at Vasco da Gama's expedition to the Indian Ocean in the fourth book of this work. If the Messianist streak is more evident in the writings of some other courtiers of Dom Manuel than those of Barros, who wrote after all in the reign of Dom Manuel's successor, Dom João III, we can nevertheless catch an echo of it in his writings.

It is only natural that Portuguese Messianism has been thought of by modern-day historians as a phenomenon to be associated more or less exclusively with Dom Manuel's great-grandson, Dom Sebastião, the focal point of a highly celebrated Messianic cult after his death in North Africa in 1578 while fighting the 'Moors'.[38] The Sebastianist cult found much later echoes in even nineteenth-century Brazil (the Canudos rebellion, for example) and twentieth-century Portugal, influencing such writers and ideologues as Fernando Pessoa. Alternatively, some recent writers have stressed the Messianism inherent in the claims of Dom Manuel's great-grandfather, Dom João I, on assuming the throne, and especially in the later chronicle of Fernão Gomes which justifies his claims.[39] But this should not blind us to the existence of a Messianism of a somewhat different sort in the court of Dom Manuel, sandwiched squarely between these two better-known cases. The proximate psychological reason for this Messianism was the monarch's rather improbable accession to the throne, after what was for him the successive deaths of a large number of persons who preceded him in terms of succession; the idea of impending contact with 'lost Christians' at the other end of the world also played no mean role. Furthermore, in his early life, Dom Manuel's Franciscan tutors almost certainly seem to have communicated to him the very influential religious ideas derivative from the celebrated Joachim of Flora, whose works were very much in the air in the late fifteenth century. As is well known, Joachite philosophical and eschatological speculation centred around some key ideas, such

[38] It is my impression that this issue is insufficiently developed even in the relatively sophisticated analysis of Sebastianism by Lucette Valensi, *Fables de la Mémoire: La glorieuse bataille des trois rois* (Paris: Editions du Seuil, 1992).

[39] Luís de Sousa Rebelo, 'Millénarisme et historiographie dans les chroniques de Fernão Lopes', *Arquivos do Centro Cultural Português* 26 (1989), 97–120. Also the more wide-ranging study by Margarida Garcez Ventura, *O Messias de Lisboa: Um Estudo de Mitologia Política (1383–1415)* (Lisbon: Edições Cosmos, 1992).

as the literalist belief that one had to wrestle with the letter of the scriptures to get at the spirit, a trinitarian approach to history, and an apocalyptic vision which made use of such central texts as the Book of Daniel.[40] Portuguese royal millenarianism of the sixteenth century shared certain common traits and themes with other millenarian movements of the same period, in both the Christian and the Islamic worlds, from Istanbul to India. The Book of Daniel, the interpretation of Nebuchadnezzar's dream, and the notion of the four empires (culminating in the Roman), with the fifth millennial empire to be awaited, was certainly a staple of a certain set of late fifteenth-century Iberian theologians, in particular those who were highly concerned with the fall of Constantinople to the Ottoman ruler Sultan Mehmed in 1453. Taken together with Jewish millenarian expectations in the same area, they combined to produce what was at times a potent and heady mixture, as we see from the writings of a fifteenth-century Spanish Franciscan, Alonso de Espina (himself said to be a *converso*), who claimed that Jews were waiting in the Carpathians, between the palaces of Gog and Magog, in expectation of the Anti-Christ. The reading here was less from Ezekiel, 38–39, where Gog was the ruler of the Kingdom of Magog in the 'extreme north', and rather more from the Apocalypse, 20, 8, where Gog and Magog are the turbulent nations at the end of time. Millenarianism and anti-Semitism thus could be combined to form a political programme; Iberian Jewish eschatology for its part read the fall of Constantinople quite differently, seeing in it a hopeful sign.[41]

These ideas, which Dom Manuel probably held already before 1495, were further fostered by some of those in his council after his accession to the throne including his wives (two of whom were, significantly, the daughters of the Catholic monarchs). They permitted Dom Manuel at times to act in a highly autocratic fashion (as Barros indirectly notes), since he believed that he was

[40] Thomaz, 'L'idée impériale manueline'; for a useful summing up of recent research, Delno C. West and Sandra Zimdars-Swartz, *Joachim of Fiore: A study in spiritual perception and history* (Bloomington: Indiana University Press, 1983). Interestingly, it has been pointed out that Columbus too was influenced by Joachite thinking; cf. Abbas Hamdani, 'Columbus and the Recovery of Jerusalem', *Journal of the American Oriental Society* 99, 1 (1979), 39–48; also John L. Phelan, *The millennial kingdom of the Franciscans in the New World* (Berkeley: University of California Press, 1970), pp. 135–6.

[41] Jacqueline Genot-Bismuth, 'Le Mythe de l'Orient dans l'eschatologie des juifs de l'Espagne à l'époque des conversions forcées et de l'expulsion', *Annales ESC* 45, 4 (1990), 819–38.

directly inspired by the Holy Spirit, who had chosen him – humble and insignificant – to confound the mighty and powerful. Further, the Joachite influence on the ruler also brought with it the effect we have noted above, namely a preoccupation with the recapture of Jerusalem, a preoccupation Dom Manuel may have shared with other contemporaries like Columbus. It has been argued that the conquest of the Holy Land, and the destruction of the pilgrimage centres of the Hijaz (notably Mecca), came to be seen as the potential climax of overseas expansion, indeed the crowning achievement that would enable Dom Manuel to claim the title of Emperor of the East.

Enthusiasm for this project, which was not given great publicity in any event, was not widely shared in Portuguese court circles. By all accounts, the Jerusalem enterprise was one that was peculiarly associated with Dom Manuel, for neither Dom João II nor Dom João III appear to have partaken of it. But the means to be used preserved a certain continuity with earlier policy. Up until 1516–17, when the Ottomans took Egypt, Dom Manuel's plan was to launch a two-pronged assault on the Mamluk Sultanate of Egypt, which writers like Duarte Galvão identified with the Biblical Babylon. This would mean, on the one hand, reactivating a major offensive on the North African front, and hence a diversion of resources from mercantile activity. On the other hand, the Asian strategy was a combination of the military and the commerical, a blockade that would damage the commercial economies of Egypt and Venice, and at the same time drastically reduce the customs-revenues available to the Sultans from the spice trade.

It was clear that the Portuguese could not engage in this acitivity alone, but European allies were seen as neither desirable (for it would mean sharing the glory in the *Respublica Christiana*) nor available. Rather, given the view that a lost body of Christians existed elsewhere, the support of other Asian or African powers was deemed more sound. Dom Manuel and his coterie clearly believed the number of Christian kingdoms in Asia to be far larger than was in fact the case; and, eventually, this led them to develop a strategy centred around an alliance with Ethiopia, which had already been an object of much curiosity from earlier times, in its guise as the state ruled over by the fabled Prester John. It was this alliance that was deeply opposed by other parties, both in Portugal and in Portuguese Asia, during the reign of Dom Manuel.

Gesta proxime per Portuga
lẽses ĩ India. Ethiopia. ⁊ alijs
oriẽtalibus terris.

Plate 8 Title page of *Gesta Proxime per Portugalenses in India, Ethiopia et aliis orientalibus terris*, a text of Manueline propaganda.

THE GAMA HERITAGE

What was the role of Vasco da Gama then in all this? We may note that practically nothing is known of the first quarter-century of his life, and that the idea which is sometimes defended that he was born in 1460 is based on no more than vague inference, with many authors (including compilers of standard genealogies of the nobility) proposing a more likely date as late as 1469. The late nineteenth-century researches of A.C. Teixeira de Aragão and J.I. de Brito Rebelo have served at least to distinguish him from some *doppelgängers*, contemporary petty noblemen with the same name.[42] One of these, a squire (*escudeiro*) to the kings Dom Duarte and Dom Afonso V, lived in Elvas, and had property in Olivença, but died in 1496. At least two others of the same name can be found in Olivença in the 1480s, besides one other in Elvas and one in Évora.[43]

Our Vasco da Gama, it has been established more or less conclusively, was the grandson of another Vasco da Gama, who lived in Olivença and was married to a certain Dona Teresa da Silva. From this union were born four children, of whom the oldest Estêvão da Gama was a *cavaleiro* in the household of the powerful Duke of Viseu, Dom Fernando, in the 1460s, and was rewarded with the post of *alcaide-mor* and captain of Sines, as well as a small revenue from taxes on soap-making (*saboarias*) in Estremoz, Souzel and Fronteira. Very much a member of the *noblesse de service* then, Estêvão da Gama accumulated a series of rights in Sines and its surroundings (notably Colos), where he seems by the 1460s to have been based: in 1478, for example, he was substituted in the post of *alcaide-mor* of Sines but was instead granted the right to collect taxes there on octroi and registration, by virtue of services rendered to the Order of Santiago, and in the next year was granted the revenues due from two Jewish residents of the settlement of Santiago de Cacém. He also held the *comenda* in the late 1470s of Cerqual for the Order of Santiago.[44]

Estêvão da Gama thus fits in well with our general profile of the activities of the Order of Santiago in the area ranging south from the Setúbal peninsula to the Algarve. Our knowledge of the economic history of the region in the late fifteenth and early sixteenth centuries is admittedly uneven at present.[45] An examination of a corpus of eighty documents of the Order, dating from

[42] More recently, also see Humberto Baquero Moreno, 'Vasco da Gama, alcaide das sacas de Olivença', in *Encontros: Revista hispano-portuguesa de investigadores en Ciencias Humanas y Sociales* 1 (1989).

[43] Teixeira de Aragão, *Vasco da Gama e a Vidigueira*, pp. 3–5.

[44] Documents published in J.I. de Brito Rebello, 'Navegadores e exploradores portuguezes até o XVI século (Documentos para a sua história): Vasco da Gama, sua família, suas viagens, seus companheiros', *Revista de Educação e Ensino* (Lisbon) 13 (1898), 49–70, 124–36, 145–67, 217–30, 274–85, 296–313, 366–70, 473–5, 508–22; 14 (1899), 560–5; 15 (1900), 28–32, 90–2, as Docs. 5 and 6, from a total of 93.

[45] But see, for the Algarve, Joaquim Antero Romero Magalhães, *Para o estudo do Algarve económico durante o século XVI* (Lisbon: Edições Cosmos, 1970). Unfortunately from our point of view, Romero Magalhães provides little information on the economic role of the military orders in the area.

1482 to 1523, and concerned with property transactions, reveals a picture of extensive penetration by the Santiaguistas into rural life.[46] The Order of Santiago in this area appears not only as a proprietor, but very like a form of state itself. It appears that many of these quasi-state powers date to the time of the king Dom Fernando, who is known to have issued an *alvará*, granting the Master of Santiago 'jurisdiction over the criminal and the civil' in Setúbal, Sesimbra, Palmela and Alcácer.[47] Several of the documents examined are land-settlement papers, or *aforamentos* and *cartas de sesmaria*, issued by men like Dom Pedro de Noronha, *mordomo-mor* of the king Dom João II, *comendador-mor* of the Order of Santiago in 1482, who by virtue of this post in the Order held the *comenda* of the *vilas* of Canha and Cabrela, the first worth 150,000 *reais* per annum and the second 130,000 *reais*.[48] Some concern olive-groves, vineyards and houses, others are sale-deeds (*cartas de venda*), where the Order of Santiago certified and registered the transaction, at times before a notary (*tabelião*), collecting a fee for the service. Besides the *comendadores* who appear frequently in the documents (for example, Jorge de Sousa, *cavaleiro* of the Order of Santiago, who appears in papers from 1493 to 1495), we also see the presence of a number of other officials of the order, ranging from its inspectors sent out from Palmela to the outlying regions, the *almoxarife* of the order, to agents of the *comendadores* (who could be termed *feitor e procurador*, that is 'factor and agent').[49]

It would appear that by the early 1480s, Estêvão da Gama had begun to rise somewhat in the Order's complex hierarchy, without ever attaining exalted heights. Thus a document, probably dating to 1481, lists those present at a formal meeting of the Order in Santarém. At the head of the list are the Dom Prior and the Comendador-Mor, followed by the Thirteen (seven seated on the right, and six on the left). Eleventh on the left side is Estêvão da Gama, who follows João Afonso de Aguiar, Pero Jaques and

[46] AN/TT, Corporações Religiosas [Old Colecção Especial 142/1356], Ordem de Santiago, Maços 4 and 5, each containing 40 documents. Those of Maço 4 run from 1482–1505, those of Maço 5 from 1505–23.

[47] AN/TT, Corporações Religiosas, Ordem de Santiago, Maço 5, Doc. 6, on parchment, dated 1 March 1508. The document is a copy (*treslado*) of an *alvará* of Dom Fernando, made at the request of Dom Jorge.

[48] AN/TT, Corporações Religiosas, Ordem de Santiago, Maço 4, Doc. 1, *carta de sesmaria* dated 30 April 1482.

[49] AN/TT, Corporações Religiosas, Ordem de Santiago, Maço 4, Doc. 11, *carta de novação* dated 28 November 1491, involving Francisco de Miranda *comendador* of Elvas, his *feitor* Nuno Vaz, and Mose Paby, a Jewish resident.

Afonso de Carvalho. The same document lists among those who received the Habit of the Order at the meeting, in the sixth place, Vasco da Gama.[50]

The father Gama's marriage with Dona Isabel Sodré, daughter of a certain João Sodré (also known as João de Resende), produced a good number of children. The family's profile in 1480 can be reconstructed through the papers of the Bishop of Safim, who conducted a visit to Sines in November that year, and oversaw the ceremony of first tonsure (*prima tonsura*) in the Gama household. Among those listed as having undergone tonsure are, probably in order of age, Paulo da Gama, another son called João Sodré (who had evidently taken his mother's name), Vasco da Gama and Pedro da Gama. No mention is made of Aires da Gama, who served two stints in India in the 1510s, and was the inheritor of his father's fiscal accumulations; it may be that he was too young to have undergone tonsure in 1480. There is also naturally no mention of the daughters, in view of the character of the document; but we do know however that Vasco da Gama had at least one younger sister, Teresa da Gama. The real surprise in the document is the emergence of yet another *doppelgänger*: this is a *second* Vasco da Gama, who is listed as having been born of Estêvão da Gama when he was still single, and of a single mother (*de soluto genitus et soluta*).[51] Thus, Vasco da Gama had an illegitimate half-brother with exactly the same name as he! Finally, mention is made of the first tonsure in 1480, of Vasco da Gama's maternal uncle, Vicente Sodré, listed as resident in Sines, albeit a native of the parish of Madalena in Lisbon. Vicente Sodré, it is noted in the document, was the son of João Sodré and Isabel Serrão; he was to play a role of some significance in the middle of Vasco da Gama's career, on his second expedition to India in 1502–3. The Sodré family, we may note in passing, descended from a certain Frederick Sudley, of Gloucestershire, who had accompanied the Earl of Cambridge to Portugal in 1381–2, to fight for the House of Avis against the Castilians, and had thereafter settled down there. It was not only the Infante Dom Henrique then who had an English connection.

Teixeira de Aragão has argued – partly from oral tradition – that

[50] AN/TT, Ordem de Santiago, B-51-135, fos. 178–9, published in Brito Rebello, 'Navegadores e exploradores portuguezes', Doc. 41.

[51] Isaías da Rosa Pereira, *Matrícula de Ordens da diocese de Évora (1480–1483): Qual dos dois Vasco da Gama foi à Índia em 1497?* (Lisbon: Academia Portuguesa da História, 1990), pp. 140–1. Also see the editor's discussion in the Introduction, pp. 23–7.

Vasco da Gama was born in Sines itself, in a house near the church of Nossa Senhora das Salas, but that he received the greater part of his education in Évora. The former statement is probable, the latter much less certain. It has even been claimed that he had a good grasp of astronomy, partaking of the wisdom of no less a master than the celebrated Jewish savant, Abraham Zacuto. Such myths, though created already by near-contemporaries like Gaspar Correia, must be treated with caution. Nor is it at all clear, as claimed by Teixeira de Aragão, that from 1478 on Vasco da Gama was engaged in numerous services on behalf of Dom João II. If he was young enough to have his first tonsure in 1480, this seems very unlikely by all accounts. Indeed, the only explicit evidence of this which Teixeira de Aragão presents is a Spanish document from November 1478, issued by Isabel, Queen of Castile, which is a safe-conduct allowing a *certain* Vasco da Gama and one Fernando de Lemos to pass to Tangiers via Spain, despite the prevailing hostilities between Portugal and Castile.[52] Since, as we have seen, persons with the same name abound, we can by no means be certain that this Vasco da Gama who went to Tangiers was the son of Estêvão da Gama. And curiously, much hangs on this question, for the rejection of the date 1469 for his birth by some historians revolves precisely on admitting that he was engaged in royal service in 1478!

Whatever be the case, the question of where and how Vasco da Gama engaged himself in the latter half of the 1480s, when I would suggest he was still only in his late teens, remains unresolved. It is possible, though once more by no means certain, that he served in campaigns in North Africa in the latter part of this decade, since this was a pattern common enough among his contemporaries, and especially younger sons of the petty nobility. Against this hypothesis is the fact that the diligent researches of scholars in the documents pertaining to Morocco have thrown up no concrete evidence thus far. Given his family's attachment to the Order of Santiago, it is equally possible that Vasco da Gama's main sphere of activity in these years was the Alentejo itself, where the Order had a great deal with which to occupy itself.

The first clear reference to him occurs quite late, in 1492, when by our calculation he should have been roughly twenty-three years of age. A gold-laden Portuguese caravel on its return from

[52] Archivo General de Simancas, Registo General del Sello, Ano 1478, Mes de Noviembre, in Brito Rebello, 'Navegadores e exploradores', Doc. 23.

São Jorge da Mina in west Africa was captured by the French, and Dom João II retaliated by having French vessels captured at Setúbal, the ports of the Algarve, Aveiro and Oporto. The charge of the expedition to Setúbal and the Algarve was given to Vasco da Gama, who carried out his instructions successfully. Our source for this is Garcia de Resende, who in his *Crónica de Dom João II* thus writes:

> [Dom João] at once sent in a great hurry, with great provisions and powers to Setúbal and the Kingdom of the Algarve Vasco da Gama, *fidalgo* of his household, who was later to be Count of Vidigueira and Admiral of the Indies, a man whom he trusted and who had served in armadas and affairs of the sea, to do the same [capture] to those [ships] which might be there, which he did with great brevity.[53]

Whether Resende, writing with the benefit of hindsight in the early 1530s, attributed a closer link between Gama and Dom João II, as also greater experience in maritime affairs to the future Admiral than was in fact the case, may be debated. It is interesting to note, nevertheless, that Vasco da Gama does appear here as a trusted agent of Dom João II, as a *fidalgo* of the royal household, and as someone with a certain maritime experience. If we stay close to the letter of the statement, we cannot argue – as Armando Cortesão has done – that Vasco da Gama had already *commanded* numerous long-distance maritime ventures on which no other documents exist; this is to abuse Resende's text. If we do take Resende at his word, we may surmise that Gama had served earlier on fleets sent to the Atlantic islands, or to Guinea, or even to Flanders, where the Portuguese king maintained an active commercial interest in these years. Another document from the same year, 1492, is however noteworthy – for it also relates to Setúbal, where we know Vasco da Gama was in that year. Issued in the name of Dom João II, from Lisbon on 22 December 1492, it relates how a certain Diogo Vaz, an *escudeiro* (squire) of the royal household and resident of Setúbal, and Vasco da Gama were walking one night on the streets of the town, when they ran into the *alcaide* of the place João Carvalho, and were challenged by him, as Vasco da Gama was concealed by a cape and hence mistaken for an evildoer (*malfeitor*). An altercation ensued, and various other officials came to the aid of the *alcaide*, who subsequently lodged a complaint against Vaz and Gama. In his

[53] Garcia de Resende, *Crónica de Dom João II*, Ch. cxlvi, p. 213.

letter of December 1492, Dom João, 'wishing to do him grace and mercy', pardoned Diogo Vaz for his violent resistance, made him pay a small fine of 1,000 *reais*, and also revoked an order for his arrest.[54] The small incident underlines some of what we later come to know of Gama's psychological makeup; violent of temper, and quick to react to perceived insults, he seems scarcely the ideal man to send to make contact with new continents and cultures. The same incident was, it is amusing to note, characteristically exaggerated and distorted by Gaspar Correia subsequently: in Correia's version, the attack was on a judge in Setúbal who was gravely wounded (perhaps even killed), and carried out not by Vasco but his elder brother Paulo da Gama. He even claims that on the eve of his departure for India, Vasco da Gama asked for and received from Dom Manuel a pardon for his brother.[55]

No more is heard of Vasco da Gama for a few years, but he then surfaces in 1495 in quite significant circumstances, immediately after the death of Dom João II. We have already noted the succession struggle that ensued after the death of the heir-apparent Dom Afonso in 1491, with Dom João II trying his utmost to promote the candidature of his illegitimate son, Dom Jorge. We have noted too that his main supporters in this enterprise were the Almeidas, who, we shall see below, were linked to Vasco da Gama in a number of ways. The Queen, Dona Leonor, supported her own brother Dom Manuel, who was clearly the legitimate successor, once Dom João II's attempt to have the Pope legitimise Dom Jorge had failed in 1494. By the time of the King's death, then, it was clear that Dom Jorge was not a viable candidate. The succession, formally opened by Rui de Pina himself, is described by him thus:

> It was found that the King had declared the Duke of Beja as the one who by right was his legitimate heir, and successor to his Kingdoms, and he commended to him with words of great love, and of greater obligation, his son Senhor Dom Jorge, whom he also left with the Dukedom of Coimbra, and the Lordship of Montemor-o-Velho with all the *vilas* and lands that had been held by the Infante Dom Pedro, his great-grandfather, and he also commended the Duke [Dom Manuel] that he should give him

[54] AN/TT, Chancelaria Dom João II, Livro 7, fo. 141, reproduced in Teixeira de Aragão, *Vasco da Gama e a Vidigueira*, doc. 6, pp. 215–6, and in Brito Rebello, 'Navegadores e exploradores', Doc. 8.

[55] Gaspar Correia, *Lendas da Índia*, ed. M. Lopes de Almeida, 4 vols. (Oporto, 1975), Vol. I, Ch. 5, p. 13.

> [Dom Jorge] all the things that he had possessed as Duke, in which he included the Mastership of the Order of Christ, and the island of Madeira. And though the King Dom Manuel gave him the title of Duke with many of these things during his reign, he excused himself from some of them, and it is believed that it was not for lack of love, and good will that he had for him, but because of the [financial] straits of the Kingdom, and the great needs of the Royal Crown, and the expectation that if he had sons they might thus ask it of him.[56]

Dom João II thus wished to leave Dom Jorge with enormous powers, as Duke of Coimbra and Master of all three military orders, which would surely have placed him in a position of great political significance. The comparison chosen by the king himself, with the Infante Dom Pedro, though based in part on the fact that the Infante had been Duke of Coimbra, must have sent out curious echoes, for it suggested that Dom Jorge was intended to play the role of a rival (and potentially destabilising) centre of power to the Crown. This may appear to square rather ill with what we know of Dom João II during his lifetime: why would this king, usually seen as an inveterate and rather ruthless centraliser, indeed the first absolutist monarch of Portugal, wish to leave behind such a divided legacy? It would seem that the apparent paradox can be resolved if one is willing to consider the matter from a late fifteenth-century perspective, sweeping away certain obvious anachronisms.[57]

In the first place, while some version of the medieval European notion of the king's two bodies (as set out by E.H. Kantorowicz and others) may be found in Portugal, it is not obvious that this captures anything more than some very broad ideas of institutional continuity.[58] The practice of centralised rule in the time of Dom João II was too recent (and too fragile) perhaps for it to have a specific and well-defined 'political theology', and thus

[56] Lopes de Almeida (ed.), *Crónicas de Rui de Pina*, pp. 1032–3.

[57] There is a brief discussion of the succession after Dom João II, in Martim de Albuquerque, *O poder político no renascimento português* (Lisbon: Instituto Superior de Ciências Sociais e Política Ultramarina, 1968), pp. 305–8, but it appears somewhat unsatisfactory for our purposes. It does however contain some illuminating anecdotes from Garcia de Resende. For another discussion, see Manuela Mendonça, *D. João II: Um percurso humano e político nas origens da modernidade em Portugal* (Lisbon: Editorial Estampa, 1991), pp. 449–70.

[58] Ernst H. Kantorowicz, *The King's Two Bodies: A Study in Mediaeval Political Theology* (Princeton: Princeton University Press, 1958); also the interesting discussion of sixteenth- and seventeenth-century France in Arlette Jouanna, *Le devoir de revolte: La noblesse française et la gestation de l'état moderne, 1559–1661* (Paris: Librairie Arthème Fayard, 1989), 282–90.

researchers who approach the issue from a primarily juridical viewpoint are unlikely to advance very far. It appears from an examination of Dom João II's actions and testament that his conception of power was not merely centralised but specifically *personal*, and thus not generalisable into a notion of 'kingship' in the abstract as centralised. Thus, the fact that he himself had centralised power and exercised it as such did not mean that Dom João considered such a political dispensation appropriate *in general*; certainly, he must have regarded it as rather inappropriate to concentrate power to such an extent in the hands of someone whom he thought as incompetent as he did Dom Manuel. The solution then was to divide power between Dom Jorge and his supporters the Almeidas on the one hand, and Dom Manuel on the other, to institutionalise and inscribe in stone, as it were, a state of endemic factional struggle.

We may return, after these reflections, to Vasco da Gama, and to a consideration of his place in the factional politics of the early Manueline court. In this context, it is interesting to note that the first documents that refer to him after the accession of Dom Manuel are two grant-letters to him by Dom Jorge, and dating some months after the death of Dom João II, namely to 17 and 18 December 1495. These letters have been published on a number of occasions, by Brito Rebelo, Teixeira de Aragão and most recently by Armando Cortesão; they grant Vasco da Gama, who is termed a '*fidalgo* of the household of the King my lord and a knight of the Order of Santiago', the *comendas* of Mouguelas and Chouparria, with all their benefits, rents, tithes, rights and tributes, a privilege which we know was worth 80,000 *reis* (60,000 from Mouguelas and 20,000 from Chouparria) in the early sixteenth century. The *comendas* were strategically located in the Setúbal area, very close to the seat of the Order of Santiago in Palmela. In the grant-letters, there is mention of services rendered by Vasco da Gama in earlier years, which Cortesão interpreted as further evidence of mysterious expeditions around the Cape of Good Hope, 'a voyage of discovery and exploration beyond the Rio do Infante to the east African coast, with caravels, about which no information has survived'. The precise phrase from which he drew such inferences is in both these letters, which mention 'the many services that Vasco da Gama *fidalgo* of the household of the King my lord . . . has rendered and I hope will render in future to the King my Lord and father whose soul is with God, and to the King my Lord, and to me'.[59] From our perspective, we may see these letters quite differently. They suggest that Vasco da Gama, as a member of the Order of Santiago, continued to be attached to Dom Jorge, and thus – in view of our earlier discussion – is very likely to have been a member of the loose grouping against Dom Manuel that was organised around the Almeidas and the Baron of Alvito.

At first this interpretation seems only to confound the confusion. Why, after all, should Dom Manuel choose a member of the opposing party, as it were, to lead the first expedition into the Indian Ocean? A little reflection shows the plausibility of the idea. First, if – on Barros's evidence – the decision to send the expedition was against the majority of the king's council, a measure of compromise is likely to have been struck on the choice of captain. Second, there was the issue of risk, and the fact that only a small fleet (with a minimal crew of under two hundred) was eventually sent out. Far better to have a man identified with the opposition lead such an expedition than one of Dom Manuel's hand-picked nobles; at least in this way, the burden of failure (if failure was indeed the outcome) could in part be passed on. Third, we have the fact that none of the chroniclers provides us a convincing explanation of Gama's selection. Barros and Castanheda suggest that Dom João II had already chosen Vasco's father, Estêvão da Gama, for the post, and on the latter's death the honour passed on to his sons. Castanheda adds further that by rights the older son, Paulo da Gama, should have led the expedition, but that he yielded the post to his younger brother for reasons of health, and instead agreed to captain one of the vessels![60] Despite some supporting circumstantial evidence – Paulo da Gama died on the return voyage from India, suggesting ill-health – the last version is not free of problems. Had Paulo da Gama refused entirely to go, the matter would have been somewhat different. Further, if it was a matter of an inherited right,

[59] Grant-letters 17 and 18 December 1495, from AN/TT, Chancelaria da Ordem de Santiago, Livro IV do Supplemento, fos. 34v–35v, reproduced with a facsimile in Cortesão, *The mystery of Vasco da Gama*, pp. 178–83, with English translations on pp. 184–5.

[60] For the chroniclers' views, see Barros, *Da Ásia*, Década I/1, Book IV, Ch. 1; Correia, *Lendas da Índia*, Vol. I, p. 12; Fernão Lopes de Castanheda, *História do Descobrimento e Conquista da Índia pelos Portugueses*, ed. M. Lopes de Almeida, 9 books in 2 vols. (Oporto: Lello e Irmão, 1979), Vol. I, pp. 10–11; Damião de Góis, *Crónica do felicíssimo Rei Dom Manuel*, 4 vols. (Coimbra: Imprensa da Universidade, 1949–55), Ch. XXIII; see also M.N. Pearson, *The Portuguese in India* (Cambridge: Cambridge University Press, 1987), p. 19, who largely accepts Correia's account.

why was Vasco chosen over his other brothers? The suspicion is thus that Vasco da Gama was chosen not merely as his father's son, but as himself.

A third chronicler, the resolutely unofficial and gossipy Gaspar Correia, offers us a wholly different version. According to him, Dom Manuel chose Gama – who was attached to his household – more or less on an impulse, because he liked his looks. This version is, when taken in its entirety, wholly improbable, but contains one element that may be taken as true to the spirit of the decision: Gama was indeed picked out. To further confound the confusion, Garcia de Resende, in his *Crónica de D. João II*, insists for his part that D. João had not only chosen Vasco da Gama to lead the expedition to India, but even got the fleet ready, and the instructions (*regimentos*) finalised; in this version, D. Manuel really did no more than order the fleet to part, 'as it was, with the same crew, and the same instructions that were set out, and for Captain-Major, the same Vasco da Gama'.[61] Whatever be the case, and it may well be that none of these chroniclers has the truth of the matter, we are certain at least that in July 1497, Vasco da Gama set out for India with his fleet of three vessels, *São Gabriel*, *São Rafael*, and *Bérrio*, and a fourth supply ship.

THE PROBLEM OF DOM JORGE

The political interpretation we have suggested in the past few pages argues for a reconsideration of the role of the Order of Santiago, and also of its Master, Dom Jorge, Duke of Coimbra, in contextualising the early career of Vasco da Gama. We still lack an adequate biographical study of Dom Jorge, which would have considerably eased our task. In its absence, we may briefly outline the major elements of his career, beginning with his illegitimate birth at Abrantes, in August 1481, to Dom João II's mistress, Dona Ana de Mendonça. He was educated to the age of nine by the King's sister, Dona Joana, who though not under holy orders lived at the convent of Jesus in Aveiro, and on her death in 1490, arrived at the court. From this time, Dom Jorge appears closely associated with the Almeida family, a link that is further strengthened after 1492, on his assuming the Masterships of Santiago and Avis. His mother's family continued to play a role in his life, and

[61] Garcia de Resende, *Crónica de D. João II*, Ch. ccvi, pp. 273–4.

we know of dealings between him and his maternal uncle, António de Mendonça Furtado, resident at Lisbon, and a *comendador* of the Order of Avis.[62] The chronicles of Dom Manuel and Dom João III constitute a form of systematic propaganda against Dom Jorge (and later against his son Dom João de Lencastre, Marquês de Torres Novas from March 1520, and then made first Duke of Aveiro), which renders our task the more difficult; both father and son are accused, in varying degrees, of immorality and licentiousness in their personal conduct.

Dom Manuel – we have noted – had been asked in the will of Dom João II to hand over the Mastership of the Order of Christ to Dom Jorge, as also control of the Atlantic islands; he was further asked to name him Duke of Coimbra, and to give him a daughter in marriage. Dom Manuel dragged his feet on some of these bequests; in respect of others, like the Order of Christ, we have noted that he refused to comply altogether. Dom Jorge was named Duke of Coimbra only in May 1500, and the title was eventually confirmed only as late as May 1509, nearly fifteen years after his father's death. Further, Dom Jorge was not given a princess in marriage; instead, in May 1500, a few days after being finally named Duke of Coimbra, he contracted marriage with Dona Brites de Vilhena, daughter of Dom Álvaro de Portugal, and hence a cousin of the then Duke of Bragança.[63]

However, Dom Álvaro de Portugal was himself a person of no mean political significance and weight.[64] Son of the second Duke of Bragança, Dom Fernando, he had emerged into a position of political authority by the last decade of the reign of Dom Afonso V, and was appointed successively to the posts of, first, administrator of the *casa de suplicação*, and then *chanceler-mor* of Portugal, the latter in 1475. The great purge that accompanied the fall and execution of his older brother, the Duke of Bragança, in 1483 saw him flee to Castile, where he was received with unprecedented honours: he was granted a territorial position, made *alcalde-mayor* of Seville and Andujar, and eventually President of the Royal Council of Castile.[65] When Dom Manuel

[62] AN/TT, Corporações Religiosas, Ordem de Santiago, Maço 4, Doc. 29, *estromento de procuração*, dated 24 November 1498.

[63] AN/TT, Gavetas XVII/5–6, 'Contrato do cazamento do Senhor Dom Jorge, Duque de Coimbra', 30 May 1500; also see AN/TT, Gavetas XVII/9–1, 'Confirmação do Dote e Arras no cazamento de D. Jorge com D. Brites, filha de D. Álvaro'.

[64] Elements for his biography may be found in Anselmo Braamcamp Freire, *Brasões da Sala de Sintra*, ed. Luís Bivar Guerra, 3 vols. (Lisbon: Imprensa Nacional, 1973), Vol. I, pp. 438–41.

[65] For Dom Álvaro's sojourn in Spain, see William D. Phillips, Jr, 'Christopher Columbus

acceded to the throne, he invited him to return as part of the general amnesty to the conspirators of 1483–4, and offered to give back to him his effects (which had been confiscated), as well as a substantial cash-stipend. Dom Álvaro accepted, and continued, until his death in Toledo in early March 1504, to play a key role as mediator between Spain and Portugal. It was he, for example, who was sent as ambassador by Dom Manuel to negotiate the king's marriage with Dona Isabel, daughter of the Catholic Monarchs, and widow of the Infante Dom Afonso (who had died in 1491). Further, Dom Álvaro showed a keen interest to participate in the Indian Ocean trade, sending his factors to the ports of south-western India, and investing a part of his substantial fortune therein. The marriage between his daughter and Dom Jorge in 1500 shows that we should not adopt an overly simplistic, and binary, view of factional politics in the Portuguese court. Dom Álvaro, an opponent of Dom João II in the mid-1480s, did not consequently automatically emerge as a supporter of Dom Manuel; he represents instead a conception of trade in the Indian Ocean where the nobility had a substantial role and the monarchy a relatively reduced one, thus – in effect – an anti-Manueline conception. Further, he appears to have been allied with Dom Jorge, one of the opponents of royal power in Portugal, and a rival to Dom Manuel as claimant to the throne, and to have brought to the latter a breath of Castilian influence.

This influence is evident, indirectly, in Dom Jorge's activities in regard to the Order of Santiago. If from the point of view of historians of Dom Manuel and Dom João III, Dom Jorge appears as an unstable prince, who eventually turns into a lecherous dotard before his death at the age of sixty-nine at Palmela in 1550, chroniclers of the Order of Santiago perceive his role rather differently. From their perspective, Dom Jorge appears as a great administrator and reformer, who helps reorganise the internal structure of the order, acting on a smaller scale, but in a similar fashion, to Dom Manuel, who centralised affairs on a nationwide scale with the *Ordenações Manuelinas* and other legislation.[66] The

in Portugal: Years of Preparation', *Terrae Incognitae*, Vol. XXIV (1992), pp. 38–40. Phillips argues that Dom Álvaro probably interceded for Columbus with the Catholic Monarchs. For another discussion of the same question, which neglects the Portuguese factor, see Demetrio Ramos, '¿Quien decidiò a Fernando el Católico a aceptar el proyecto descubridor de Colón?', *Mare Liberum* 3 (1991), 184–91.

[66] Fr Jeronimo Román, *Historia de la inclita cavalleria de San Tiago en la corona de Portugal*, 13 chapters (78 ff.), dated 5 July 1591, especially Ch. 9, 'Del decimo seisto Maestre y ultimo de la Orden de San Thiago el Señor Don Jorge', fos. 43v–49; partially legible copy in the Biblioteca Nacional de Lisboa, Mss. Pombalina No. 24.

key work that is cited in this respect is the *Regras, Statutos e Diffinições da Ordem de Santiago* (1509), which emerged from a general body meeting (*capítulo geral*) called by Dom Jorge at Palmela in October 1508, but there are also a number of other pieces of internal legislation (*regimentos*) which run from roughly 1508 to the late 1520s. Of particular significance, as a recent historian of the Order of Santiago remarks, is that the norms in such works as that of 1509 are 'a translation, adaptation or reformulation of the Castilian ones, not only of the fifteenth century, but of the entire history of the Order'.[67]

A part of Dom Jorge's purpose was clearly to defend the Orders of Santiago and Avis from the more powerful Order of Christ, whose master since 1484 had been Dom Manuel. To this end, he sought and received in May 1505 a royal *alvará*, forbidding knights of the Orders under his control from moving to the Order of Christ, save with his express consent, but Dom Manuel then rapidly found other means to undermine his position. A papal letter of July 1505 gave the king the right to dispose of the possessions of all three Orders quite freely, and then, in January 1506, the Pope Julian II issued a bull, *Sincerae devotionis*, permitting the free movement of knights from other Orders to the Order of Christ. Dom Jorge still resisted and, some years later, made an issue in respect of the Order of the Hospital (St John); in fact, he confiscated the *comenda* at Sesimbra of, and also fined, Dom João de Meneses, Conde da Tarouca, for having accepted membership of the Order of St John, and for having become its head in Portugal (the title being that of Prior of Crato).[68] But the battle against royal power was, in the long term, a losing one. By 1510, Dom Jorge had been definitively marginalised, as the successive births of Dom Manuel's sons moved him further and further away from the throne. In 1516, Dom Manuel petitioned the Pope Leo X, and was granted the right to determine the successor to

[67] Isabel Lago Barbosa, 'A Ordem de Santiago', p. 31. Relevant texts include AN/TT, Livros de Santiago, B-50–137, dated 4 June 1528, 'Regimento da casa do convento de Palmela'; B–50–135, 'Livro dos Privilégios da Ordem', etc. The *Regras, Statutos e Diffinições da Ordem de Santiago* was published at Setúbal: Herman de Kempis, 1509.

[68] Biblioteca Nacional de Lisboa, Reservados, Mss. 90, No. 8, late copies of documents of the Order of Santiago, *carta de sentença*, dated 4 September 1509. For the rapid expansion of the order of Christ in the period, see Francis A. Dutra, 'Membership in the Order of Christ in the Sixteenth Century: Problems and Perspectives', *Santa Barbara Portuguese Studies* 1 (1994), 228–39, based in turn on documentation published in António Machado de Faria, 'Cavaleiros da Ordem de Cristo no século XVI', *Arqueologia e História* 6 (1955), 13–73.

Dom Jorge in the Masterships of Santiago and Avis, so that the royal shadow loomed ever larger.[69]

The rulers of the House of Avis, both Dom Manuel and especially Dom João III, seem to have relished every opportunity to demonstrate to Dom Jorge and his son, Dom João de Lencastre, their dependence on royal power. The confrontations and scandals were several. In the case of Dom João de Lencastre, the scandals centre around his opposition to the marriage of the Infante Dom Fernando (son of Dom Manuel) to the heiress, Dona Guiomar Coutinho in the early 1520s, on the grounds that he himself already had secretly married her; as a consequence, he was imprisoned in the Castelo São Jorge for several years, an episode which was used in the late nineteenth century as the subject of Camilo Castelo-Branco's play *O Marquês de Torres Novas*. As for Dom Jorge, moralising chroniclers and Whiggish historians have often delighted in the story of his passion, at the age of sixty-seven, for the sixteen-year-old Dona Filipa de Melo, daughter of Dom Fernando de Lima, whom he actually married; but Dom João III then intervened with the Papacy to annul the marriage. The fact that a similar interpretation could be given to Dom Manuel's own comportment late in life is of course another matter.

Between 1516 and the end of his life, Dom Jorge did continue though to exercise a role in politics from Palmela, and his significance for the history of art patronage in the epoch has begun in recent years to receive scholarly attention. Equally, it would appear that in the 1530s and early 1540s, he served as protector to a number of Jewish and crypto-Jewish merchants, artists and intellectuals, at times intervening to have them granted a passage to the relatively safety of India and the Indian Ocean.[70] Finally, on Dom Jorge's death at Palmela in late July 1550, Dom João III moved rapidly and decisively to centralise control of the military orders. This was effected through two papal bulls, the first *Regimini universalis ecclesiae* of Julian III, dated 25 August 1550, which granted the Master-

[69] For details, see Fortunato de Almeida, *História da Igreja em Portugal*, rev. ed. by Damião Peres (Oporto–Lisbon: Livraria Civilização, 1968), Vol. II, pp. 219–22.

[70] I am grateful for this information (drawn in part from early Inquisition records) to Dr José Alberto Tavim, which will eventually appear as part of a larger work in preparation by him on the role of Jewish and New Christian merchants in sixteenth-century Portuguese expansion in Asia. For a first approach, see José Alberto Rodrigues da Silva Tavim, 'Os Judeus e a expansão portuguesa na Índia durante o século XVI: O exemplo de Isaac do Cairo: Espião, "língua" e "judeu de Cochim de Cima"', *Arquivos do Centro Cultural Calouste Gulbenkian* 33 (1994), 137–260.

ships to Dom João III in a personal capacity; and a second and far more extensive one of 30 December 1551, conceded under severe diplomatic pressure, naming the King of Portugal in perpetuity as the Master of the Orders of Santiago and Avis, and thus bringing to fruition in Portugal what the Catholic Monarchs had already achieved a half-century earlier in Castile and Aragon, namely the union under the monarchy of the major military orders. The ceremony of takeover itself was conducted in 1552, in the Monastery of St Eloy.[71] Dom João de Lencastre, by now Duke of Aveiro, was made a member of the Thirteen, but reduced to a mere *comendador*. By the early seventeenth century, the Dukes controlled revenues of 1,470,000 *reis*, from the Order of Santiago's total revenues of 23,449,000 *reis*, or just over 6 per cent.[72] Retrospectively, then, the episode of Dom Jorge appears as a hiccup in the ongoing trend of royal fiscal centralisation. Its resonances in the context of the late fifteenth and early sixteenth centuries were, nevertheless, not insignificant.

CONCLUSION

Portuguese historiography, for long dominated by a nostalgic nationalism, has tended to ignore the extent to which the élite of that country was divided at the turn of the sixteenth century on the question of overseas expansion. Neo-Marxist historians, especially the influential school of thought developed by Vitorino Magalhães Godinho, have for their part been excessively preoccupied with the issue of class conflict in this epoch, and hence also downplayed the problem.[73] As for observers of Portuguese expansion in other historiographies (say, historians of early sixteenth-century India or East Africa), their tendency was obviously to treat 'the Portuguese' as a monolith, and then to observe this monolith undergo the cycle of rise, consolidation and

[71] AN/TT, Colecção São Vicente, Vol. II, fos. 326–47, 'Modo que se teve por El Rey Dom João o 3° quando tomou a posse dos mestrados de Santiago, e Avys, em Lixboa no Mosteiro de santo Eloy'; also Colecção São Vicente, Vol. III, fos. 475–75^{v}, 'Ordem que se devia ter no tomar a posse do Mestrado da Ordem de S. Tiago'.

[72] AN/TT, Casa da Fronteira, Vol. XXI [M-VII-21], fls. 175–7, 'Relação de todas as comendas que a Ordem de S. Tiago tem nestes Reynos de Portugal, e Algarves . . .', by the Contador do Mestrado, Diogo Rabello (late 1620s?).

[73] Godinho's approach, its strengths and weaknesses, can be gauged from a collection of dispersed articles by him; cf. Vitorino Magalhães Godinho, *Mito e mercadoria, utopia e prática de navegar, séculos XIII–XVIII* (Lisbon: Difel, 1990).

decline. Vasco da Gama is thus no more or less to them than the embodiment of Portugal.

The point of departure in this work is different from the above, and further develops a historiographical trend that has emerged in the past two decades in the study of not merely Portugal, but the early modern world more generally.[74] As royal courts emerged as the dominant arenas of political action in that epoch, the nature of factional politics changed in many states, both European and Asian. The Namierite style of analysis proves insufficient to the task, although it would obviously be useful in the Portuguese case to have a better élite prosopography for the early sixteenth century than we dispose at present. Differences were not *merely* factional; they are equally ideological, even if they cannot be closely tied down to class interest. The life of Vasco da Gama that we shall trace will see him adopt shifting positions in court politics, but for now we should retain the essentials of his origins, as the uneasy bearer of the heritage of Santiago.

We have argued in the preceding pages that the contests surrounding fifteenth-century Portuguese overseas expansion must be understood in the context of an institutional complex, involving the Crown, the nobility, and also – in a place of particular significance – the military orders of Santiago and Christ. Royal centralisation proceeded in fits and starts, and if the Portuguese Crown had to compromise in large measure with the territorial nobility in the course of the fifteenth century, its strategy with respect to the military orders was also characterised by ambiguity. The Masterships of these Orders, while vested in princes of the blood after roughly 1420, did not form a subordinate arm of royal power in a direct sense until as late as the mid-sixteenth century; and once they did so, they lost their political significance for the most part. From the 1470s, an attempt was made by the Infante Dom João (after 1481, Dom João II) to make of the military orders a part of the repertory of instruments easily available to royal power. Yet this king, though himself Master of Avis and Santiago, never gained control of the Order of Christ; and on his death left behind a curiously divided legacy as well. Partly as a consequence, the military orders could continue to play a role as a form of opposition to royal

[74] Cf. Robert Shephard, 'Court Factions in Early Modern England', *The Journal of Modern History*, 64, 4 (1992), 721–45; earlier Sharon Kettering, *Patrons, Brokers and Clients in Seventeenth-Century France* (New York, 1986).

absolutism in the early part of the reign of Dom Manuel. However, in their internal composition and political positioning, they also reflected the wider struggles that were taking place in Portuguese society to define and delimit the character of expansion overseas.

CHAPTER THREE

TO CALICUT AND BACK

Th'Age shall come, in fine
Of many years, wherein the Main
M'unloose the universal Chain;
and mighty Tracts of Land be shown,
To Search of Elder Days unknown.
New Worlds by some new Tiphys found,
Nor Thule be the Earth's farthest Bound.

Seneca, *Medea* (2nd Chorus) (translated Sir Edward Sherburne, 1648)[1]

THE RETURN OF THE ARGONAUT

IN PORTUGUESE GOA, IN MID-NOVEMBER 1599, AS THE chronicler Diogo do Couto was bringing to a close one of his more obscure works, the *Tratado dos Gama* (Treatise of the Gamas), he sought a passage from a classical text with which to dedicate it to his patron, the viceroy of the *Estado da Índia*, Dom Francisco da Gama (1565–1632), fourth Count of Vidigueira – the very same whose bones later suffered the indignity of being wrongly transferred from Vidigueira to Belém in 1880. It was almost inevitable that he should fall on Seneca's *Medea*; Lucius Annaeus Seneca was after all Iberian (born at Córdoba, to be precise), a Stoic (which Couto at times fancied himself to be), and a statesman and intellectual who was close to power, as Couto would have liked to be. The tragedy *Medea* had a good deal to recommend it from Couto's viewpoint, its link with the legend of Jason and the Argonauts not being its least attraction. As he

[1] *The Tragedies of L. Annaeus Seneca, the Philosopher*, translated Sir Edward Sherburne (London: S. Smith and B. Walford, 1702), pp. 43–4. For the text, see Humbertus Moricca (ed.), *Medea – Oedipus – Agememnon – Hercules* (*Oetaeus*) (Turin: G.B. Paravia and Co. 1925), pp. 14–15.

himself paraphrased the verse of the chorus (which he cited correctly in Latin, not always his wont), 'There will come a time, even if it be there in the very last years, in which the chains of the Ocean will be broken, to the last of the Lands.'[2]

Couto, of course, in his fixation on *Nec sit terris ultima Thule*, changed the significance of the verse somewhat to suit his purposes in his paraphrase. It is generally supposed that Thule referred to a northern land, said to be six days' sail north of Britain, and perhaps reflecting the Ancients' perception of Norway or Iceland. In the sixteenth century, many regarded the verses as a prophecy; it is interesting to note in passing that Ferdinand Columbus wrote in Latin, in the margin of his own copy of Seneca, that Seneca's prophecy had been fulfilled by his father Christopher Columbus in 1492.[3] Neither Ferdinand Columbus nor Couto pays attention to the preceding verses of the chorus, which rather inappropriately from their points of view ran (and I cite once more Sir Edward Sherburne's mid-seventeenth-century translation):

> The passive Main
> Now yields, and does all Laws sustain.
> Nor the fam'd Argo, by the hand
> Of Pallas built, by Heroes mann'd,
> Does now alone complain she's forc'd
> To Sea; each petty Boat's now cours'd
> About the Deep; no Boundure stands,
> New Walls by Towns in foreign Lands
> Are rais'd; the pervious World in'ts old
> Place, leaves nothing. Indians the cold
> Araxis drink, Albis and Rhine
> The Persians.[4]

[2] Biblioteca Nacional de Lisboa, Reservados, Códice 462, 'Tratado de todas as cousas socedidas ao valeroso capitão D. Vasco da Gama, primeiro Conde da Vidigueira . . .', fl. 3v (the text of the *Tratado* is at present unpublished, but will be edited soon by António Coimbra Martins). A more accurate prose translation is: 'There will come an age in the far-off years, when Ocean shall unloose the bond of things, when the whole broad earth shall be revealed, when Tethys shall disclose new worlds and Thule not be the limit of the lands'; cf. Samuel Lieberman (ed.), *Roman Drama*, (New York: Bantam Books, 1964), p. 360. The translations from Seneca there are by Frank Justus Miller.

[3] Seneca, *Medea*, edition and commentary C.D.N. Costa (Oxford: Clarendon Press, 1973), pp. 107–8.

[4] The prose translation by Miller in Lieberman (ed.), *Roman Drama*, runs, 'Now, in our time, the deep has ceased resistance and submits utterly to law; no famous Argo, framed by a Pallas' hand, with princes to man its oars, is sought for; any little craft now wanders at will upon the deep. All bounds have been removed, cities have set their walls in new lands, and the world, now passable throughout, has left nothing where it once had place; the Indian drinks of the cold Araxes, the Persians quaff the Elbe and the Rhine.'

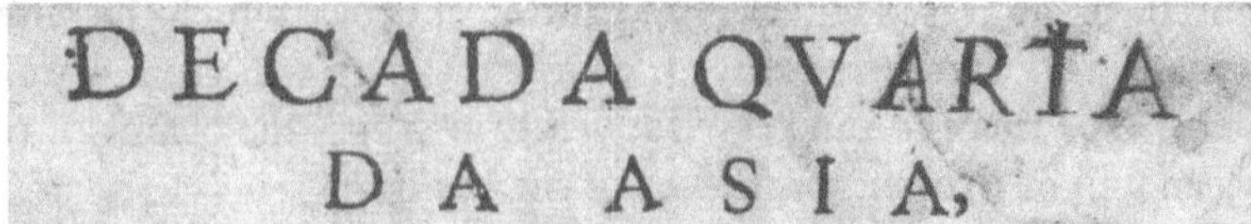

DECADA QVARTA
DA ASIA,

DOS FEITOS QVE OS PORTVGVESES FIZERAM NA CONQVISTA E DESCOBRIMENTO das terras, & mares do Oriente: em quanto gouernaraõ a India Lopo Vaz de ſaõ Payo, & par te de Nuno da Cunha.

Compoſta por mandado do inuẽciuel Monarcha de Eſpanha dom Felipe Rey de Portugal o primeiro deſte nome:

POR DIOGO DO COVTO GVARDA MOR da torre do tombo do eſtado da India.

Com licença da ſanta Inquiſição & Ordinario.

EM LISBOA.
Impreſſo por Pedro Crasbeeck, no Collegio de ſanto Agoſtinho. Anno M.DCII.

Com Preuilegio Real.

Plate 9 Title page of the first volume of the chronicle of Diogo do Couto, *Década Quarta da Ásia*, from the first edition.

There is a curious irony in this vision of Indians drinking the waters of an Armenian river, and Persians sipping from the Elbe and the Rhine (*Indus gelidum potat Araxen, Albin Persae Rhenumque bibunt*), being used to sum up Portuguese expansion into Asia or Castilian expansion into the New World. It had equally been forgotten by Couto that the import of Seneca's verse was to compare an earlier age of heroes (the Argonauts), with the present age when exploration has become so trivialised that 'any little craft now wanders at will upon the deep'. Worse still, the hero of an earlier epic, Jason, has been reduced by the end of *Medea* to declaring that there are no Gods. None of this fitted very well with Couto's own vision of the voyage of his very Christian hero Vasco da Gama from Lisbon to Calicut.

RECONSTRUCTING THE VOYAGE

But we have already moved considerably ahead of our story; let us return therefore to 1497, when Gama prepared to leave for India. He was given command, as we have already noted, of a fleet initially comprising four rather small vessels, *São Gabriel* (which he himself captained), *São Rafael* (captained by Paulo da Gama, his brother), a third known as the *Bérrio* (commanded by Nicolau Coelho), and a fourth supply-vessel which did not go as far as India; the fleet was also accompanied by a fifth vessel, a caravel which was destined for São Jorge da Mina on the African west coast, and which was captained by a certain Bartolomeu Dias – perhaps the discoverer of 1487, but perhaps not.[5] If this is indeed the case, it is significant that despite his experience, and availability, Dias himself was not chosen to perform Gama's role. The crew in the four vessels numbered between 148 and 170, depending on which of several estimates one prefers. The major roles were distributed as follows:

São Gabriel: captain – Vasco da Gama, captain-major of the fleet; pilot – Pêro de Alenquer, chief pilot; scrivener – Diogo Dias.[6]
São Rafael: captain – Paulo da Gama; pilot – João de Coimbra; scrivener – João de Sá.

[5] At least one document of mid-December 1497 suggests that by that time, Bartolomeu Dias was in Lisbon; cf. AN/TT, Chancelaria D. Manuel, Livro 31, fls. 84–84v, reproduced with a facsimile in Luís de Albuquerque *et al.*, *Bartolomeu Dias: Corpo Documental, Bibliografia* (Lisbon: CNCDP, 1988), Document 6.

[6] Diogo Dias was probably the brother of Bartolomeu Dias, the leader of the 1487–8 expedition.

Bérrio: captain – Nicolau Coelho; pilot – Pêro Escobar; scrivener – Álvaro Braga.
Supply-ship: captain – Gonçalo Nunes; pilot – Afonso Gonçalves.

Until the early nineteenth century, historians wishing to reconstruct the details of Gama's first voyage were wholly dependent on Portuguese chronicles produced several decades after the event, especially those of the official chroniclers João de Barros and Damião de Góis, and the unofficial but still highly 'respectable' Fernão Lopes de Castanheda. By following the narratives of Barros and Castanheda, not only could the broad outlines of the voyage easily be established, but a large number of interesting and lively details were available, suggesting that their writings in turn rested on another, earlier, layer of materials. Indeed, these chroniclers themselves made it clear that they had had access to the verbal or written accounts of men who had themselves participated in the expedition, and Castanheda in particular specified that one or more written accounts by an eyewitness had passed through his hands. These accounts proved elusive until the second quarter of the nineteenth century, when an anonymous one was finally found in 1834 by the historian Alexandre Herculano, and edited four years later by Diogo Köpke and António da Costa Paiva from a unique manuscript, a sixteenth-century copy of a lost original. The manuscript, comprising seventy-nine folios, had earlier been conserved in the monastery of Santa Cruz de Coimbra before being taken by Herculano to the Municipal Library of Oporto, the somewhat obscure northern Portuguese library where it now reposes.

Köpke and Costa Paiva's edition of 1838 has been followed by a number of others, in both the nineteenth century and the present one.[7] On the occasion of the fourth centenary of Vasco da Gama's arrival in Calicut, an authoritative translation into English was also published for the Hakluyt Society, by Ernst George Ravenstein. Besides, a vast critical literature has emerged around this text, addressing both its usage, and on a number of occasions the possible identity of its author, who makes it clear that he travelled for the most part during the outward voyage on board the vessel commanded by Paulo da Gama, *São Rafael*, and further, that he was one of the group that accompanied Vasco da

[7] The accepted standard edition from the twentieth century is that by António Baião, A. de Magalhães Basto and Damião Peres, *Diário da Viagem de Vasco da Gama*, 2 vols. (Lisbon, 1945).

Gama on shore on his first visit to the town of Calicut in late May 1498. Now since the chroniclers, and especially Castanheda, identify a number of those who were part of this group, several candidates for the text's authorship emerged. Of these Diogo Dias, scrivener of Vasco da Gama's ship, and Álvaro Braga, scrivener of the *Bérrio*, were eliminated on account of the fact that they were both on the wrong vessels during the outward voyage. The interesting figure of Fernão Martins, who spoke some Arabic and acted as Gama's interpreter, was equally ruled out because he is mentioned by name in the anonymous text.[8] This left three known names (besides the unknowns who accompanied Gama ashore): namely, João de Sá, who was later to be treasurer of the Casa da Índia, a certain Gonçalo Pires, and one Álvaro Velho.[9] Pires was rapidly eliminated by commentators, by collating Castanheda's account with the anonymous one: from this it emerged that Pires was involved in certain actions near Calicut in which the anonymous author was not.

The field was thus rapidly narrowed, among the known names, to two possibilities: João de Sá, and Álvaro Velho. Now, Portuguese historians came very early on (indeed from the time of Köpke and Costa Paiva) to the conclusion – and even the 'moral certainty' (*certeza moral*), as some were later to assert – that the text was the handiwork of Álvaro Velho, a rather obscure figure, but who had possibly travelled on board Paulo da Gama's ship on the outward voyage.[10] In most modern versions, in keeping with the late nineteenth-century researches of Franz Hümmerich, Velho has been identified with a personage of that name known from a few other sources, a native of the southern Portuguese town of Barreiro (on the banks of the Tejo), who moreover had spent a period of eight years, prior to 1507, on the Guinea coast and especially Sierra Leone.[11] This latter Velho, it was noted, had

[8] On Martins, also see AN/TT, Chancelaria D. Manuel, Livro 6, fl. 121, a grant-letter dated 17 December 1502, in which it is mentioned that he had gone with Gama to India, and served there as Arabic interpreter ('de lingua aráviga'); the document is published in Brito Rebello, 'Navegadores e Exploradores Portuguezes', Doc. 21, pp. 68–9.

[9] Gaspar Correia provides still other names from amongst the twelve who disembarked in Calicut: João de Setúbal, and João Palha, but the reliability of his information is open to doubt.

[10] See, most recently, the entry by José Manuel Garcia, 'Velho, Álvaro', in Luís de Albuquerque and Francisco Contente Domingues, eds., *Dicionário de História dos Descobrimentos Portugueses*, Vol. II (Lisbon: Círculo dos Leitores, 1994), pp. 1064–5.

[11] For discussions, see in particular, Abel Fontoura da Costa, *Roteiro da Primeira Viagem de Vasco da Gama (1497–1499) por Álvaro Velho* (Lisbon: Agência Geral das Colónias, 1940); also José Pedro Machado and Viriato Campos, *Vasco da Gama e a sua viagem de descobrimento* (Lisbon: Câmara Municipal de Lisboa, 1969), pp. 63–4. Machado and

in fact supplied detailed information in about 1507 on the religion of Sierra Leone to a Lisbon-based Moravian notary, printer and scholar, who went by the name of Valentim Fernandes, and who was at that time in the process of putting together a manuscript collection of travel-accounts for his correspondent Konrad Peutinger in Augsburg; such an activity would be in keeping with the alleged observant nature of the author of the text of 1497–9.[12] The identification has, on the other hand, raised a number of problems, and notably forced historians to ask why Velho did not equally divulge the text of the 1497–9 voyage to Fernandes (who would almost certainly have been avid to see it, and probably would even have wanted to print it), rather than allow it to languish in obscurity.

Some other historians, notably the translator of the text into English, E.G. Ravenstein, and more recently the philologist and historian Carmen Radulet, have instead proposed the name of João de Sá, arguing that after all Sá was a trained scrivener, and that he later held an administrative position (as 'treasurer of spices') in the Casa da Índia (India House) between February 1511 and April 1514; this would presumably make him a more likely candidate for the writing of a text, which it emerges besides is remarkably official, as well as discreet, in character.[13] Opponents of the Sá hypothesis note however that opinions are attributed to him in Castanheda's chronicle which are the opposite of those espoused in the anonymous text; and that besides, a close reading of the chronicle of João de Barros suggests that he was involved in specific actions on the east coast of Africa (the erection of a memorial pillar, or *padrão*), to which the anonymous author alludes as if he were a mere spectator. A reading of other documents apparently concerning Sá's career further confounds the issue; from one of these (a royal letter giving him the status of cavalier, in January 1512), it emerges that he probably spent some

Campos even speculate that Velho was a convict-exile (*degredado*), and that he was hence not allowed to return to Portugal in 1499, but had instead to leave ship in west Africa on the return voyage.

[12] On Valentim Fernandes, see Marion Erhardt, *A Alemanha e os Descobrimentos Portugueses* (Lisbon: Texto Editora, 1989), pp. 31–40; and for the text itself *O Manuscrito 'Valentim Fernandes'* (Lisbon: Academia Portuguesa de História, 1940).

[13] Carmen M. Radulet, *Vasco da Gama: La prima circumnavigazione dell'Africa, 1497–1499* (Reggio Emilia: Edizioni Diabasis, 1994), pp. 31–49. For elements on Sá's later career, see AN/TT, CC, I-11-101, *alvará* dated 5 February 1512, when he was already employed in the Casa da Índia; for a copy of the letter naming him to the post, AN/TT, Chancelaria D. Manuel, Livro 15, fl. 70v (copy in Chancelaria D. João III, fl. 55); also his *carta de quitação*, dated 22 August 1517, in AN/TT, Livro 6 de Místicos, fls. 149–50v.

time at Safi (in North Africa) in the years after his return from India, that he was seen as a servitor (*criado*) of the Bishop of Coimbra, D. Jorge de Almeida, and that in Safi he was closely associated with D. Pedro de Almeida, with whom he participated in the military exploits that permitted him to rise up the social hierarchy.[14] This would suggest, if nothing else, that while in India in 1497–9, he – like Vasco da Gama – was closely associated with the powerful Almeida family, and represented their interests. In any event, the debate remains open and surprisingly lively (considering its arcane character), and it is perfectly possible that the actual author of the text might have been neither Velho nor Sá, but one of the other, in the event unnamed, men who accompanied Gama ashore at Calicut. It is a remarkable fact that very few of the historians who have addressed the issue wish seriously to admit this possibility, even though the identification with Álvaro Velho does not particularly render the text more intelligible, or authoritative. Since Velho himself is a more or less anonymous character, the mere giving of a name to the Oporto manuscript's author does not in fact greatly aid the historian's task.

Vasco da Gama's fleet eventually left the estuary of the Tejo on 8 July 1497, as we learn from the anonymous text, as well as from a number of other sources. Let us begin therefore with a consideration of the fractured syntax of our anonymous text's somewhat neutral and enigmatic beginning:

> In the name of God. Amen.
> In the year 1497, the King Dom Manuel, the first of this name in Portugal, sent out to discover, four ships, which went in search of spices, of which ships there went as Captain-Major Vasco da Gama, and of the others, of one of them Paulo da Gama his brother, and of another Nicolau Coelho.

A deceptive start, with the curious but altogether characteristic use of the verb 'discover' without a stated objective (*mandou . . . a descobrir*). What was to be discovered by the four ships? New lands? New routes? Or was 'discovery' an activity that explained itself without further reference to a purpose or destination, closer to the notion of 'uncovering' the hidden or concealed? We are told then later in the same phrase, in an agglutinative fashion, that

[14] AN/TT, Chancelaria D. Manuel, Livro 7, fl. 10, grant-letter dated in Lisbon, 30 January 1512; also see the later grant-letter, elevating him to the status of *fidalgo*, Chancelaria D. Manuel, Livro 29, fl. 99v. It is of course possible, though unlikely, that this João de Sá was a homonym of the man who went to India.

Plate 10 The Torre do Belém, Lisbon, from which ships set out for India.

the ships went in search of spices. Three captains are mentioned, and the fourth ship is passed over for detailed mention: it was, after all, a mere supply ship (*nau dos mantimentos*). The author's tone is pious (or at any rate officious), even in the next passage.

> We left Restelo on a Saturday, which was the eighth of July of the said Year of 1497, on our way, which may God Our Lord allow to be completed in his Service. Amen.

The author has now identified himself to the reader as one of those who made the voyage, and suggests moreover that the voyage is not complete at the time of the writing. He now moves rapidly into the practical details, starting with the next Saturday (15 July 1497), when the small fleet passed the Canaries, then to sight the west African coast south of Cape Bojador. The seamen occupied themselves with fishing for a couple of hours on the 17 July, but then encountered the first minor problem, as the fleet accidentally came to disperse during the night. We now gather directly that the anonymous author was aboard Paulo da Gama's vessel, for he notes that his ship lost sight of the ships of the Captain-Major, and of Nicolau Coelho. But we learn too that the captains were well prepared for this eventuality, and the ships hence all made for a pre-determined rendezvous in the Cape Verde islands. Within about a week, that is by 26 of July, the other ships had all met up with the *São Rafael* on which the anonymous author was. In particular, the sighting of the *São Gabriel* with Vasco da Gama, which joined the others last of all, was greeted with the celebratory firing of bombards, and trumpet blasts. This minor detail suggests a certain residual anxiety already at this early stage of the voyage: sailing even this part of the Atlantic was by no means a matter of certainty.

The four months between the departure from Restelo (the 'suburban' settlement that was downriver from Lisbon proper), and the sighting of the Cape of Good Hope on 18/19 November 1497 occupy a mere six folios of the manuscript. We pass quickly from the vicinity of the Cape Verdes (and the island of Santiago, where the fleet arrived in late July to take on water and meat), to a long period in the south Atlantic, without sight of land. After the mast of the flagship *São Gabriel* had broken, and been repaired in mid-August, the fleet made a complex and much-debated wide loop in the Atlantic, coming very close as it turns out to Brazil.[15] In these months between August and early November, laconically described by the anonymous author in a mere half-folio, whales and various other sea-creatures and birds were observed by the fleet, until finally the south-west coast of Africa was sighted once more on 4 November. As one might have expected, the sight of land was greeted with relief and celebration; the fleet had arrived at the Bay of St Helena, not to be confused with the island of the same name in the Atlantic.

However, caution was exercised, for the anonymous author notes, 'we had no knowledge of the land'. After sailing about to explore the bay on 5 and 6 November, on the 7th Vasco da Gama sent out a ship's boat (*batel*) with the pilot Pêro de Alenquer, to ensure that the place was a sound anchorage, and sheltered from the winds. It was only on 8 November that the fleet anchored in the Bay of St Helena, to spend eight days there cleaning the vessels, and mending the sails. This period was also used to explore the region around, and the Portuguese quickly came across a river to the south-east, which was given the name

[15] For the Atlantic passage, see the paper by Carlos Viegas Gago Coutinho, 'Discussão sobre a rota seguida por Vasco da Gama entre Santiago e S. Brás', *Anais da Academia Portuguesa da História*, 2ª Série, Vol. II (1949), pp. 99–131, which argues that Gama took a sensible zig-zag route in the southern Atlantic, suggesting that the Portuguese had some experience already in navigating these waters.

Santiago. A preliminary naive ethnography was also attempted, of the natives who ate seals (*lobos marinhos*), roots and deer-meat, were covered with skins, and who went about accompanied by dogs. Indeed, on the 9th, Gama and several others went on land, and immediately captured an autochthone who was gathering honey, who was taken back to the *São Gabriel*. Here, he was placed at the same table as Gama himself, and ate whatever the others ate; the following day, he was dressed up by the Portuguese, and sent back on land. This experiment had the desired consequences, for the very next day, the man returned with fourteen or fifteen others, to whom Gama showed off his trade-samples (cinnamon, cloves, gold, and seed-pearls) to see if they had any of the same. However, when this produced no signs of recognition, Gama left them with some odds and ends of no great value. In turn, a day later, a larger group of forty and fifty returned, with whom some petty exchange turned out to be possible, giving them copper coins (*ceitis*) for conch-shells, and other goods.

One of Gama's men, a certain Fernão Veloso, now decided to make more bold, and asked the Captain-Major for permission to return with the Africans to their place of residence. This eventually led to a minor misunderstanding, for having taken him along a small distance and given him something to eat, the Africans then indicated to Veloso that he should turn back to the ship. He began calling out to the ships; some of the Portuguese who had gone back to the ships at once got on board a small sail-boat (*barca a vela*) and made for the land. Seeing them approaching in haste, the blacks (*os negros*) now began to run along the beach, throwing spears (*zagaias*) at them, in which process Gama and three or four others suffered slight injuries. The incident left a bad impression on the Portuguese, for the anonymous author assures us that it was the result of being too trusting, and of the fact that they were insufficiently armed.

The Portuguese thus returned to their ships, and set sail on 16 November, having been assured by their experienced pilot Pêro de Alenquer that they were not far from the Cape of Good Hope. Indeed, two days later, on the afternoon of 18 November, they sighted the Cape, heading south-southeast. Several days of struggle with the winds followed, until they were finally able to double the Cape, which eventually happened only about midday on the 22nd. Thereafter, passing a large curvature in the coast, they arrived on the 25th at the Bay of São Brás, today Mossel

Bay, where the fleet remained for all of thirteen days. The fourth ship, the supply-vessel that had accompanied them, was dismantled at this spot, and the supplies themselves divided amongst the other vessels. This strategy was evidently the result of the small size of the three main vessels, which could not thus carry adequate supplies for such a long voyage on board.

Nearly a week passed without any further contacts with the local inhabitants. Then, on 1 December, a group of about ninety men, who appeared similar to those whom the Portuguese had earlier seen at St Helena, arrived on the beach. The Portuguese sent out some ship's boats, and Gama and others approached the shore. The first contacts were positive, for the men on ashore actually approached the Portuguese, whereas (our author notes) 'when Bartolomeu Dias was there, they ran away from him, and would not take anything that he gave them, but instead one day when he was taking water from a source, which is very good and close to the sea, they prevented him from doing so by showering stones from a hill that overlooks the source'. At this, Dias is reported to have fired on them with a crossbow (*besta*), killing one of them.

Why this 'changed' attitude then, on part of the blacks? To begin with, it is not in the least obvious that the group that met Gama was the same as that which had encountered Dias a decade earlier. The Portuguese speculated – somewhat improbably – that it might be because the blacks had been in communication with the others, from St Helena, who had told them that they were harmless. Nevertheless, Gama was suspicious, and noting that there was a wooded area just beyond the beach, decided not to land there; he instead went to another spot that was more open, and where there was less room for concealment. Those on the strand followed them, and Gama and the other captains now went ashore accompanied by men carrying arms (*bestas*). He also signalled that only one or two blacks should approach at a time, and exchanged his goods against their ivory chains, which were rather more to the Portuguese taste than the conch-shells and worked metal of St Helena.

Contacts continued the next day. Thus:

> On Saturday, some two hundred blacks both large and small arrived, bringing along about a dozen cattle, both oxen and cows, and four or five sheep. And as soon as we saw them, we at once went ashore, and they began to play four or five flutes, and some

> played high and some low, in a fashion that was very well attuned for blacks, of whom one does not expect music, and they danced like blacks. And the Captain-Major ordered that our trumpets be played, and we in our boats danced, and the Captain-Major too, along with us.

One finds here, then, the reverse of the later stereotype of the 'naturally' musical African. Once these festivities were over, the Portuguese exchanged a black ox against three chains, and slaughtered it for Sunday dinner. On that very Sunday, as mutual suspicions were somewhat diminished, a group of cattle-herders arrived, this time including the women and children; and they too played music and danced, although it is noted that the children were left in the woods with the arms. Gama now sent a certain Martim Afonso, who had some experience in west Africa, to see if he could communicate somewhat better, in order to buy another ox. This man was taken by the Africans to the water-source, and given to understand that the Portuguese should not take the water from there; further, the cattle were driven into the woods, and exchange refused. Gama became suspicious, thinking that some 'treason' was being armed by the Africans, and withdrew his men onto the ships. They were then ordered to put on armour, and return on shore fully armed. This was done, writes our author, 'to show them that we had the power to harm them, but that we did not wish to do so'. In an ambience of suddenly heightened tension, the cattle-herders began to run about in a panic, and the Portuguese for their part once more withdrew to a safe distance; however, two shots of a bombard were then fired from the prow of one of the boats. The shots had a dramatic effect. Those who were seated near the woods now ran pell-mell, dropping their arms and the animal skins with which they were covered. Driving their cattle before them, they made for a hill-range that was within sight of the shore.

A brief description now follows in the text of the local fauna, both the seals in a nearby island, and the birds, besides the cattle of course, of which our author is quite admiring (indeed rather more so than he is of the inhabitants). As often enough in his account, his comparisons take him to the southern part of Portugal (thus, 'the oxen of this land are very large, like those of the Alentejo'), suggesting that he was from that region. The contacts with the Africans having ended badly, we hear little of them, save for a minor, but still significant event. When, on 5

December 1497, the fleet was ready to leave São Brás, they placed a commemorative pillar (*padrão*), and a cross in the ground. But the very next day, as they were leaving, 'around ten or twelve blacks arrived, who, before we left there, broke both the cross and the pillar'. As with the water-source, then, Portuguese intentions and local notions were very evidently at cross-purposes.

The fleet then continued to follow the south-eastern African coast, in the face of weather that was less than ideal. On one occasion, the other ships lost sight of Nicolau Coelho's vessel in a violent storm, but they soon found one another. Finally, on 16 December, they passed the last *padrão* left by Bartolomeu Dias, thus at last coming to sight land that they felt they were truly 'discovering'. Complications followed both with the wind and with the currents just off the coast; for some days, the fleet was forced backwards rather than being able to make further headway. It was also becoming increasingly difficult to continue on account of the shortage on board of fresh water, and food. At last, on 10 January 1498, the fleet arrived at a small river (identified as the Inharrime by modern historians), thinking to put in there for water and food.

The account describes this new land and its people in terms that are wholly opposed to those of São Brás. That the land was dubbed *Terra da Boa-Gente* ('Land of the Good People') is already indication enough. The same Martim Afonso, who had earlier been in west Africa, was sent ashore, and managed to make contact with someone who appeared to be locally a man of influence (*um senhor entre eles*), whom the Portuguese promptly presented with red trousers, a coat, and some trinkets. Communication of some sort was established (how, it is not clear, given the linguistic differences between the Congo and this region), and Martim Afonso and another Portuguese then went along to a nearby village. Here, they were given cereal-cakes and chicken, and inspected all night long by men and women who were apparently curious about them. The next day, the two returned to the ships with chickens, and accompanied by as many as two hundred men. This local notable is also alleged to have said that he would show the gifts he had received to 'a great lord whom they had, and who it seemed to us was the King of that land'.

The language of the account here is distinctly more favourable, also in the sense of attributing to the inhabitants the familiar qualities of courtesy, hierarchy, and certain familiar skills too. Thus, when the man to whom the trousers and jacket were given

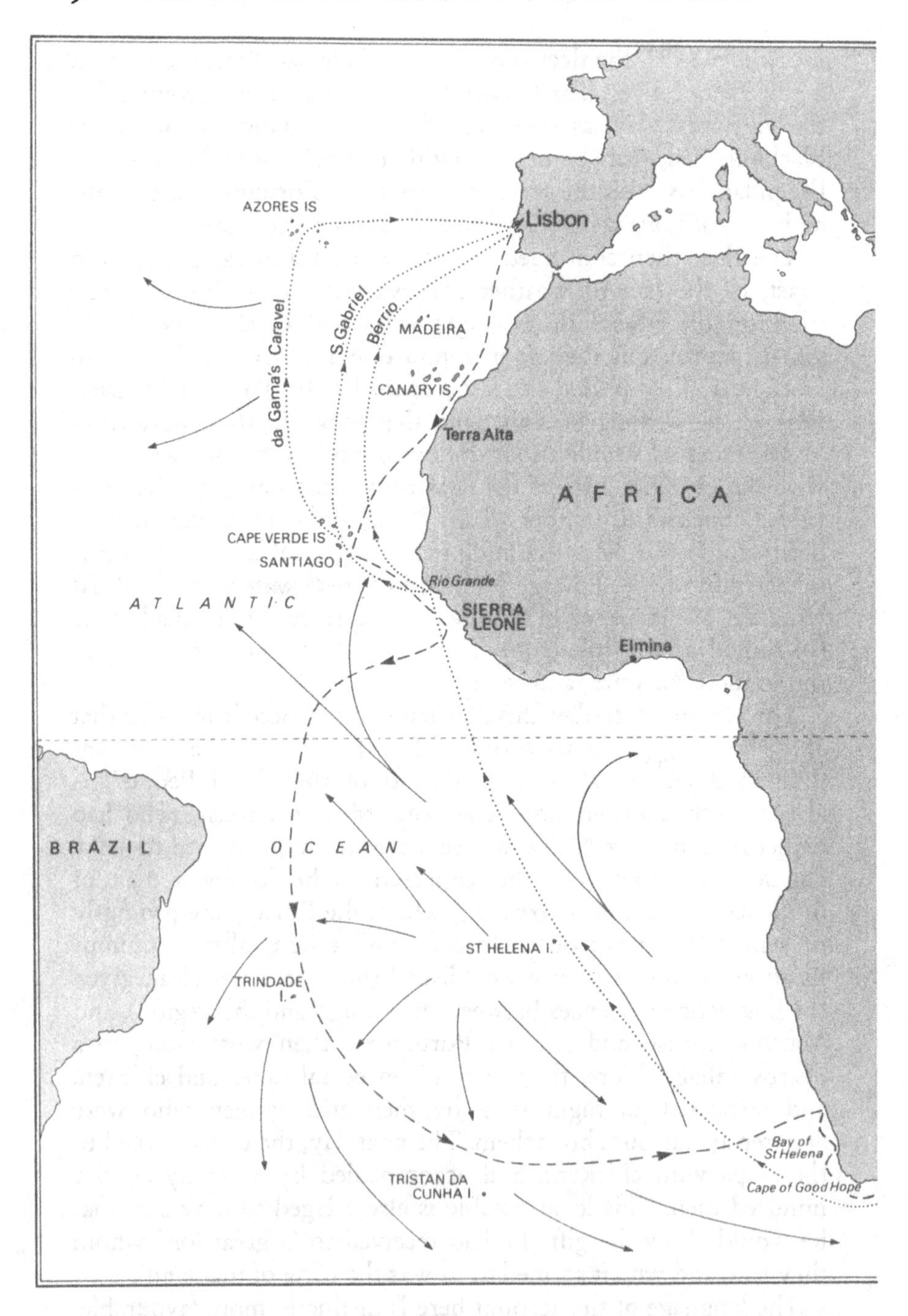

AZORES IS
Lisbon
da Gama's Caravel
S.Gabriel
Bérrio
MADEIRA
CANARY IS
Terra Alta
AFRICA
CAPE VERDE IS
SANTIAGO I
Rio Grande
SIERRA LEONE
ATLANTIC
Elmina
BRAZIL
OCEAN
ST HELENA I.
TRINDADE I.
Bay of St Helena
TRISTAN DA CUNHA I.
Cape of Good Hope

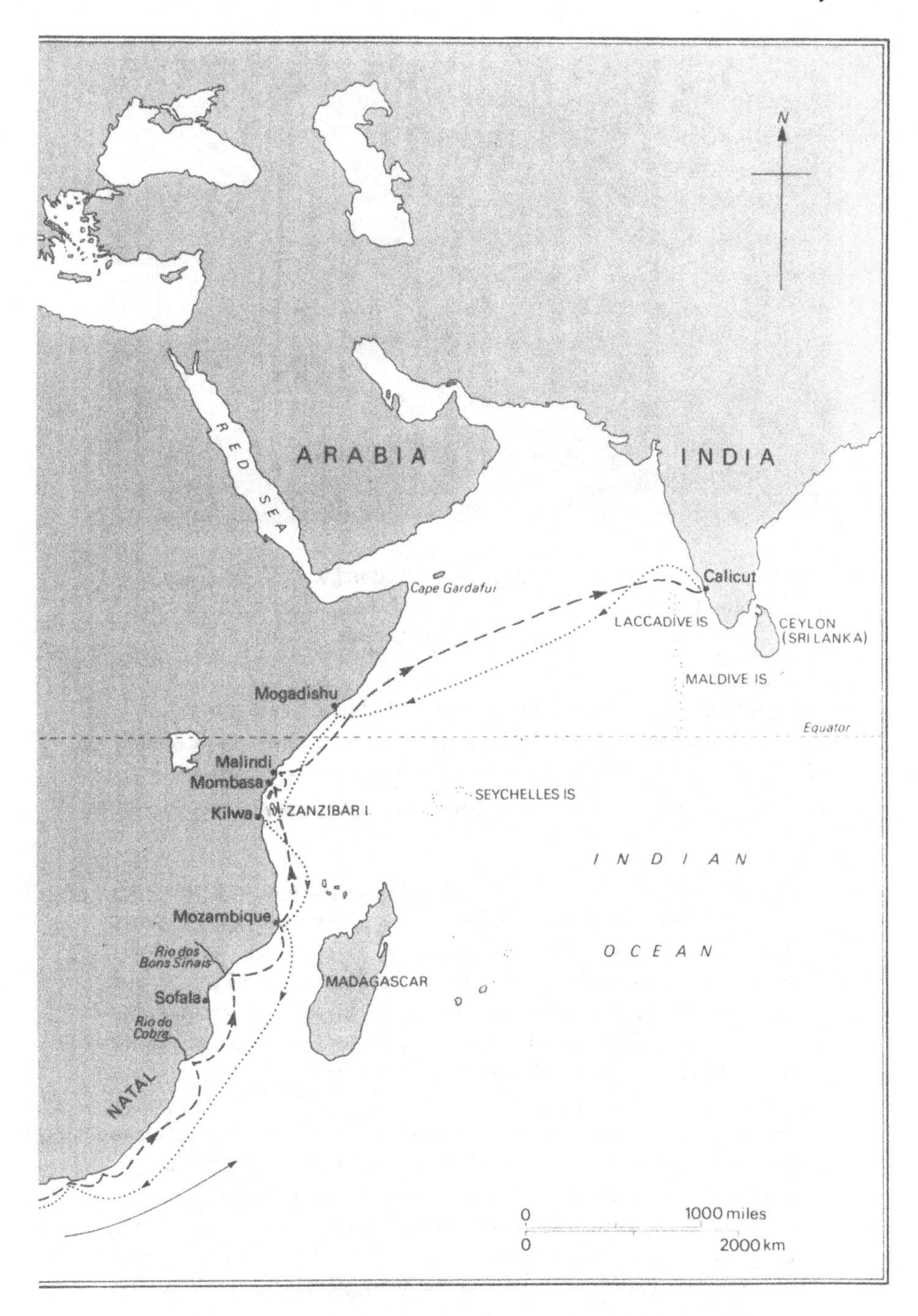

Map II Route taken by Vasco da Gama, 1497–1499

Plate 11 Drawing of a set for *L'Africaine* (Act III) by Charles-Antoine Cambon and Joseph Thierry (1865), showing a storm at sea in the Indian Ocean.

was going back homewards, it is stated that he said proudly to all those whom he passed: 'Do you see what they gave me?', to which the others responded by 'clapping their hands as a courtesy'. Or again, we may take the description of the region in its entirety:

> This land, so it seemed to us, is heavily populated and there are many lords in it, and it seemed to us that there were more women than men, because where twenty men came, there were forty women along. And the houses of this land are made of straw, and the arms of these people are very large bows, and arrows and spears of iron. And it seemed to us that there was a great deal of copper in this land, which they wore on their legs, and arms, and twisted in their hair. There is also tin in this land which they use to decorate the scabbards of their dagger (*guarnições de punhais*), and their handles (*bainhas*) are made of ivory. And the people of this land greatly prize linen, and they gave us much copper in exchange for shirts, if we were willing to part with them. These people bring along certain large containers in which they carry back salt-water from the sea inland, and they throw it in pits in the ground, and make salt from it.

This broadly favourable tone is carried over into the next port-of-call, for which they departed after a five-day halt. In this new land, further optimistic signs were found, notably the first set of boats other than their own that the Portuguese had seen since leaving west-central Africa. The fleet had now arrived, unknowingly of course, at one of the distributaries of the Zambezi, an area with which the Portuguese were to have a long and complex relationship over the next five centuries! Fresh water was made available to them, and a halt of over a month (thirty-two days to be precise) could be made at the river's mouth, during which time the *São Rafael*'s mast was repaired. At the same time, disquieting signs of illness appeared amongst the mariners, with swellings in the hands and feet, and a severe inflammation of the gums (characteristic of scurvy). It was here that the Portuguese began to sense that they were at last on the fringes of the commercial world that they sought, as we see from the following passage:

> And after we had been here two or three days, two lords of this land came to see us, who were so proud that they prized nothing that we could give them. And one of them had a cap on his head with tassels embroidered with silk, while the other had a head-dress (*carapuça*) of green satin. And together with them there came a youth who, so they said, was from another land a good distance from there and who said that he had seen ships as large as those that we had brought, at which signs we rejoiced greatly for it really seemed to us that we were coming closer to where we wanted [to go].

With these men, whom the text directly calls *fidalgos*, the Portuguese traded over a week, at the end of which they departed up-river in their boats. The Portuguese then decided to leave a *padrão* there, and to call the place the River of Good Omens (or *Bons Sinais*), before leaving there on 24 February. On 1 March, they then sighted some islands, and the very next day were accosted by some sailing vessels that seemed to belong to the settlement there, and who signalled to them that the Portuguese could pass between the island and the continent, and put in to port by following them. The captains consulted, and it was finally decided to let Nicolau Coelho go first; the anonymous author notes that at this time, he himself was on the *Bérrio*. Despite breaking his rudder, Coelho managed to make it to port, and dropped anchor a short distance from the main settlement. There was an immediate recognition on the part of the Portuguese that they had reached a considerable commercial centre: in fact, they were on Mozambique island.

> The men of this land are dark and well-built, and of the sect of Muhammad (*Mafamede*) and speak like Moors; and their clothes are made of cotton and linen, very fine and multi-coloured, and striped, and they are rich and embroidered. And all of them have caps on their head, with silk tassels (*vivos*) embroidered with gold thread. And they are merchants, and trade with white Moors (*mouros brancos*), of whose ships there were four in this place, which used to bring gold, silver and cloth, and cloves, and pepper and ginger, and silver rings (*aneis*) with many pearls and seed-pearls, and rubies, and all of these things are also brought by the men of this land. And it appeared to us, from what they said, that all these things came there [overland] by cart, and that those Moors brought them, except for the gold, and that further down where we were headed there was a good deal [of gold] and that the stones, and the seed-pearls, and the ivory were in such plenty that it was not necessary to barter for them, but simply to gather them up in baskets (*apanhá-la aos cestos*). And all of this was heard by a sailor whom the Captain-Major had brought along, and who had been a captive of the Moors, and so understood these [Moors] whom we found here.[16]

The investment in linguistic expertise was beginning to pay off, for here at last was an Arabic-speaking settlement. This was a reality at once familiar and disquieting, and the thought seems at once to have struck the Portuguese that the Eastern Christians they sought must be somewhere in the vicinity. They were at once reassured by the traders of the port, who assured them that further along the coast was an island 'in which there were half Moors and half Christians, the which Christians were at war with the Moors'. Besides, 'they told us that Prester John was close by there and that he had many cities on the sea-shore, and that the residents there were great merchants and had many large ships, but that the Prester John was deep in the interior, and that one could only go there on camels'. Hearing this, our author reports, many of the Portuguese were so happy that they wept with joy, and praised God.

THE WESTERN INDIAN OCEAN, C. 1500: A DIGRESSION

Let us leave the anonymous text here for a time, now that we have been over roughly a quarter of its narrative, and turn to some

[16] Álvaro Velho, *Roteiro da Primeira Viagem de Vasco da Gama*, ed. Neves Águas (Lisbon, 1987), p. 36. For an earlier English translation, E.G. Ravenstein, *A journal of the first voyage of Vasco da Gama, 1497–1499* (London, 1898).

considerations of a more general nature, to place in context the episodes that follow. What was the nature of the commercial system that the first Portuguese maritime expedition into the western Indian Ocean found itself confronted with? And to what extent did this system differ, first, from what the Portuguese expected to find, and second, from the type of commerce to which they were habituated in the Mediterranean and the Atlantic?[17]

Now, it has often been stated by historians that the commercial expansion of the early centuries of the second millenium in the Indian Ocean was part of a phase of 'Arab dominance', and even more generally that the area was a 'Muslim lake', or that the years after about 750 AD saw the formation of an 'Islamic world-economy' in the Indian Ocean.[18] The key question obviously lies in the strict association of trade networks with Islam, to the extent that a historian could write anthropomorphically as late as the 1960s that on the arrival of the Portuguese in the Indian Ocean, 'Islam took up arms against the invader'.[19] Now much commercial history tends to treat both Islam and Muslims in a largely monolithic and undifferentiated fashion, and historians are often strikingly reticent both on questions of ideology and on the social and economic competition and conflict between different groups operating in the Indian Ocean at this time.

It seems clear enough that the history of Indian Ocean trade in the centuries after 750 AD is far too complex to fit the set of straitjackets that have often been imposed on it in the literature. We are dealing with a political and commercial network that was poly-centric in its organisation; and there was no single epicentre that generated a pulse to which the entire 'system' responded – even in the western Indian Ocean. Those who argue for the existence of an 'Islamic world-economy' in the Indian Ocean in this period are simply unable, for example, to explain the trade of other groups than the Islamic ones – save as minor actors

[17] For a more detailed discussion of these questions, see Sanjay Subrahmanyam, 'Of *Imârat* and *Tijârat*: Asian merchants and state power in the western Indian Ocean, 1400–1750', *Comparative Studies in Society and History* 37, 4 (1995), on which the following paragraphs draw in part.

[18] For the most recent formulation of this type, see André Wink, 'Al-Hind: India and Indonesia in the Islamic world-economy, c. 700–1800 AD', *Itinerario* 12, 1 (1988); and by the same author, *Al-Hind: The Making of the Indo-Islamic World*, Vol. I (*Early medieval India and the expansion of Islam, seventh to eleventh centuries*) (Delhi: Oxford University Press, 1990). Similar views are echoed in Geneviève Bouchon, *Albuquerque, le lion des mers d'Asie* (Paris: Editions Desjonquères, 1992), pp. 30–7.

[19] M.A.P. Meilink-Roelofsz, *Asian trade and European influence in the Indonesian archipelago from 1500 to about 1630* (The Hague: Martinus Nijhoff, 1962).

operating essentially within the monopolistic framework of this Islamic world-economy. But this is surely less than accurate, for two reasons. First, other non-Muslim Asian merchant groups participated in this trade on a large scale as well – and often wholly or largely on their own terms. These included Gujarati *vaniyas*, Tamil and Telugu Chettis, Syrian Christians from south-western India, Chinese from Fukien and neighbouring provinces, and others besides the Jews studied by S.D. Goitein on the basis of the Geniza papers. Second, even the Muslims, to exercise a *monopoly*, must have acted *en bloc*, something that is simply undemonstrable. The very strong connotation of a term like 'monopoly' suggests a certain caution in its usage, before creating the image of a monolith where none existed.

It remains true that the fourteenth and fifteenth centuries saw the expansion of a variety of forms of Islam on the shores of the Indian Ocean, and the growing presence of Muslim mercantile communities, whether in East Africa, India, or Southeast Asia. But this is not the same process as that envisaged by the constructs of a 'Muslim lake' or an 'Islamic world-economy', for we should bear in mind that this Islam was as often heterodox as orthodox. We see this for example in the disdain expressed by a celebrated fifteenth-century author, to whom we shall return at far greater length below, namely the Arab navigator Ahmad ibn Majid, in his discussion of Malay Muslims:

> These are bad people, who do not know any rule; the Infidel marries the Muslim, and the Muslim the Infidel woman; and when you call them 'Infidels', are you really sure that they are Infidels? And the Muslims of whom you speak, are they really Muslims? They drink wine in public, and do not pray when they set out on a voyage.[20]

No doubt to some later European observers (and notably to some sixteenth-century Portuguese) the entire process appeared one of the spread of the 'law of Muhammad', but it seems more than a little problematic to equate Mappilas from Kerala and Maraikkayars from the Tamil coast to Bohras and Khojas from Gujarat, and to assume that all of these were bound up tightly in one social and economic system. What seems more reasonable as an assertion is that the expansion of Islam, and to a lesser extent

[20] Cited in Luís Filipe F.R. Thomaz, 'Malaka et ses communautés marchandes au tournant du 16e siècle', in Denys Lombard and Jean Aubin (eds.), *Marchands et hommes d'affaires asiatiques dans l'Océan Indien et la Mer de Chine, 13e–20e siècles* (Paris: Editions EHESS, 1988), p. 42.

Theravada Buddhism – religions that laid particular stress on individual salvation – may indirectly have had an influence on the social perception of the role of trade and profits. The Christianity that the Portuguese brought to Asia therefore competed with other religions freshly on the ascendant, which were being spread not merely by conquest but by acculturation and trading contacts. It will not do however to speak of the European monolith confronting an earlier, equally monolithic, structure – for this does scant justice to the history of both the pre-1500 and the post-1500 periods.

A brief *tour d'horizon* of the western Indian Ocean may be useful here, in order to set the record straight. In the course of the fifteenth century, a set of port-based city-states grew up, and came to dominate the maritime world from Melaka in the east to Aden in the west. Their rise was associated with what has recently been called an 'Age of Commerce', in the context of Southeast Asia, a process that the limited records that we have at our disposal from the fifteenth century only dimly allow us to comprehend. Nevertheless, these sources, which include Chinese travelogues, Persian chronicles and memoirs, as well as European accounts from Italian and other merchants, do permit the delineation of a network of interlinked ports, and trader communities, concerning which the Portuguese then provide us more detailed information after 1500.

An instructive case with a somewhat mixed profile from the fifteenth century is located on the very western fringes of our region: this is the Mamluk state of Egypt, of crucial significance in political terms, as the centre of the surviving structure of the Caliphate. There are numerous suggestions in the texts of the fifteenth century which hint at a shadowy suzerainty exercised by the Mamluks over some states of the western Indian Ocean; the Sultans of Gujarat, for example, seem formally to have had their successions legitimised by embassies to Cairo.[21] Nevertheless, the Mamluks intervened to a rather limited extent in the actual conduct of trade in the western Indian Ocean, contenting themselves with their control over the Red Sea, and the Hijaz in particular. Direct control over the Hijaz remained vested, nevertheless, in a lineage of Sharifs, Shafi^{c}i Muslims, who drew revenues from ships wrecked in the Red Sea, from a share of the

[21] It is instructive in this context to read Gaston Wiet, trans., *Journal d'un bourgeois du Caire, chronique d'Ibn Iyâs*, 2 vols. (Paris: Librairie Armand Colin, 1955–60), Vol. I, pp. 176–7, 269, *passim*.

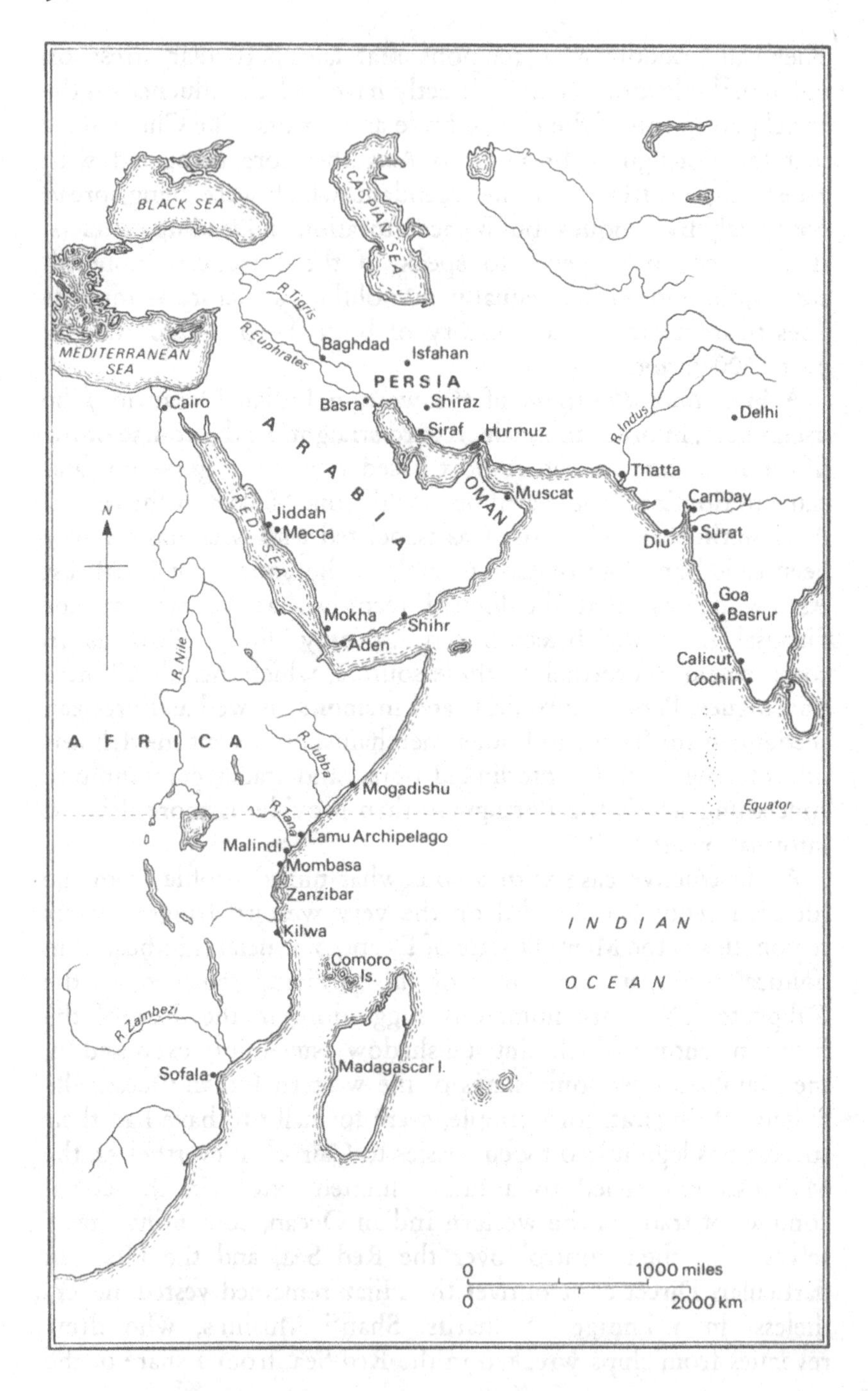

Map III The Western Indian Ocean in about 1500

gifts sent into Mecca, and a 10 per cent levy on imports into Jiddah.

It is clear that the Mamluks were to an extent exemplars, looked up to by some of the larger states of the Indian subcontinent at least, and to this extent what they represented is significant for our purposes. To understand what underlay the manner in which the Mamluks approached trade in the fifteenth century, we should recall that the idea of seapower as an instrument of statecraft was by no means alien to these rulers of Egypt. This was because their kingdom was vulnerable to attack from both the Mediterranean and the Red Sea. During the period of the Circassian Mamluks (1382–1517), the use of maritime power was required at various points, to defend the kingdom's economic interests, and never did the Sultans show signs of thalassophobia. Barsbay (r. 1422–38) in the 1420s built coastal defences as well as a fleet, to defend his kingdom's northern shores from the depredations of Christian pirates and privateers, and it is also in the reign of this Sultan that Mamluk policies towards the Red Sea trade witnessed significant changes. A conscious policy was adopted of developing the trade of Jiddah, to the detriment of Aden, since the former port was far more closely linked to Cairo fiscally than the latter. In Barsbay's period, a contemporary source reported that Jiddah alone yielded a revenue of 200,000 *dînârs* on average to the treasury each year.[22]

The commercial policy of earlier Sultans, articulated for example by the contemporary writer al-Qalqashandi (who in his encyclopaedic manual *Subh al-aᶜshâ* speaks of the need to receive merchants kindly 'and treat them justly, for the profits accruing from them . . . are very great'), was reversed to an extent in Barsbay's time. A series of commodity monopolies was created, first on local products like sugar, and later even on imported goods such as pepper.[23] In the late 1420s and early 1430s, the Sultan forbade the sale of pepper to Europeans by anyone save his own official apparatus, and this policy was periodically revived by later rulers like Khushqadam (r. 1461–7) and Qa'it Bay (r. 1468–96).

Lest the case of the Mamluks seem to us to be wholly unique, we should equally recall that the southern Indian kingdom of

[22] Ahmad Darrag, *L'Egypte sous le règne de Barsbay, 825–841/1422–1438* (Damascus: Institut Français de Damas, 1961), p. 222.

[23] Walter J. Fischel, 'The Spice Trade in Mamluk Egypt', *Journal of the Economic and Social History of the Orient* 1, 2 (1958).

Vijayanagara, while far from being a port-state, was nevertheless interested in maritime trade already in the fifteenth century. We see this for example from the account of Kamal al-Din ᶜAbd al-Razzaq Samarqandi (1413–82), who between 1441 and 1444 acted as the ambassador-at-large of the Timurid ruler Shah Rukh in the Persian Gulf and India.[24] ᶜAbd al-Razzaq's travels took him to Hurmuz, from where he embarked for Calicut in a horse-trading vessel in 1442–3. While in Calicut, he was invited by the Vijayanagara ruler Deva Raya II to his court, and eventually spent several months there. The account of the ambassador from Herat is important for several reasons. First, while it is apparent from his relation that the control of the ports of the west coast by Vijayanagara was somewhat tenuous (save perhaps for Mangalore), there nevertheless was a considerable west Asian trading community at Vijayanagara. Several references, mostly uncomplimentary, are made to the Hurmuzis, and on his departure the ambassador set out accompanied by two Khorasani merchants Khwaja Masᶜud and Khwaja Muhammad. Second, he stresses the key role played by imported commodities, coinage metals and war-animals, in motivating Deva Raya's external policies. In the case of horses, the Vijayanagara rulers in the fifteenth and sixteenth centuries seem to have followed a semi-monopsonistic policy, which was further strengthened from about 1480, when the key port of Bhatkal fell under their control. Trade between Bhatkal and Hurmuz, shared by the Hurmuzis, Mappilas, Navayats and Saraswats, was particularly flourishing in the late fifteenth and early sixteenth centuries. In turn, the overland route between Bhatkal and the city of Vijayanagara was jealously guarded by Vijayanagara rulers like Krishnadevaraya (r. 1509–29).

Besides these states with an interest in trade, we are aware that, in about 1500, a number of commercial nodal points existed in the western Indian Ocean, some embedded in far larger political structures (the case of Khambayat, or Bhatkal mentioned above), others themselves the *centres* of political formations. In the latter category, we could treat as prototypical the celebrated instance of Melaka in the Malay peninsula, but there are also other clear instances where one equally finds political power seeking expression in the form of relatively small and compact trading states not dissimilar to Melaka. The two clearest examples are Yemen in the

[24] See Wheeler M. Thackston, trans., *A Century of Princes: Sources on Timurid History and Art* (Cambridge, Mass.: The Aga Khan Program for Islamic Architecture, 1989), pp. 299–321.

Red Sea, and Hurmuz in the Persian Gulf, both of which also controlled key nodes in Indian Ocean trade – respectively the ports of Aden and Jarun.

Considering the case of Aden first, it emerges into importance as a centre in the late ninth or early tenth century, several centuries before the rise of Melaka.[25] Together with the ports of Zafar and al-Shihr, Aden remained one of the centres where ships from India and Southeast Asia customarily made first landfall on the Arabian coast. Under the rule of the Ayyubid dynasty, the port duties (*ᶜushûr*) at Aden were first codified in the twelfth century by a certain Khalaf al-Nahawandi, and these rulers also instituted a system of galleys to patrol the coast and the mouth of the Red Sea in order to protect merchant shipping from pirates. This coastal fleet (*al-ᶜasâkir al-bahriya*) survived in some form into the early sixteenth century, and was used to defend Aden from attacks by the rulers of Shihr and other neighbours on at least two occasions between the 1450s and 1490s.[26] The consolidation of Aden's fortunes is usually dated to Ayyubids in the twelfth century; but the structure of trade and state in the area, somewhat obscure in the period, becomes clearer during the rule of the Rasulid dynasty that succeeded. Of particular importance as evidence is a text of 1411–12, the *Mulakhkhas al-fitan*, which sets out at some detail the administration of the port, the nature of the communities resident there, the extent of revenues, and so on.[27] From this work, it emerges that the Aden port-duties were the major source of revenue for the Rasulid Sultans resident at Taᶜizz.

It is natural enough, then, that from the early thirteenth century, the Rasulid Sultans should have taken great interest in developing the port, building a reputation of such dimensions that, in 1374–5, the *qâzî* of Calicut (himself of the Shafiᶜi school of jurisprudence) brought a letter to the ruler al-Malik al-Ashraf, requesting that the *paradesi* (Iranian and Rumi) Muslims of this Malabar port be permitted to read the *khutba* in the name of the

[25] Salomon D. Goitein, 'From the Mediterranean to India: Documents on the trade to India, South Arabia and East Africa from the eleventh and twelfth centuries', *Speculum* 29, 2, 1 (1954), 181–97; also R.B. Serjeant, 'Yemeni Merchants and Trade in Yemen, 13th-16th Centuries', in Lombard and Aubin, *Marchands et hommes d'affaires asiatiques*, pp. 61–82.

[26] Cf. Lein Oebele Schuman, trans., *Political History of the Yemen at the Beginning of the 16th Century: Abu Makhrama's account of the years 906–927 H. (1500–1521 A.D.), with annotations* (Groningen: Druk V.R.B. Kleine, 1960), pp. 12–14, 69.

[27] Claude Cahen and R.B. Serjeant, 'A Fiscal Survey of the Mediaeval Yemen', *Arabica* 4, 1 (1957), pp. 22–33.

Rasulids. The regulations of the port speak of a certain degree of bureaucratic orderliness; all vessels entering Aden were to carry a manifest, in the hands of the ship's scrivener (*karrânî*), all goods were to be examined in detail, body-searches of passengers were to be conducted and so on. Many of these practices, described in the case of Aden by Ibn al-Mujawir in his *Târîkh al-Mustabsir*, were to be found in later centuries in ports like Mughal Surat.[28]

But evidence also exists of conflict between the Rasulid Sultans and merchants, especially the substantial magnates (*a^cyân al-tujjâr*) of Aden, and in particular the so-called Karimi merchants. These conflicts arose largely because the Sultans, like their counterparts at Melaka, were themselves engaged in commerce, and also because – as was to occur later in the Sultanates of Aceh and Banten, and *unlike* what usually obtained in Melaka – they at times acted as monopsonists or monopolists.[29] The royal trading establishment (*al-Matjar al-Sultânî*) was probably founded in the mid-fourteenth century, in imitation of the Fatimid practice at Cairo. Engaged in the trade in madder, pepper, aromatics and so on, the headship of the establishment appears to have been a hereditary post. It is likely that the Sultan also owned ships, though one has no evidence that they traded over long-distance routes. For the most part, the trade of the *Matjar* seems to have been limited to the Arabian peninsula, the Red Sea littoral, and Egypt; this would also explain why the merchants who were most often in conflict with the Sultan were the Karimis, rather than, say, the Gujarati *vaniyas*, whose quarter at Aden (the so-called *hâfat al-bâniyân*) is to be encountered already in the late fourteenth century.

Fifteenth-century Aden bears more than a passing resemblance to Melaka, although some differences also existed between the two. It also resembles its nearer neighbour, Hurmuz (or rather Jarun) in the Persian Gulf in more ways than one. Jarun, unlike either Aden or Melaka, was an island, with two port-sites on it, one for small vessels, the other for large ones. Founded in around 1300, the island-city was virtually impregnable from land, and succeeded quickly in superceding the centre of Qays that had earlier dominated Persian Gulf trade for two centuries. Hurmuz,

[28] R.B. Serjeant, 'The Ports of Aden and Shihr', in *Les grandes escales* (Brussels: Société Jean Bodin, 1974).

[29] Denys Lombard, *Le Sultanat d'Atjéh au temps d'Iskandar Muda (1607–1636)* (Paris: EFEO, 1967), pp. 101–26; Claude Guillot, 'Libre entreprise contre économie dirigée: Guerres civiles à Banten, 1580–1609', *Archipel* 43 (1992).

like Melaka and Aden, was organised in terms of quarters, where different communities resided: these included the Gujarati *vaniyas*, Iranian merchants, and also a substantial community of Jewish traders in the fifteenth century. More than the rulers of Melaka, the Shahs of Hurmuz appear to have run a semi-tributary, semi-trade-based state. They controlled a number of islands in the Persian Gulf, of which the most important – Qishm – was also the major supplier of agricultural products to feed Jarun. Other islands had a more clearly strategic function, such as Kharg (associated in legend with the figure of Khwaja Khizr) which guarded the entrance to the Shatt-al-Arab waterway at the interior of the Gulf. Besides these islands, Hurmuz's rulers also collected revenues from their dominions on the mainland, on both the Iranian and the Arabian sides, with the latter being more significant from the viewpoint of revenue yield.

Thus, despite the apparent similarities, Hurmuz provides us with a case distinct from Melaka and Aden at one level, being somewhat closer perhaps to the situation in the south-western Indian kingdom of Calicut, where the Samudri Raja (or Zamorin) derived a great part of his revenues from taxing trade, but still did not run a fully fledged mercantilist state. In Calicut, as in certain other small kingdoms with access to the seaboard like Kotte in western Sri Lanka, the rulers left trade in the hands of specific communities – be they indigenised Muslims, like the Mappilas of Malabar, and foreigners (*paradesis*) from as far as Bahrain, Baghdad, Kazerun and Shiraz, or Chettis from the Coromandel coast, and traders from Gujarat.[30] Apocryphal, but nevertheless evocative, local histories of the kingdom, such as the *Keralolpatti*, propose an elaborate chronology of establishment and consolidation, which can be verified against inscriptions; these suggest that the kingdom would have been consolidated in the eleventh century, and then commenced a process of expansion that extended from the mid-thirteenth century to the third quarter of the fifteenth century, reaching its apogee during the reign of Manavikrama (r. 1466–74), a great patron of the arts. This expansion created an admittedly fragile sovereign umbrella, under which the rulers of Calicut attempted to bring together the congeries of petty chieftains (*nâḍuvaḻis*) whose domains extended over more or less extensive territories in central and northern Kerala. The

[30] Geneviève Bouchon, 'Les Musulmans du Kerala à l'époque de la découverte portugaise', *Mare Luso-Indicum* 2 (1973), 3–59; also Bouchon, 'Un Microcosme: Calicut au 16e siècle', in Lombard and Aubin, *Marchands et hommes d'affaires asiatiques*, pp. 49–57.

Calicut state itself, or Eralanadu, by the late fifteenth century appears to have recognised only two sovereigns of broadly equal status in Kerala: the Kolattiri Rajas of Kolathunadu (whose domains centred on the northern port of Cannanur), and the Rajas of Venadu, whose major centre appears to have been further to the south, at Kollam.[31] Now, contrary to what has sometimes been suggested in recent literature, all evidence points to the fact that trade was never far from the Samudri's conception of statecraft. Thus, arriving there after an eighteen-day voyage from Qalhat in 1440s, the Timurid envoy ᶜAbd al-Razzaq was rather unimpressed by the person of the ruler, or Samudri ('whom I found to be as naked as other Hindus'), but nevertheless has a number of positive things to say concerning the development of the port itself as a site for fair trade. To be sure, it was a 'city of infidels and therefore . . . in the *dâr al-harb*', but as against this there was a resident Muslim population with two congregational (*jâmaᶜ*) mosques, who were well protected. And above all, the conduct of trade itself was assured.

> In that city, security and justice are such that wealthy merchants who sail the seas bring many goods there from Daryabar. They unload them from the ships and store them in lanes and the bazaar as long as they wish without having to worry about guarding them. The *dîwân* watchmen keep guard and patrol them day and night. If they make a sale, one-fortieth is taken in alms; otherwise no duties are imposed on them. It is the custom of other ports to seize as windfall and plunder any ship headed for one port but driven by God's destiny to take refuge in another. However, in Calicut, no matter where a ship is from and where it is headed, if it docks there, they treat it like any other ship and subject it to no more or no less duty.[32]

The idea that the Samudri's reputation depended crucially on the security he was able to afford merchants is confirmed by an anecdote that serves as a sort of foundation-myth in the *Keralolpatti*: Ambaresan, a Chetti merchant from Coromandel, on his return from a trading voyage to the Red Sea, is supposed to have

[31] For a discussion of the early medieval situation in Kerala, from which these three states emerged, see Kesavan Veluthat, *The Political Structure of Early Medieval South India* (New Delhi: Orient Longman, 1993), pp. 113–21.

[32] Translation in Thackston, *A Century of Princes*, p. 304. Compare the earlier translation in R.H Major, *India in the Fifteenth Century: Being a Collection of Narratives of Voyages to India* (London: The Hakluyt Society, 1857), p. 14. For Calicut in the mid-fourteenth century, see Mahdi Husain, trans., *The Rehla of Ibn Battûta (India, Maldive Islands and Ceylon)* (Baroda: Oriental Institute, 1976), pp. 188–95.

had a ship so overloaded with gold that it was in danger of sinking. He hence left a large treasure-chest in the safe custody of the Samudri (in a stone cellar dug by him for the purpose), and sailed way, none too optimistically. To his astonishment, when he returned, he found that the whole treasure had been preserved intact, and in gratitude he offered the Samudri a half of it. The latter refused, stating that he had done no more than what was expected of him as ruler, and the merchant for his part went on to found the bazaar at Calicut.[33] Thus, we see the Samudri not as trading-monarch, but as a protector-monarch, who sees his interests and those of the merchants as intimately associated. But this did not preclude attempts at territorial aggrandisement, both along the coastal plain, and into the interior domains of other chiefs. This image is not entirely at odds with the vision that the Portuguese sources present us with.

Even closer than Calicut to Melaka and Aden in some respects was the East African Sultanate of Kilwa, ruled over from the late thirteenth century by a family of Yemeni *sharîfs*, the Mahdali.[34] In the case of Kilwa, or its northern neighbours and rivals Mombasa and Malindi (which emerged into prominence in the fourteenth century), we are not aware of any data on the fiscal foundations of the state, though the links between prosperity of these states and the Indian Ocean commercial triangle of Gujarat–Red Sea–East Africa is often asserted (as is the importance in the case of Kilwa of control of the gold trade).[35]

While our discussion so far has taken us from Melaka at the eastern end to Aden at the western end of the geographical space that we are considering, it has largely ignored the area that can rightly speaking be termed the key or 'hinge' of western Indian Ocean trade – namely Gujarat.[36] It may be partially true that the trade between Malabar and Kanara (ports such as Calicut, Can-

[33] The story is summarised in K.V. Krishna Ayyar, *The Zamorins of Calicut (From the earliest times down to A.D. 1806)* (Calicut: Norman Printing Bureau, 1938), pp. 85–7.

[34] For a general survey, see Randall L. Pouwels, *Horn and Crescent: Cultural Change and Traditional Islam on the East African Coast, 800–1900* (Cambridge: Cambridge University Press, 1987); but see also Gabriel Ferrand, 'Les Sultâns de Kilwa', in *Mémorial Henri Basset: Nouvelles Etudes Nord-Africaines et Orientales* (Paris: Librarie Orientaliste Paul Geuthner, 1928), pp. 239–60, and the survey by G.S.P. Freeman-Grenville, *The Medieval History of the Coast of Tanganyika* (London, 1962).

[35] For a sense of the history of this centre, based largely on archaeological evidence, see H. Neville Chittick, *Kilwa: An Islamic Trading City on the East African Coast* (Nairobi: British Institute in East Africa, 1984), 2 vols.

[36] Cf. in this context Michael N. Pearson, *Merchants and Rulers in Gujarat: The response to the Portuguese in the sixteenth century* (Berkeley: University of California Press, 1976).

nanur and Bhatkal) and the Red Sea and the Persian Gulf could be conducted autonomously of Gujarat. Precious metals and horses from the north-west could be exchanged for pepper, spices, sugar and rice from the south-eastern fringe of this maritime area. And it is also true that the Portuguese, when they first appeared on the western Indian Ocean scene, concentrated on the one hand on the Red Sea, Persian Gulf and East Africa, and on the other hand on Malabar and Kanara, ignoring Gujarat to a large extent.

The Gujarati mercantile network at the close of the fifteenth century has been called a 'forgotten thalassocracy'. Whether in the ports of Burma (Martaban and Dagon), or in Melaka, or even in Bengal, the Gujarati presence was a significant one. Closer at hand, on the Malabar coast and in the Maldives, in the Konkan ports of Chaul and Dabhol, in the Persian Gulf, Red Sea and East Africa, they had to be reckoned with in greater force. The Gujaratis in question were often termed *baniyas* in the epoch; the term *baniya* today refers to a set of Hindu and Jain merchant communities, resident over a great part of northern and western India. However, there has been some debate in recent times over the accuracy with which this term was used by early modern observers. Some historians have claimed that the term was extended to include practically any merchant group in western India, and particularly those who performed the function of intermediaries. This may have been the case in the late eighteenth century, when the English in particular used the term to speak of such dependent 'comprador' merchants.

Two other merchant groups operating out of Gujarat merit some attention. The first were the Parsis, of whom a rare earlier mention is in the thirteenth-century Persian account of Sadid al-Din Muhammad ᶜAwfi, who accuses the Zoroastrian merchants of Khambayat of having incited the local residents against the Muslim merchants of the port, resulting in the death of eighty of them, and the destruction of a mosque and its minaret.[37] But it is only from the late seventeenth century that one encounters the great Parsi merchant figures of the early modern period, such as the renowned Seth Rustamji Manakji and his family. These merchants, operating first out of Surat, and then later from Bombay, came to enjoy a place of major prominence by the late eighteenth century. The second group, far more visible in the

[37] Iqtidar Hussain Siddiqui, *Perso-Arabic Sources of Information on the Life and Conditions in the Sultanate of Delhi* (New Delhi: Munshiram Manoharlal, 1992), pp. 24–6.

sixteenth and seventeenth centuries, are the Isma^cilis, usually divided into two groups – the Khojas and the Bohras. The Isma^cili Bohras in turn can be divided into two, the Sulaimanis and the Da'udis, with the latter being largely Shi^ci migrants from Yemen who settled in Gujarat in the first half of the sixteenth century, and who were in the late seventeenth century to suffer persecution at the hands of Aurangzeb. But there also existed a third group of Bohras, the so-called Sunni Bohras, concentrated at Rander but also with some important representatives at Surat.

While the *baniya*s originated largely from western and northern India, their trade oriented them towards settling in urban nodes which were often far distant from their land of origin. At least on the face of it, they appear then to correspond quite well with the idea of the 'diaspora', and indeed have been used by Philip Curtin as an illustration of the notion in the context of Indian Ocean trade.[38] One finds them, after all, from Melaka to the Red Sea, and in the seventeenth century, even in the small (ostensibly Portuguese) settlements of the Zambezi valley in East Africa. Nowhere do they show a particular hunger for political power, nor do they seem to wish to shape the policies of continental states in a way that might be advantageous either to trade in general, or to their community in particular. On the other hand, not all traders – even those based in Gujarat – shared the same mentality. Earlier historians have examined the careers of such entrepreneurs as Malik Ayaz and Malik Gopi, and the seventeenth and early eighteenth centuries provide us other examples. The governor of the port of Diu in the early sixteenth century was Malik Ayaz, himself a great trader, and a manumitted slave of the Sultan of Gujarat, who may have been of either Slav, Turkish or Persian origin. In Surat, renowned as a centre of *baniya* trade, the richest merchant in the early sixteenth century was a certain Malik Gopi (or Gopinath), a Brahmin and not a *baniya*, described by one contemporary as 'richer than all the men of the Orient'. Gopinath traded extensively in other ports like Khambayat too, operated a trading fleet of thirty vessels, patronised poets, and was accused by the author of the near-contemporary chronicle *Mirât-i Sikandarî* of tyrannising the local Muslim population using his access to the state machinery.[39] For besides being a great trader,

[38] Philip D. Curtin, *Cross-cultural trade in World History* (New York: Cambridge University Press, 1984).

[39] Cf. the discussion in Jean Aubin, 'Albuquerque et les négociations de Cambaye', *Mare Luso-Indicum* 1 (1971), 9–11.

Malik Gopi was a major actor in the politics of the Gujarat Sultanate in the epoch. On the death of the Sultan Mahmud I, he is known to have played a decisive role in arranging the succession of Muzaffar II.

The will to political power here was expressed by seeking a slice of the action within a larger political structure – the Sultanate of Gujarat for Ayaz and Gopi, the Mughal state for later merchants in the same region. Elsewhere, mercantile-minded elements had tried other means, even seeking to shape the very structure of the state to their own ends. An important and relatively little-known example of a sort of 'merchant republic' form, in the fifteenth and sixteenth centuries, can be found at Basrur on the Kanara coast, south of Goa. The dominant trading community here were Saraswats, a caste of open status, which at times claimed Brahminhood but more usually was identified with mercantile activity (the Portuguese usually term them *chatins*, from *chetti*).[40] In the middle years of the fifteenth century, with the exception of a small stretch of coast near Mangalore, most of the region was free of Vijayanagara control. It was thus under the control of various minor rulers, as part of whose chieftaincies the ports held out an independent character. In the fourteenth century, Ibn Battuta found Honawar, for example, under the rule of a certain Sultan Jamal al-Din Muhammad ibn Hasan; further south, in Barkur, he encountered a Hindu ruler with a fleet of thirty warships, engaged in raids on merchant ships. The Moroccan traveller also visited Basrur, the great centre of Saraswat trade, but found little there to interest him.[41] Two centuries later, the Portuguese paid far greater attention to the town.

The case of Basrur is a relatively little-known one, and brings out the existence of surprising forms of mercantile political autonomy in the western Indian Ocean in the fifteenth and sixteenth centuries. Nevertheless, the Basrur merchants' situation was quite different from that of another semi-autonomous mercantile community located somewhat further to the south, the Mappilas of Malabar. Mappila settlement concentrated largely on two nodes, one to the north near Cannanur, and another further south in the vicinity of Ponnani. By the late fifteenth century, Mappilas were extensively involved in the trade to Melaka and

[40] Cf. the discussion in Sanjay Subrahmanyam, *The political economy of commerce: Southern India, 1500–1650* (Cambridge: Cambridge University Press, 1990), pp. 260–5.
[41] Cf. Husain, trans. *The Rehla of Ibn Battuta*, pp. 180–5.

the Coromandel coast of eastern India; however, their strongest zone of operation was an area defined by the west coast of Sri Lanka (the ports of the kingdom of Kotte), Malabar, the Maldives and the Laccadives. Thus, they do not correspond all that well to the notion of a diaspora community, being somewhat localised in their activities. The cohesiveness and militancy of the Mappilas in the sixteenth and seventeenth centuries are testified to by numerous European accounts, and were celebrated by Shaikh Zain al-Din Maᶜbari, writing his *Tuhfat al-Mujâhidîn* in the 1570s.[42]

The reader would have immediately discerned from the discussion above (however digressive its apparent character) that the world of trade in the western Indian Ocean should not have been an entirely bizarre one from the Portuguese point of view. Here, as in the Mediterranean, one found states operating on different scales, with a lesser or greater degree of integration with the interests of merchant communities. Religious identities played some role in determining the nature of solidarities and networks, but were by no means the sole factor to be taken into consideration. If the 'Christian' powers operating in the Mediterranean were unable to reach a common understanding in dealing with Mamluk Egypt, nor was Islam in the Indian Ocean a binding force that fully determined commercial or political strategies. Other rivalries and attempts to build spheres of influence cut across the ostensible lines of solidarity, and also provided fault-lines that the Portuguese could at times exploit, on the occasions that they themselves were not too divided internally.

But surely, it may be argued, there was *one* major difference between what Vasco da Gama left behind in the Atlantic, and what he found in the western Indian Ocean, namely the systematic use of violence on the sea in the one case but not in the other. It has after all long been a tenet of historians that this difference constituted a major rupture in the 'rules of the game' as they existed in the trading world of the Indian Ocean. This is not a statement concerning the inherently non-violent character of Asian history, which might be a difficult proposition to defend as such. But violence in pre-Vasco da Gama Asia, it would seem from a reading of a number of authors from K.M. Panikkar to M.N. Pearson, was confined to the *land*, where large armies were mobilised and brought to battle, substantial urban populations

[42] For the Arabic text, and a Portuguese translation, see David Lopes, *História dos Portugueses no Malabar por Zinadim* (Lisbon: Imprensa Nacional, 1899).

decimated (the celebrated practice of the *qatl-i ʿâmm*, or 'general massacre' being but one amongst many), and tracts of agricultural land devastated.

Now, two types of counter-objection could be raised to this formulation. First, the existence of populations of pirates and/or corsairs is a recurrent feature of the history of the Persian Gulf, the Indian west coast (we have noted Ibn Battuta's encounter with them), in eastern Bengal and northern Burma, and in Southeast Asia. But second, and more to the point perhaps, a number of scattered references to maritime violence as part of routine statecraft in fifteenth-century Asia have often gone relatively unnoticed. We have mentioned the system of galleys that were used by the rulers of Aden both for patrolling, and more offensively. The Geniza papers suggest that this was a tradition on the South Arabian coast of somewhat longer standing than merely the fifteenth century. Thus, from 1135, we have the following account in a letter sent from Aden to the merchant Abraham Yaju in India:

> This year, at the beginning of seafaring time, the son of al-ᶜAmîd, ruler of Kish, sent an expedition against Aden demanding a part of the town, which was refused, whereupon he sent this expedition. It consisted of two big *burmas*, of three *shaffâras* and ten *jâshujiyas*, altogether manned with about 700 men. They remained in the haven (*makalla'*) of Aden attending the [incoming] ships, but did not enter the city. The people of the city were very much afraid of them, but God did not give them victory and success. Many of them were killed and their ships were thrust with spears, and they died of thirst and hunger. The first of the [merchants'] ships to arrive were the two vessels of the shipmaster (*al-nâkhudâ*) Râmshat. They attacked them, but God did not give them victory. As soon as the ships entered the port (*bandar*), they were manned with a great number of regular troops, whereupon the enemy was chased from the port and began to disperse on the sea.[43]

An even more significant event in this respect is reported from the early 1490s, a few years before Gama's fleet arrived in the Indian Ocean. During the early part of the reign in the Deccan of the weak monarch Mahmud Shah Bahmani (1482–1518), the area on the western seaboard appears to have been largely controlled by a notable of Iranian origin, a certain Bahadur Khan Gilani. Bahadur

[43] See S.D. Goitein, 'Two eyewitness reports on an expedition of the King of Kîsh (Qais) against Aden', *Bulletin of the School of Oriental and African Studies* 16, 2 (1954), 247–57.

Khan is reported to have used a fleet to seize control for a time of the ports of Dabhol and Goa, and sent one of his Abyssinian slaves, a certain Yaqut, with twenty warships to take over Mahim (near the site of modern Bombay). In the course of these activities, Bahadur Khan is equally alleged to have seized a number of richly laden vessels belonging to Gujarati merchants, who complained to the Sultan of Gujarat, Mahmud Shah Begarh. He, in turn, both sent envoys to activate the Bahmani court, and sent out a fleet of his own, under the command of his admiral, one Safdar al-mulk. The maritime expedition turned out badly. Caught in a gale off the Konkan coast, the bulk of the Gujarati fleet was stranded; the crews were then massacred by those on land on Bahadur Khan's orders. The admiral himself suffered humiliation and capture. Eventually, Bahmani forces had to intervene, attacking Bahadur Khan from the rear; he was killed in battle near Kolhapur. Interestingly, his fleet (some twenty ships in all, including the captured merchant vessels) is alleged to have been handed over to the Gujarati admiral, who returned home consoled.[44] Further east, it is equally clear that, by the early sixteenth century, several states of the Malay-Indonesian world had a tradition of war-fleets, in which large jongs as well as smaller vessels could and did participate. Most telling of course are the very well-known Ming maritime expeditions of the years 1403 to 1433, which must be construed as representing a show of force by maritime means. Even if earlier accounts of maritime expeditions by the Cholas and the rulers of Srivijaya are exaggerated, there is no doubt that Asian war-vessels were known from the early centuries of the second millenium; the iconographic evidence from the Konkan suggests much the same for that area. Indeed, even the sixteenth century evidence might be re-read in part, since not all Asian maritime violence in that period comprised a mere 'reaction' to the Portuguese, or was based on an adoption of their methods. The huge war-fleets of the rulers of Arakan after 1530 were, it is clear, only based to a limited extent on the use of imported weapons, or strategic conceptions borrowed from the Portuguese.

It is nevertheless the case that some differences existed between these instances and the Portuguese way of going about matters. In the first place, we do not have clear evidence of the use of cannon

[44] The account here derives from the early seventeenth-century chronicle of Muhammad Qasim Ferishta, *Gulshan-i Ibrâhîmî*, translated by John Briggs, *History of the Rise of the Mahomedan Power in India till the Year 1612*, 4 vols., reprint (New Delhi: Atlantic Publishers, 1989), Vol. II, pp. 332–5; Vol. IV, pp. 43–4.

or firearms on the fifteenth-century Asian vessels (though Chinese vessels may have carried firearms by the early sixteenth century). Second, in many of the instances the use of maritime force was supplementary to a campaign conducted by land (the case of Mahmud Begarh in the 1490s, for example). Third, force was used in these cases either to define a rather localised sphere of hegemony (as with corsairs), or to defend a maritime zone contiguous with an already existing territorial space. What was fundamentally new about the Portuguese in the Indian Ocean was thus not the *fact* that they used force on water: it was the degree of expertise with which they did so, the fact that they did so over such large maritime spaces, separated moreover by such a distance from anything that could be thought of as their home territory, and the relatively systematic effort they brought to bear in this sphere.

THE SEARCH FOR ALLIES

Instances of Portuguese maritime violence can be found already during their stay on the African east coast, where we left Vasco da Gama and his expedition in early March 1498. We have already seen that the Portuguese had realised very rapidly that Mozambique island was dominated demographically by Muslims, that Arabic was spoken there, and that a rich trading network linked these dark Muslims with other 'white' ones, who arrived from the exterior in ships. Now, very quickly, relations developed between the Portuguese fleet and the local ruler (or Sultan), who is reported to have come to visit the Portuguese vessels on several occasions. The Portuguese tried to present him with hats, coral and the like, but he for his part is said to have been 'so proud that he despised everything that was given him and asked that he be given scarlet (*escarlata*) which we had not taken along'.

It is of some significance at this stage that Vasco da Gama used these contacts to persuade the Mozambique ruler to 'give' him two pilots to take along. This is, in the first place, a recognition on the part of the Portuguese that they had entered a world of maritime skills, where it made more sense to use local expertise rather than to experiment in the interest of 'discovery'. The Sultan is reported to have agreed to this request, provided the Portuguese paid the pilots well; Gama hence decided to give them 30 *misqâls* of gold, and some cloaks, binding them to the condition that one of them had to stay aboard the fleet at all times. Soon after, relations began to deteriorate. The fleet had decided to anchor now at some distance from the settlement, off a small island, apparently 'to celebrate Mass on Sunday, and to confess'; the implicit sense is that they did not wish the local inhabitants to be able to observe their religious practices. While Vasco da Gama and Nicolau Coelho were out in the ships' boats, they were accosted by a group of five or six local boats, with men armed with bows and arrows, who asked them to return to the main settlement. Gama not only refused, but ordered his men to fire on the boats. Hearing the sound of shots, Paulo da Gama arrived on the *Bérrio*, and the local boats fled landwards in all haste.

Some days earlier, the Sultan and Nicolau Coelho had had a rather curious encounter. The former asked the Portuguese captain's men to accompany him to his house, and an exchange of gifts took place, of cloth and prayer-beads, and also of conserves. The anonymous account continues: 'And that was so long as it seemed to him that we were Turks or Moors from some other part, because they asked us if we came from Turkey, and that we should show them the arks (*arcos*) of our land, and the books of our religion (*lei*).' In brief, the Portuguese, realising that Mozambique was inhabited by a Muslim population, had adopted a strategy of dissimulation, and did not wish their interlocutors to know that they were Christians. This explains too why they went off to a nearby island to celebrate Mass, as also why they implicitly allowed the Sultan to proceed with gifts of a certain religious significance, such as black prayer-beads (*contas pretas*). The Portuguese attitude is interesting, given the fact that, after all, they knew that Christian Italians traded with Muslims in the Levant, and were also aware that the same could not have been unknown to the Arabs trading in Mozambique.

At any rate, whether through the pilots who now co-habited with the Portuguese or by some other means, the Sultan seems to have discovered the Christian identity of the Portuguese. Once they suspected that their masquerade was seen through, the Portuguese began to fear an adverse reaction; the anonymous writer thus notes: 'And after they knew we were Christians, they ordered that we should be captured and killed by treason, but their pilot whom we took along with us, revealed everything that they had ordered should be done to us, if they could put it into effect.' Thus, believing the veracity of this rumour, the fleet left Mozambique and went further down the coast, but was forced back to the island where Mass had been celebrated, by the weather and winds.

It is altogether unfortunate that we do not have any texts available from the Mozambican side, to provide us a perspective on how this curious behaviour on the part of the Portuguese was perceived. Let us nevertheless imagine how it must have appeared: the arrival of three ships, rigged and outfitted in an unusual style, with fair men on board, of whom a few speak Arabic, and who begin to make enquiries as if they are innocent of the local trade-routes and goods. They demand two pilots, and keep one of them constantly on board (almost as a hostage for the good behaviour of the other), and rather than staying in the port, choose to anchor some distance away in an island. When challenged, and asked to return to the port, they fire with arms that were still unusual in the region (but in use by the Ottomans already, hence possibly their identification with Turks). When asked for details of their religion, they are evasive. And finally, they sail away without the usual formalities. All of this must have added up to a rather suspicious picture, the key to which was the Portuguese desire to conceal their real identity.

Diplomatic means were hence broached when the Portuguese returned, forced to do so – as we have seen – by the adverse winds. A 'white Moor', who had the title of *sharîf* (and who was reputedly a great drunkard) was sent by the Sultan to the Portuguese vessels, to 'make peace with us and be our friend', which suggests that the atmosphere was already somewhat unfriendly. Nothing much seems to have come out of it. Another Muslim, who stated that he was a pilot from the Red Sea, also came aboard with his young son, and offered to show the Portuguese where they could get fresh water; he may of course have been sent to gather information. In search of fresh water, the Portuguese sent out boats after midnight (a rather significant hour for such activity) to the port with this so-called pilot, but he failed to show them the water-source. After spending the next day aboard the ship, the boats were again sent out under cover of darkness. This time, they found the water-source defended by twenty men armed with spears, but Gama fired on them with bombards, and managed to get the water he desired.

We are now in the last week of March 1498, and the affair of the fresh water has truly become contentious. Gama had decided that a show of force was necessary so that the local inhabitants would become aware 'how much harm we could do them if we wanted'. He hence took out the ships' boats, armed them with bombards, and approached the water-source, where the 'Moors' for their part had improvised a wooden stockade to defend it. With a whiff of bombard-shot, the Portuguese cleared the beach of the first set of defenders, and a minor battle commenced that lasted three hours. Two local men were killed, and the Portuguese also managed to capture two boats, taking several prisoners back to their vessels. One of these boats belonged to the *sharîf* who had earlier visited the fleet as ambassador; the booty, mostly cotton cloth and some meal, was distributed amongst the sailors, while Gama himself kept some books ('of their law') that he found, to take back to Dom Manuel.

With a spirit of triumph then, on Sunday 25 March, the Portuguese went to supply themselves with water without any resistance, and then went and fired off some bombards in front of the main settlement, where 'the Moors remained inside their houses for they did not dare come out on the strand'. Then, on the 27th, satisfied that they had quelled the local resistance, the three ships went back to the island where Mass had been read (*ilhéus de S. Jorge*), and departed from there on the 29th, with a poor wind, further north along the coast.

Their main informants at this stage appears to have been the Muslim pilot, and the other man who had come aboard of his own volition with his son. The pilot was, however, viewed with some suspicion, and even whipped by Vasco da Gama on 1 April for telling him wrongly that some islands that were in view were in fact the mainland. The account tells us moreover that 'to the first of the said islands we gave the name of the Flogged Man's Island (*ilha do Açoutado*)', to mark this event. The fleet seems to have been in search of the island inhabited by Christians, which they had been assured at Mozambique existed somewhere along the coast (and which it is sometimes believed was a reference to Kilwa). Eventually, on 4 April, on being informed that they had already gone past it, they decided to make for Mombasa. But, meanwhile, having passed Zanzibar (which they mistook for the mainland), the ship *São Rafael* ran aground on some shoals, and had to await the high tide to continue its voyage. Passing Pemba, the fleet finally arrived before Mombasa on 7 April. This was with some relief, as a great part of the crew was still rather ill, and the Portuguese believed that in Mombasa at least they would find themselves amongst Christians. Thus, the text tells us:

Plate 12 Drawing of Vasco da Gama from Sir Richard Fanshawe's translation of *Os Lusíadas* (1655).

> There [in Mombasa], we rested with great pleasure, for it seemed to us that the next day [Sunday 5 April] we would go and hear the Mass on land with the Christians who we were told were here, and who lived separately from the Moors, and had their own head (*alcaide*). The pilots whom we had brought said that on this island of Mombasa there existed and lived Moors and Christians, and that they lived separately from each other, and that each of them had their head (*senhor*), and that when we reached there, they would do us great honour and that they would take us to their residences.

A prospect that was radically different then to what the Portuguese had left behind in Mozambique. But the very next day, the Portuguese began to have doubts. First, there arrived a boat carrying a hundred men, of whom Vasco da Gama allowed only four or five on board, fearing that otherwise they would take his ships by force. After a discussion with these men, they returned ashore, and the ruler sent a sheep and some fruit as an offering, asking the Portuguese to come ashore. But after his experience in Mozambique, we see a change in Gama's attitude. The arrival of two men ('who said they were Christians, but who did not seem so to us') did not reassure him, and nor did the presence of four of the 'most honoured Moors'. Eventually, the next day, two Portuguese (probably convict-exiles, or *degredados*) were sent to the King's palace, where they were received with ceremony; they also visited the town, and the house of two 'Christian merchants', who showed them some texts referring to the Holy Spirit. Samples of ginger, pepper and cloves were also sent aboard.

Two main contrasts need to be developed between the comportment of Gama's fleet earlier on and in Mombasa. First, we should note that the Portuguese do not seem to have attempted to conceal their identity as Christians in Mombasa, but this was perhaps because they believed (wrongly) that there was a substantial Christian population there. But, secondly, it is noticeable that Vasco da Gama and the other captains did not go ashore in Mombasa (unlike in São Brás), leaving this task to the expendable convict-exiles, who had been brought along for such dangerous tasks in the first place. Their suspicions were confirmed at the following incident, which took place on Tuesday 10 April, the day after the two convict-exiles had visited the town:

> And in the ships there were Moors with us, who having seen that we were not leaving, collected together in a boat (*zavra*), and as they were passing by the prow, the pilots who came with us from

> Mozambique threw themselves in the water, and those in the boat picked them up, and when it became night the captain tortured (*pingou*) two Moors from amongst those who were [still] with us, to make them tell us if some treason was planned. They said that when we went in [on land], orders had been given to capture us and take revenge for what we had done in Mozambique. And just as another was to be tortured, he threw himself into the sea with his hands tied, and another threw himself in the *quarto de alva*.

Hostilities thus commenced and, later the same night, two boats (*almadias*) approached the Portuguese ships, and several men swam across in an attenpt to cut the cordage of the *Bérrio* and *São Rafael*. However, they were foiled, and our account notes: 'These and other evil deeds were conceived by these dogs, but Our Lord did not permit that they should happen, for they did not believe in Him'! The Portuguese on the fleet thus seem very largely to have seen the conflict in Manichaean terms: Christians opposed to the Muslim 'dogs' (*perros*). The description of Mombasa is interesting in this respect, for it states:

> This city is large and is located on a height above the sea, and it is a port where many vessels enter every day, and at the entrance there is a pillar (*padrão*). And the part of the settlement (*a vila*) close to the sea has a low fortress. And those who went on land told us that they saw many prisoners walking about this quarter in irons, and these, it seemed to us, must have been Christians, because the Christians in this land are at war with the Moors.

Thus, the implicit point of departure is that the prisoners and subject peoples 'must have been Christians', and it is noted a little further down the text that even the Christian merchants were greatly oppressed by the Moor King. Now despite the relatively healthy air and water of the city (as a result of which symptoms of scurvy diminished dramatically), the fleet could hardly stay on in a city once they 'knew of the malice and treason that these dogs wished to bring about against us'. The fleet hence proceeded down the coast, and took another vessel belonging to an old and 'honoured Moor' and carrying him on board with his young wife and seventeen other men, as well as a cargo of gold, silver and foodstuff.

We have seen that the pilots from Mozambique had been lost at Mombasa. The Portuguese thus cast about for other pilots, and indeed this is described as their main motive in taking the vessel mentioned above; however, they were disappointed in respect of pilots. Thus, on the advice of some of the Muslim prisoners they still held on board, the fleet decided to put in at Malindi (described as thirty leagues further up the coast from Mombasa) on 14 April. In particular, Gama had heard that there were four Christian ships at Malindi, belonging to Indians (here *índios*, and not *indianos*), and thought that these might be able to give him Christian pilots, as well as water, food and other supplies. At the same time, much mutual suspicion is evident in the Portuguese fleet's dealings at Malindi. When the Portuguese anchored off the port, no boats showed up to greet them, since – our anonymous author himself notes – 'they had already been advised and knew that we had taken a vessel with Moors'.

Thus, after a wait, on the day after Easter (which fell in that year on 15 April), the old Muslim shipowner was set ashore, and sent to the ruler of Malindi with a message saying that Gama wished to enter into peaceful relations with him. He returned later that day, with presents, and a message from the ruler, saying that he was willing to receive Gama, and to give him the goods *and* the pilots he wanted. But the latter showed a certain reluctance to set foot on land, and instead asked the ruler to come out in a boat and meet him off shore. When the two vessels (Gama's *batel*, and the *zavra* of the king) were side by side, the king entered Gama's boat and a long conversation followed. The king offered to visit the Portuguese ships, if Gama were willing to visit his palace; the latter refused, stating that he had orders from his King against this. However, since the atmosphere seemed largely propitious to the Portuguese, Gama released the Muslim prioners on board as an act of goodwill, and also had the cannons fire a salvo to honour the king. At last, after about three hours, the latter departed, taking back with him the former captives, and two Portuguese (to whom he wished to show his palaces); however, as hostages against these two Portuguese, he left behind one of his sons and a *sharîf*.

Mutual suspicions were thus somewhat (if not wholly) dissipated, so that the next day Gama and Nicolau Coelho went to inspect the city from closer up; two of the king's cavalrymen put up a show for them on the strand. The king redoubled his request that Gama come ashore, but the latter steadfastly refused. It was here that the Portuguese had their first encounter with Indian merchants, more concretely with those on board the four vessels mentioned above, who may have been St Thomas Christians from Kerala. The description in our text thus insists that these men carried Christian icons, worshipped Christ and the Virgin, and said Christian prayers. We note besides that they were dark, wore few clothes, and had long beards and long hair; further, that they did not eat beef, spoke a language that was quite different from 'that of the Moors', but nevertheless could communicate crudely in Arabic.

A curious aspect of the description of their ships is the repeated mention of firearms there; thus, when they held a celebration for the Portuguese, it is noted that 'they fired many *bombardadas*, and let off fireworks (*foguetes*), and gave great cries'. These very same men are reported to have told Gama not to step ashore, as the local men were not to be trusted, as they 'did not speak from their hearts, nor of their will'. Taking them at their word, he seized one of the king's favourites (*um seu privado*) a few days later (on 22 April), and imperiously demanded that he be given pilots as promised. The text continues: 'And when he got the message, the King at once sent a Christian pilot, and the Captain [Gama] released the *fidalgo* whom he had kept on board. And we rejoiced greatly with the Christian pilot whom the King sent us.' Finally, after nine days at Malindi (an urban centre that is compared by our author to the Portuguese town of Alcochete), on 24 April, the fleet departed for 'a city which is called Qualecut, of which city the King [of Malindi] had notice'. While the disappearance of the *regimento* carried by Gama once more hampers us, it is possible to interpret this phrase as meaning that the choice of Calicut as destination was one that was improvised by Gama and his captains, once on the African east coast.

We may pause here for a moment to reflect briefly on the experiences of Gama's fleet in the area, and also to locate them in the context of the area's political history. The fleet, we have noted, while sailing from south to north, had put in at Mozambique, Mombasa and Malindi in that order; rumours were also gathered concerning the significance of Kilwa and Pemba, but these centres were not actually visited. Now all of these cities are referred to by Arab geographers and travellers from at least the twelfth century (for instance by al-Idrisi, Yaqut, Ibn al-Mujawir, and Ibn Sa^{c}id), and in particular Kilwa and Mombasa are singled out for attention. Kilwa, we may note, was visited by Ibn Battuta in the 1330s, and described by that time as having surpassed Mombasa as the dominant centre of the *Bilâd al-Zanj* (as the Swahili coast was known to the Arabs). However, by the late fifteenth century, sharp trading and political rivalries of some importance existed between these centres; if, in the Somali coast (Arabic, *Bilâd al-Barbar*) to the north, Muqdisho (Mogadishu) preserved its pre-eminence, the dominant position of Kilwa further south was under challenge by both Mombasa, and the ruling family at Zanzibar.[45] In order to understand the relatively warm welcome afforded to the Portuguese at Malindi, it is also useful to bear in mind that the Sultan there too was struggling at this time to shake off the tutelage of Kilwa.

The Portuguese did not immediately comprehend this situation. But we should nevertheless insist here that, contrary to what a number of historians implicitly assume, this East African sojourn was crucial in defining Portuguese conduct once in Kerala. It is a fundamental error to have a vision of Gama arriving in Calicut, fresh out of Portugal, so to speak. We note the growing shadow of a form of extreme suspicion in Gama's attitude towards the commercial centres, and rulers that he comes across, as he proceeds from Mozambique to Mombasa, and finally to Malindi. By the end of the stay in East Africa, a pattern may be discerned; the Portuguese captains wait for vessels to approach them, and typically take hostages before setting foot on shore. The three captains (the two Gamas and Coelho) rarely if ever set foot ashore themselves, even in circumstances as favourable as those in Malindi.[46] Concessions are extracted by the strategic use of force, and one of these concessions was – as we have noted – the 'Christian' pilot whom Gama acquired in Malindi.

THE 'MYSTERY' OF IBN MAJID

Now, the wisdom of hindsight was to show that the Portuguese fleet of Gama, which set out from Lisbon in early August 1497, had in fact set out much too late in the year to benefit from the weather patterns along the route. In the sixteenth century, it became usual therefore for vessels to leave Portugal considerably earlier in the year, typically in February or March. Further, and aggravating the situation, was the fact that by the time the fleet

[45] Cf. the useful discussion in Jean-Louis Triaud, 'L'expansion en Afrique', in Jean-Claude Garcin, *et al., Etats, Sociétés et Cultures du Monde Musulman Médiéval, Xe–XVe siècle* Vol. I (Paris: Presses Universitaires de France, 1995), pp. 420–9.

[46] By 1505, this approach had crystallised into a policy, as we see from the *regimento* given to D. Francisco de Almeida, where he was instructed that 'in no place (. . .) never at any time should you go out on land (. . .); and if you need to come and speak to some king, it shall be on the sea'; cf. R.A. de Bulhão Pato (ed.), *Cartas de Afonso de Albuquerque*, Vol. II, pp. 314–15.

had made the slow and long trek up the Swahili coast, it was somewhat late in the season to cross the western Indian Ocean, in view of the impending arrival of the south-west monsoon. A dim awareness of this may have lain behind Gama's anxiety to find a local pilot at Malindi, once those from Mozambique had left him so summarily. But far more important, on balance, was the fact that he needed to have the services of someone to advise him of where the fleet was when landfall was made on the Indian west coast. After all, unlike in East Africa, Gama had a specific task in India: namely to make contact with not a region but a commercial city, that of Calicut. These pressing and concrete reasons, rather than the abstract need to find someone to 'unlock the mystery of monsoons', appear to have been what lay behind the recourse to the aid of a pilot from the region. Besides, the search for, and the use of, local pilots is a recurring theme once the fleet entered the trading zone of the Indian Ocean, indeed from Mozambique onwards.

It seems clear enough that the Portuguese knew that between the African east coast and India lay a somewhat large stretch of water (*huma muito grande enseada*), which they had to traverse from west to east; and, on the face of it, traversing this stretch was hardly the most major of technical problems, when compared to the far more problematic navigation of the south Atlantic. Of some significance is their recognition very early on (by the time they reached Mozambique), that they were in a region with its own maritime knowledge and skills, which could be put to use to facilitate their task, even if this diminished their pretensions to 'discovery'.

Now, much has been made of the identity of the pilot who was sent over to Gama by the ruler of Malindi. The text that we have cited claims that he was 'Christian'; but few modern historians have been willing to accept this claim, especially since Barros and Castanheda are unanimous in terming him a 'Moor'. Instead, there has been a persistent (and, over time, even popular) belief in the twentieth century that Gama had at his disposal none other than *the* most celebrated Arab navigator and theorist of navigation of the fifteenth century, Shihab al-Din Ahmad ibn Majid al-Najdi.[47] Where do the origins of this theory lie, and what are its

[47] For a discussion of this personage and his work, see G R. Tibbetts, *Arab navigation in the Indian Ocean before the coming of the Portuguese, being a translation of the 'Kitâb al-Fawâ'id fi usûl al-bahr wa'l-qawâ'id' of Ahmad b. Mâjid al-Najdî* (London: Oriental Translation Fund New Series, 1971).

ramifications? To address the latter question first, the theory that Ibn Majid was 'the pilot of Vasco da Gama' opens up several pleasing possibilities for romantically inclined historians of various stripes. In the first place, it permits – by sheer coincidence – the meeting of two 'celebrities' of the past from two quite different domains, as it were, somewhat like the poet Hafiz Shirazi and the conqueror Timur.[48] This meeting, which is entirely in keeping with the hagiographic logic of so many accounts of Vasco da Gama, thus adds, interestingly enough, to the lustre of Vasco da Gama in the eyes of Portuguese nationalist historians. For was he not thus initiated into the mastery of the Indian Ocean by the greatest Master of the epoch? Besides, reading his voyage as a mystic one (a tradition which, as we shall see below, dates to Camões's *Lusíadas*), the meeting with a great master (and a consequent 'initiation') should naturally form a part of the structure of the voyage. At the same time, Asian nationalists can read the matter in a different but perfectly complementary fashion: for what greater contrast than between the greed and rapine of the Portuguese, and the scientific generosity and openness of spirit of Ibn Majid, who let the Portuguese share his knowledge, even as he had a premonition that his world would thus be destroyed![49] Or, taking still another tack, one could read this as an example of how Europeans used some Asians against others, exploiting the knowledge of the great Ibn Majid, to prise open the Indian Ocean to their own advantage.

In fact, the identification of the pilot (specifically termed a 'Moor from Guzerate' by both Barros and Góis, and equally called a Gujarati by Castanheda) with Ibn Majid – who was reputedly an Arab from Oman – was largely the work of the well-known French Orientalist, Gabriel Ferrand, writing earlier in this century.[50] Ferrand's view, which he developed in his authoritative biographical article on Ibn Majid, in the *Encyclopédie de l'Islam* (1927), was based in turn on a particular Arabic text of the late

[48] The difference, of course, is that Hafiz was sought out by Timur on his conquest of northern Iran; for an early seventeenth-century reflection on this 'paradigmatic' meeting, see *Khâtirât-i-Mutribî Samarqandî (being the Memoirs of Mutribi's sessions with Emperor Jahangir)*, ed. Abdul Ghani Mirzoyef (Karachi: Institute of Central and West Asian Studies, 1977), Account of the First Meeting.

[49] On the occasion of the fifth centenary of Gama's voyage in 1998, at least one well-known centre of academic counter-culture in Delhi had been planning until recently to organise a programme around the figure of Ibn Majid, as symbolising non-western knowledge incorporated and crushed by the hegemonic expansion of western culture!

[50] Gabriel Ferrand, 'Le pilote arabe de Vasco da Gama et les instructions nautiques arabes au XVe siècle', *Annales de Géographie* 172 (1922), 289–307; also see Ferrand, *Introduction à l'Astronomie Nautique Arabe* (Paris, 1928), p. 186, *passim*.

sixteenth century, the chronicle by Qutb al-Din Muhammad al-Nahrawali al-Makki (1511–82), entitled *Al-Barq al-Yamânî fî al-Fath al-ᶜUṣmânî*. Coincidentally, Nahrawali was himself of Gujarati origin, but lived in Mecca, where he authored several chronicles. The text in question was commissioned after the Ottoman reconquest of the Hijaz by the governor of Yemen, Çighalazade Sinan Pasha, and was thus meant to celebrate the Ottoman achievement; it seems to have been written about 1565.

Now the relevant passage of the text, from the second chapter of the first book, would run somewhat as follows:

> At the beginning of the tenth century [AH], among the astounding and extraordinary occurrences of the age was the arrival in the lands of al-Hind of the accursed Portuguese (*al-Fartaqâl*), a nation of Franks (*al-Firanj*) – may they be cursed! One of their bands had embarked in the straits of Ceuta, and entered the Sea of Darkness [the Atlantic], and passed around the White Mountains (*al-Qumr*) where the Nile originates. They went east, and passed close by the coast through some straits, on one side of which is a mountain, and on the other the crested waves of Darkness. This place is so tempestuous that the ships of the Franks could not cast anchor, and were shattered to pieces, and none of them survived. Thus they persisted for a time, perishing each time at that spot, and none of the Franks managed to arrive in the Indian Ocean (*al-bahr al-Hind*), until one of their galleys managed to escape and make its way towards al-Hind. However, they were unable to gather knowledge of this sea, until it was given to them by an experienced pilot, whose name was Ahmad ibn Majid, with whom one of the Frankish captains, called Admiral (*almilandî*) came to be acquainted. After giving him wine to drink on a number of occasions, the pilot in a state of drunkenness showed him the way, and said to him: 'In this place, do not follow the coast, and make for the open sea; then turn and approach the coast, and do not fear the waves.' Once they did so, their ships avoided frequent wrecks, and the Franks became numerous in the Indian Ocean.[51]

Here then is an unmistakable finger pointed at Ibn Majid, who – in a state of inebriation – is supposed to have shown the

[51] This passage from the text was first published, with a translation, in David Lopes, *Extractos da História da Conquista do Yaman pelos Othomanos: Contribuições para a história do estabelecimento dos Portugueses na Índia* (Lisbon: Imprensa Nacional, 1892), pp. 11–12 (text), 39–40 (Portuguese translation). For a modern edition, see Qutb al-Din Muhammad ibn Ahmad al-Nahrawali al-Makki, *Al-Barq al-Yamânî fî al-Fath al-ᶜUṣmânî*, ed. Ashraf ᶜAli Tabye and Hamd al-Jasir (Riyadh, 1967), pp. 18–19. My translation differs slightly from the English version (based on Ferrand's French text), in A. da Silva Rego and T.W. Baxter (eds.), *Documents on the Portuguese in Mozambique and Central Africa, 1497–1840*, Vol. I (Lisbon: CEHU, 1962), pp. 33–5.

incompetent Franks (otherwise destined to suffer shipwreck after shipwreck) the way to navigate to India. His advice, as we can see, was rather simple: to leave the African coast behind, and sail on the high seas, until they made landfall! Let us note that this rather befuddled (indeed, altogether naive) passage in Nahrawali's text does not claim that Ibn Majid ever sailed with the Portuguese, nor does it identify him specifically with the 'Gujarati' pilot from Malindi. This connection was one made by Ferrand.

The next step in the identification had to await the discovery by the Russian Orientalist, I.J. Kratchkovsky, of three rutters (or sailing manuals) by Ibn Majid in the form of poems (*arjûzas*), in a manuscript collection in St Petersburg, in the 1920s. After a number of vicissitudes, these were eventually published with a Russian translation by his student T.A. Chumovsky in the late 1950s; the texts at once attracted the attention of Portuguese scholars, so that a translation into that language from Russian was undertaken on the occasion of the grand celebrations orchestrated by Salazar for the Fifth Centenary of the Death of the Infante Dom Henrique (1960).[52] Portuguese interest obviously stemmed from Ibn Majid's association with Vasco da Gama, and not really from the remaining information contained in the three rutters in question, which dealt respectively with Sofala, Melaka, and the Red Sea. It was clear from the start that the texts had had an initial prose redaction, and that they were subsequently put into verse by someone who may or may not have been the pilot himself. Particularly close attention was paid by Portuguese scholars to the Sofala *arjûza*.

In this section of the text, there are several mentions of the Franks, and their navigation in the Indian Ocean. One of these refers to their arrival in Calicut, their oppression of local peoples, and their obstruction of trade to the Red Sea. Still another notes that some Portuguese vessels were wrecked near Sofala on the 'feast of Michael' (*fî ᶜîd mikal*), something that does not appear in Portuguese texts. A third, extremely cryptic, passage is the most significant of all for the purposes of the Ibn Majid–Vasco da Gama association, and runs:

[52] T.A. Chumovsky, *Três Roteiros Desconhecidos de Ahmad ibn-Mâdjid, o piloto árabe de Vasco da Gama*, tr. Myron Malkiel-Jirmounsky (Lisbon: Comissão Executiva das Comemorações do V Centenário da Morte do Infante D. Henrique, 1960); Russian edn, Moscow–Leningrad: Academy of Sciences of the USSR, 1957. A Portuguese translation of the Sofala rutter had already been made by S.S. Costa Brochado in 1959.

The Franks' ships went to Wazah in 900 A.H. [1494–5], o my brother.
They oppressed it two full years, then they proceeded forward to India certainly.
Whoever tries to go to China (*Sîn*) minds the dangers of his destination; otherwise he would have expected the fulfilment of his desires.
The Franks returned from India to Zanj by this road; later on in 906 A.H., they went to India, my brother.
They bought houses and settled; they had friendly relations with the Zamorin and trusted him.
Meanwhile, people tried to guess their plans; they thought they might be conquerors or foolish thieves.
The Franks used to coin currency in the harbour of Calicut between their voyages.
I would have liked to know what they were going to do, while all men were amazed by their actions.[53]

Consider the peculiarities of this passage (to say nothing of its obscurity), namely its claim that the Portuguese were already in Sofala in 1494–5, that they coined money in the port of Calicut, that their power extended to the Egyptian Nile, and so on. Yet, many Portuguese historians in their enthusiasm brushed these aside. To Armando Cortesão, the references to the voyages of 1494–5 were signal proof of his hypothesis, that there had been Portuguese voyages to the Indian Ocean before 1497–8. And to many others, the last-but-one line was virtually an admission of guilt by Ibn Majid; had he but known how it would all turn out, he would never have let the Portuguese into the Indian Ocean! Thus, wrote two official Portuguese historians in the context of the 1969 celebrations: 'Ibn Mâjid in the *arjûza* does not admit his intervention in the voyage of Vasco da Gama, but it is obvious from the fact that the pilot confesses his regret here.'[54]

The counter-offensive against this interpretation was mounted, in the first place, by the Syrian historian Ibrahim Khoury, then by Cortesão's former collaborator Luís de Albuquerque.[55] Khoury argued that the texts of the rutters preserved in the Oriental

[53] Cf. Ibrahim Khoury, *As-Sufaliyya, 'The Poem of Sofala', by Ahmad ibn Mâğid* (Coimbra: Junta de Investigações do Ultramar, 1983), p. 93. For an earlier translation into Portuguese, see Chumovsky, *Três Roteiros Desconhecidos*, p. 48; Machado and Campos, *Vasco da Gama*, pp. 51–2.

[54] Machado and Campos, *Vasco da Gama*, p. 52 n. 2.

[55] Khoury, *As-Sufaliyya, 'The Poem of Sofala'*, pp. 17–25; Luís de Albuquerque, 'Madjid, Ahmad ibn', in Albuquerque and Domingues, eds., *Dicionário de História dos Descobrimentos Portugueses*, Vol. II, pp. 639–40.

Institute in Leningrad (St Petersburg) were clearly corrupt, and had suffered from a number of, at times rather clumsy, emendations and additions by the hand (or hands) of copyists. This had led to stylistic changes, but also to changes in content; further, some strange digressive passages had been introduced. Amongst these were the references to the Portuguese in the Sofala rutter. He established from internal evidence that the text should have contained 701 verses, whereas the extant copy had as many as 807. Further, Khoury was able to demonstrate that, by the late 1480s, the pilot considered himself old, and unable to navigate competently any more. More telling still, internal textual references demonstrated that the 'Poem of Sofala' was composed originally before the mid-1470s, thus nearly two and a half decades before Gama's voyage, and was at best revised in the 1480s. References to the Franks therein are hence, to put it mildly, somewhat anachronistic. Drawing on this work, and that of some other authors (who had argued that the Portuguese shipwrecks in the *arjûza* dated to after 1500), Luís de Albuquerque for his part stressed the highly improbable nature of the identification of the pilot, called 'Malemo Cana' by João de Barros, and 'Malemo Canaca' by Damião de Góis, with the celebrated Arab navigator. Nevertheless, doubts remained, largely on account of the rather explicit statement in al-Nahrawali's chronicle, which despite its rather later composition, was after all an authentic 'Oriental' source.[56] It was thus a matter of some significance that the scholarly argument could finally be closed in the mid-1980s, although the popular legend of Vasco da Gama guided by Ibn Majid continues to grow and spread. The last nail in the 'Ibn Majid hypothesis' was driven in by the discovery of some Italian letters written from Lisbon to Florence, immediately after Gama's return. Here, there is mention of how Gama had brought back the Moorish pilot (who apparently could speak some Italian!) with him to Lisbon, where the Italian representatives of the Sernigi-Marchionni combine were able to question him.[57] Thus,

[56] Cf. the recent, rather confused, discussion in Joseph Chelhod, 'Les Portugais au Yémen d'après les sources arabes', *Journal Asiatique* 283, 1 (1995), 1–3. Chelhod demolishes with much sound and fury his chosen straw man, the Moroccan historian Abdelhadi Tazi, who had claimed for his part that Ibn Majid could not have been drunk when Gama was at Malindi, since it was the month of Ramadan!

[57] Biblioteca Riccardiana, Florence, Codex 1910, fl. 66r (b), 'Copia della siconda lettera dipoi [che] venne il pidotto', in which he is described as 'el Moro che fu per piloto de Malinde, terra di mori, in fino a Chalichut'. A later letter from Guido di messer Tommaso Detti, Codex 1910, fl. 70r (a), dated 10 August 1499, mentions that the pilot spoke Italian; this confirms the statement in Sernigi's letter of 10 July 1499, in Codex

unless we wish to argue that Ibn Majid had in his dotage taken on a Gujarati guise, visited Lisbon, and then said nothing of it in his writings, we shall have to bury the hypothesis once and for all, the more so since there is no indication that Gama's (probably Gujarati) pilot ever returned to Asia. As for al-Nahrawali, writing under Ottoman patronage, we may speculate that this accusation was part of a more general attempt to discredit the Arabs, and portray the Ottomans as being at the forefront of the anti-Portuguese struggle.

MOORS AND SPICES: CALICUT, 1498

We have left the Portuguese fleet in late April, en route from Malindi to the Indian west coast. The traversing of the western Indian Ocean took a mere twenty-three days' sailing, which are passed over quickly by our anonymous author, even if they receive more elaborate treatment at the hands of Barros and later chroniclers. On 18 May, land was sighted (a *terra alta*, possibly Mount Eli, in northern Kerala), and the pilot then advised them to follow the coast so that he could be sure of where they were. On a closer inspection of the Indian Western Ghats, which were visible with some difficulty on account of squalls and showers, the pilot then declared that this was the land they had wished to attain. The ships anchored on Sunday 20 May, just north of Calicut, between that city and the settlement of Pantalayini-Kollam ('Pandarane' to the Portuguese). Since they were a league and a half distant from the shore, boats approached them that evening from the land, to ask who they were; they pointed the Portuguese in the precise direction of Calicut. Thus it was on 21 May 1498 that the first Portuguese was sent ashore with these local boats.

Now the man sent on shore by Gama was a convict-exile (or *degredado*), possibly called João Nunes, not someone of authority in the fleet's hierarchy. Strongly marked by his experiences on the African east coast, we note that Gama allowed the local inhabitants to take the first initiative (that is, by sending out boats), and also did not risk either his own person or that of the other captains or well-born persons on board at first. The *degredado*,

1910, fl. 61v (a), though Sernigi also adds, on fl. 63 v (a), that the pilot was brought back by Gama against his will (*contro a sua voglia*). For an edition of these texts, see Radulet, *Vasco da Gama*, pp. 169–98. Radulet insists, like Ravenstein, that the pilot could not have spoken Italian, and that the texts must in fact refer to Gaspar da Gama. This is an unwarranted conclusion, with no clear basis in documentation, as is Radulet's notion (p. 179, n. 40), that the pilot was Arab.

we are told, was taken to a place where there were 'two Moors from Tunis, who knew how to speak Castilian and Genoese', one of whom is later referred to variously as 'Monçaide' or 'Bontaibo' (perhaps Ibn Tayyib); this appears to suggest, once more, that the Portuguese were initially identified as Muslims from the Levant or Maghreb. The convict-exile, and not Vasco da Gama, is thus the real protagonist of the following celebrated scene:

> And he was taken to a place where there were two Moors from Tunis, who knew how to speak Castilian and Genoese. And the first greeting that they gave him was the following:
> – The Devil take you! What brought you here?
> And they asked him what he had come to seek from so far; and he replied:
> – We came to seek Christians and spices.
> And they said to him:
> – Why do the King of Castile and the King of France and the Seignory of Venice not send men here?
> And he replied that the King of Portugal did not permit them to do so. And they said that he did well.
> Then they welcomed him and gave him wheaten bread with honey, and when he had eaten, he came back to the ship. And one of those Moors came back with him, who as soon as he entered the ships, began to say these words:
> – *Buena ventura, buena ventura* (Good fortune, good fortune)! Many rubies, many emeralds! You should give many thanks to God for having brought you to a land where there are such riches!
> We were so amazed at this that we heard him speak and we could not believe it – that there could be anyone so far away from Portugal who could understand our speech.[58]

Let us insist once more, then, that contrary to what is often stated by modern historians therefore, Islam and Christianity did not confront each other directly at the moment of Vasco da Gama's arrival in Calicut. On the contrary, the curious unity of western Mediterranean culture is affirmed, where Tunisian Muslims spoke Arabic, Castilian and Genoese, and where the Portuguese could actually have a sense of relief at seeing such 'familiar' faces. Further, let us note that the search for 'Christians' is in direct continuation of the attempts on the African east coast, where news of Prester John is sought anxiously, and evidence of a Christian presence at Kilwa gathered with relief. 'Christians' here

[58] Velho, *Roteiro*, pp. 54–5. The identity of the convict-exile (*degredado*) is sometimes given as João Martins, and he is also sometimes identified as a converted Jew (*cristão-novo*).

does not mean the desire to proselytise, but rather the search for eastern allies in Dom Manuel's plans. As for the Tunisian traders, they did not see in the small fleet of vessels from distant Portugal a particular threat. There were riches enough in Calicut to be shared, it seemed at first.

We directly come across a problem though in the Portuguese conception of the town, brought about by their poverty of categories. Faced with the 'primitive' bushmen and pastoralists of southern Africa, the problem of religious identity had not struck the Portuguese as significant. Once into the city-states of East Africa, 'Moors' and 'Christians' were the two prime categories for distinction, with some further refining of course ('white' and 'black' Moors, those who were treacherous and those who were not). On the Indian south-west coast, faced with a dense, relatively sophisticated, network of small urban trading settlements, the anonymous author proceeds directly as follows:

> This city of Calecut is of Christians, who are dark men, and some of them go about with large beards, and long hair on the head, and others have shaven heads, and still others cropped. And they have certain topknots on their crowns, as a sign that they are Christians, and moustaches with their beards. And they have their ears pierced, and in the holes they wear much gold, and they walk about naked from the waist up, and below they wear certain very delicate cotton-cloths. And those who go about dressed like this are the most honoured folk, and the others wear what they can. The woman of this land are in general ugly, and have small bodies, and they wear much gold jewelry around their necks, and many chains on their arms, and on their toes they have rings with precious jewels. All these people are well-mannered and suave, so it would seem. And [But?] they are men who, on going beyond first impressions, know little and are very greedy.

These impressions were, however, the product of a stay in Calicut that lasted all of three months, for Gama's fleet left the port definitively only on 29 August. Returning to the situation in late May, Gama's next step was to send two men ashore, to seek out the King of Calicut wherever he might be, and to inform him that 'an ambassador from the King of Portugal was there and that he brought letters from him'. These men made their way south to Ponnani (where the ruler was), were rather well received by him, and were told that he would return shortly to Calicut. In the meantime, the Portuguese were advised to take their ships to Pantalayini, which was a good port unlike Calicut itself; a pilot was sent along to guide them. Gama agreed to this, but decided – we note – not to enter the roadstead, presumably out of suspicion. By the time the three vessels had anchored, a message arrived from the king, to the effect that he was now in Calicut; further, he sent his representative, called 'Bale' in the text (and described as 'like an *alcaide*') to Pantalayini, to invite Gama for an audience.[59]

This activity had taken nearly a week. Thus, it was only on 28 May that Gama actually set foot on land, taking along about a dozen men (including the anonymous text's author), and carrying flags, trumpets and bombards on their boats. On reaching land, he was at once met by the *Bale*, with an armed retinue including men with naked swords; Gama was placed in a palanquin, and the party left for Calicut by way of a place called 'Capua' (Kappatt). Here, halting midway, Gama and his men entered the house of a local notable (*um homem honrado*), and ate a meal of rice and fish. They embarked immediately thereafter in two boats that were tied together, and descended the backwaters towards Calicut itself. The Portuguese party seems to have been quite a spectacle, for the anonymous writer notes that an 'unending' number of people crowded to see them go by. Passing a spot where ships and boats were dry-docked, the party eventually disembarked, still surrounded by large crowds of people, including 'women who came out of their houses with their children in their arms'.

The Portuguese were now taken to a large structure, which they thought was a church, described as being the size of a monastery, and with a pillar at its entrance with the figure of a bird on it, 'which seemed to be a cockerel' (*galo*). Seven small bells hung before the principal door. The 'church' had a large tank attached to it, and there were many others in its vicinity. Entering the structure, the Portuguese found themselves at another smaller door halfway into the body of the building, with stone steps leading up to it. 'Inside', writes our author, 'was a small image, which they said was Our Lady'. The Portuguese were not allowed to enter the inner part of the structure, and hence Gama and the others had to rest content with saying their prayers at the outer enclosure. They received a sprinkling of holy water (*água benta*), and a white ash 'which the Christians of this land have the habit of putting on their foreheads, and bodies, and around the neck and along their upper arms'. Gama took the ash, but said

[59] The 'Bale' or 'Baile' of this text, is the same as the 'Catual' of the chronicles and of Camões. 'Catual' is the Portuguese rendering of the Perso-Arabic *kotwâl*, meaning magistrate or warden.

Plate 13 Exterior of a medieval Hindu (Saiva) temple at Kottakkal, not far from Calicut.

that he would put it on later; the anonymous author noted that the whole ceremony was in the hands of certain men wearing cords of thread across their upper bodies, who were called *quafees*. While leaving, our author noticed other curious aspects of this 'church': the many saints painted on its walls, with diadems, some with 'teeth that were so large that they came out of their mouths a bit', and most with 'four or five arms'. Before entering Calicut, they also visited another such 'church', which, it is reported, 'had the same things described above'.

Our author's powers of observation and description have thus not deserted him; contrary to what has often been claimed, he did carefully note the religious practice and peculiarities of this 'Eastern Christianity'. He was, of course, obviously dealing with a temple, and the bird's figure on the *stambha* outside suggests moreover that it was a Vaishnava temple (and not a Kali temple, as is often imagined), with the god's vehicle Garuda maintaining the customary vigil outside. We note, besides, that the Portuguese did not in fact get an opportunity to inspect the figure of 'Our Lady' closely, and took the word of their local interlocutors for what it was meant to be. Now, two obvious questions arise in this context. Given that none of the Portuguese spoke Malayalam (or any other Indian language), the communication must have been in

Arabic, through Fernão Martins, or one of the other Arabic speakers. It was obviously *theoretically* possible in this case to distinguish between Christianity and local forms of Hinduism, since early sixteenth-century Arabic vocabulary in India clearly allowed this. Indeed, the word used for the temple-priests (*quafees*) is very probably a Portuguese rendering of an amalgam of the Arabic terms *qasîs*, and *kâfir* ('unbeliever', hence also Hindu), which suggests that the Portuguese discussed who these men were with an Arabic speaker. Thus, our first question: Why were the Portuguese taken into the temple, and what were they told which they translated as *Nossa Senhora*? Again, given the presence in Kerala of St Thomas (or Syrian) Christians of long standing, were their local interlocutors really unable to comprehend that the Portuguese were Christians? Now, the fact that the Portuguese were not allowed into the inner enclosure (the *garbha gṛha*, or *sanctum sanctorum*) suggests that there were some doubts about their status, if not the outright certainty that they were Christians. The problems raised above can probably be reduced to two simple propositions. *First*, the Portuguese expected to see 'lost' eastern Christians, whose practices would differ widely from their own; they were thus willing to see in any structure that was not obviously a mosque, a church of some sort. *Second*, we should not forget that the Portuguese of the expedition had learnt their Arabic in the Maghreb, and we may imagine that there were significant differences of dialect, and oral expression, between that Arabic, and the version spoken in Kerala.[60] We need not appeal to complex philosophical propositions concerning 'cultural translation' to see how Fernão Martins and his companions may have been led astray. Since they were convinced that they were in the land of some sort of deviant Christians, anything that was not explicitly Islamic appeared, residually, to be Christian.[61]

As the Portuguese party advanced further into the town, the curious crowds that gathered to see them did not diminish. At one stage, notes the account, they had to be closeted briefly in a

[60] This point has been made by Geneviève Bouchon, 'L'interprète portugais en Inde au début du XVIe siècle', in *Dimensões da alteridade na culturas da lingua portuguesa – O Outro* (1° Simpósio interdisciplinar de Estudos Portugeses – Actas) (Lisbon: Universidade Nova de Lisboa, 1985), Vol. II, pp. 203–13.

[61] Here, the contrast is marked with the early Spaniards in Mexico, who at times described Aztec religious structures as 'mosques' (*mesquitas*); cf. Carmen Bernand and Serge Gruzinski, *De l'idolâtrie: Une archéologie des sciences religieuses* (Paris: Editions du Seuil, 1988), pp. 11–22.

house, while the crowds dispersed. Then, accompanied by the brother of *Bale*, who was 'a great lord in this land', and preceded by drums and pomp, and also by a firearm (*espingarda*) that was fired from time to time to clear the route, they made their way to the palace. Our account reflects a certain claustrophobia, in the face of the multitudes that accompanied the Portuguese party and watched them from nearby houses and rooftops: 'the people were so many that they could not be counted'. By the time they arrived at the palace, it was an hour before sundown; at its entrance, they were met by 'very honoured men and great lords', and accompanied through an open courtyard through four doors, with their escort pushing and beating the crowds to make way. At the inner door, they were met by a short, old, man, 'who is like a Bishop, and the king refers to him in church matters', who embraced Gama, and led him inside, still struggling to elbow away the crowds of curious spectators.

We are now in a key scene of Portuguese national mythology: Vasco da Gama before the ruler of Calicut. This is a scene that is celebrated by Luís de Camões in the *Lusíadas* (Canto VII, 57–65), and which has been painted, engraved on blue-tiles (*azulejos*) and drawn innumerable times, with the gallant, bearded captain in his European dress, gesturing grandly with broad sweeps of his arms, while before him reclines the sinister and half-naked ruler, surrounded by his even more sinister (and often vaguely Islamic) henchmen, with a part-malevolent and part-awed expression on his face. From the pen of our anonymous author, though, the scene has a somewhat different flavour:

> The King was on a platform, reclining on a couch, which had these things: a green velvet cloth underneath, and on top a very fine cover (*colchão*), and on top of the cover, a cotton cloth that was very pure and delicate, more so than any of linen. And he also had pillows of this sort. And to the left-hand side, there was a very large golden cup, the size of a pot of half an *almude*, and its mouth was two palms wide, which appeared to be very thick, into which pot he spat the liquor of some leaves that the men of this land eat for its soothing effect, which is called *atambor*. And on the right side was a basin of gold, the width of which could be attained by a man's two arms, in which those leaves were kept, and many silver vases (*agomis*), and the roof was all gilded above.

Gama had been informed to a minimal extent about court etiquette, and seems to have known for example that in speaking before the king one had to bend forward, and hold one's hand in front of the mouth. As he entered, the King folded his palms high in salutation ('as Christians are accustomed to do before God'!), then signalled to him that he should bow down before his raised hands, which Gama did without approaching too close (since he knew that no one was to approach the king save the betel-leaf bearer). Gama then seated himself in the first rank of courtiers, while the rest of his entourage was given water to wash their hands, and fruits to eat. The king and his betel-bearer are reported to have been highly amused at their fashion of eating the fruits (probably bananas, and jack-fruits).

Gama was now instructed by the King to tell the courtiers who were seated around him of why he had come there, and what he wanted; but he refused, saying that he was an ambassador, and wished to speak to the King. About sundown, Gama was thus taken inside an interior room (*câmara*), where the King reclined anew on another couch covered with gold-embroidered cloth, and asked him to say his piece. We may presume that interpreters were present, though the account is not explicit as to this; Gama seems to have gone into the interior room alone (except, that is, for Fernão Martins), and thus what follows is obviously what was communicated to the anonymous author afterwards. Gama is thus supposed to have made a little speech, presenting himself, Portugal and Dom Manuel.

> And the Captain told him how he was the ambassador of a king of Portugal, who was lord of much land, and who was richer in all things than any other king of those parts. And that for sixty years, the kings who were his predecessors had sent ships towards those parts every year to discover (*a descobrir*), for they knew that in those parts there were Christian kings like them. And that it was for that reason that they had ordered this land to be discovered, and not because they needed gold or silver, for they had those in such abundance that they did not need them from this land. That those captains had been going back and forth for a year or two until they ran out of supplies, and without finding anything, returned to Portugal. And that now a king, called Dom Manuel, had ordered these three ships to be made, and had sent him as their Captain-Major, and had told him that he should not return to Portugal until he had discovered this King of the Christians, and that if he should return [unsuccessful], he would have his head cut off. And that if he should find him, he should give him two letters, which he would give him the next day, and that he had thus ordered him to say verbally that he was his brother and friend.

How much of this could be communicated through the Arabic interpreter (especially the notion of 'discovery') we cannot say. It is nevertheless remarkable that Gama is explicitly reported here as stringing together a whole series of half-truths and falsehoods: that the Portuguese king was more powerful than the rulers of Spain or France, that there was so much gold and silver in Portugal that they had no need of any, and that his own head would be cut off if he returned unsuccessful. He also asked the Calicut ruler for return ambassadors, whom he could take back to Portugal; the latter replied that he would arrange for them.

This discussion is reported to have lasted several hours into the night, and at its end, the Calicut ruler asked Gama whether he wished to spend the night with the Moors or with the Christians. To this Gama, always suspicious, replied that he wished to have a separate resting place, apart from everyone else. Thus, he broke with the usual trading custom, which was to place visiting foreign merchants in one or the other ethnic quarter in the port city; however, the King agreed, and Gama rejoined his men, who by now were out on a verandah, in the light of a large metal lamp (*castiçal de arame*). It had been raining heavily, and the streets were flooded; Gama had thus to be carried on the backs of his men, in order no doubt to preserve his dignity.

After being carried thus through the slushy streets, Gama grew impatient, and complained to one of the accompanying Muslims (reputedly the 'factor of the King'), who took him to his own lodgings, and offered him a horse to continue on his way. The Muslim's house, with its carpets and large metal lamps, is briefly described. But since there was no saddle, Gama refused to mount the horse; the party thus wearily continued on its way as before, and eventually reached their lodgings (*pousada*), where a number of Portuguese awaited them already with a bed for Gama, and other goods that were meant to be given as presents to the king of Calicut.

On the next day, 29 May, Gama began to get his gift ready, which included some cloth, a dozen coats, six hats, some coral, six basins, a bale of sugar, and two barrels each of butter (probably rancid from the long journey) and of honey. This notion of making a gift to the ruler or local governor was a fairly common one in Indian Ocean ports of the period, and formed part of the accepted etiquette, corresponding probably to something like the Perso-Arabic *nazr*, an honorific gift to signify at least temporary subordination to authority. The text translates this transaction as 'rendering service' (*fazer serviço*) to the King, probably in order to avoid too strong a connotation of subordination. It was equally normal to have the present inspected beforehand, and in this instance the inspection was conducted by the same Muslim merchant (the King's *feitor*), and by the *Bale*. On seeing the gift proferred by the Portuguese, these men are reported to have burst out laughing, and to have said that 'it was not something to send to the King, that the poorest merchant who came from Mecca or from the Imdios [*sic*] gave him more than that, and that if they wish to render service to him, that they should send him some gold, for the King was not about to accept that [which they had]'. Gama on hearing this is said to have become 'melancholy', and to have stated by way of defence that he was in fact not a merchant but an ambassador, and also proferred the excuse that these goods were his own and not from Dom Manuel. When Dom Manuel would send his gift, it would of course comprise 'many other things, and much richer ones'. Besides, Gama is reported to have threatened in a huff that if the King *Samorim* (this is the first use of the title in the text) would not accept the gift, he would simply leave. His interlocutors for their part refused to give the gifts to the King; and several other Muslim merchants then visited the Portuguese residence to add their sneers to the earlier ones at this paltry gift.

Gama now asked for a further audience with the King, so that he could go back to his ships thereafter, and the Muslim merchants promised to arrange it later that day. However, by that evening, nothing had transpired, and the Gama temper (which we shall have further occasions to remark) began to stir; he is reported to have become 'enraged (*apasionado*) at seeing himself among men who were so phlegmatic and of such little certainty'. He hence thought at first to go on his own to the palace, but then decided on reflection to wait; meanwhile, the rest of the party for its part decided on a little light entertainment, sang, danced and played music, as if everything was in order. The next day, Gama and his party were actually taken to the palace, but made to wait several hours before the door. At last, a message came from the King, asking that Gama enter with just two men; he decided to take the Arabic-speaking Fernão Martins, and his scribe Diogo Dias. The others waited nervously, fearing that this separation would finish badly.

In the event, Gama appeared before the King, and was surprised to find himself reproached for not having appeared the previous day. Once more, he saved face by claiming that 'as he was tired from the journey, for that reason he had not come to see him'; obviously, he did not wish to mention the vexed question of the *serviço*. The King now began to probe him. If indeed he came from such a rich kingdom, why had he brought nothing? Where was the letter he was to have brought? Gama retorted once more that 'he had brought nothing because he had only come to see and discover and that when other ships would return, he [the King] would see what they would bring'. He then said he would hand over the letter promptly. But the King was not done. 'What had he come to discover', he asked ironically, 'stones or men?' And if the latter, why had he brought nothing along? He also mentioned (as a broad hint) that he had heard that the Portuguese had an icon of the Virgin made of gold, to which Gama replied that, first, it was not made of gold, and second, that he could not part with it in any case, since it was a talisman which had kept him safe on his voyage. The issue of the *serviço* rested there, at this point of stalemate.

The letter was now produced, and linguistic complications began. Gama noted that he had two letters, a Portuguese original (which he could read, but nobody in the court could), and an Arabic translation (that he himself could not read). The Arabic letter would thus have to serve, but Gama stated bluntly that since the Muslims at the court were against him, he did not wish them to translate the letter, and confound its sense. He hence asked for an Arabic-speaking Christian, and one (or at least a non-Muslim youth) was produced. However, the problem was that this youth could *speak* Arabic, but not read it! Hence, in the final analysis, it had to be handed over to four Muslims, who read the letter, and communicated the contents (probably in Malayalam) to the King.

Further discussions ensued, as the King asked Gama to describe the chief products of Portugal, and the latter spoke of much wheat, textiles, iron, brass, and other products (though not gold and silver!). He also stated that he had brought along some samples (*amostra*) of these, and asked for permission to go back to his ships to fetch them, while leaving four or five men in the residence (*pousada*). However, the King insisted that he go back at once with all his men, unload his goods, and sell them as best as he could. Gama was summarily given leave, and left to gather together his remaining men, so as to be able to leave the next day. It was the last meeting that Gama was to have with this Samudri Raja (literally, 'King of the Sea'), or *Samorim*, as the Portuguese called him. By the time of the next Portuguese voyage, that of Cabral in 1500, another monarch would be sitting on the Calicut throne.

THE MAKING OF HOSTILE TRADE

The narrative in the text is about to make a transition now, to detail what may be seen as the first instance of more or less 'hostile trade' conducted by Europeans in Asia. Many elements of this encounter are paradigmatic in terms of European dealings with Asian monarchs in the sixteenth and seventeenth centuries: confusion over etiquette and the form of gifts, deep-rooted suspicions on the part of Europeans concerning the machinations at the Asian court, nervousness about the comportment of the Europeans (and the internal inconsistencies of their claims) on the part of the Asians. Various institutional arrangements would be found over the next two centuries to deal with trade in this Age of Contained Conflict; we may continue with that found by Gama and his interlocutors.[62]

On the morning of 31 May, Gama was brought still another horse without a saddle, and offered its services; he refused it once more, preferring a palanquin. This was eventually arranged through the intercession of a very rich Gujarati merchant (our anonymous author insists that the merchant's name was *Guzerato*!), and the party set out northwards once more, to Pantalayini, where the ships were anchored. However, Gama's palanquin advanced very rapidly; the rest of the party was left behind, and briefly even got lost, before finding the right route. En route, the party was passed by the Bale (or *Baile*), who was in haste to overtake Gama; the rest of the group thus caught up at last with both Gama and the Bale in Pantalayini, in a wayside rest-house (*estau*), meant for 'travellers and voyagers who shelter themselves there from the rains'. They at once sensed that something was wrong, and Gama is variously described as being *doente* (literally 'ill'), and 'melancholy', which evidently indicates that he was in a foul temper. An argument was under way, for Gama wished to have a boat (*almadia*) to regain his ship, and Bale and his men insisted that since it was already sunset, they would do best to wait until the next day. Gama protested, and appealed

[62] I return here to a concept developed in Subrahmanyam, *The Political Economy of Commerce*, pp. 252–75.

to their finer feelings, for was he not 'a Christian just like them'! Eventually, the party was taken to the strand, and told they would be given thirty boats; but this sudden turn of affairs made Gama suspicious. He hence sent out three men along the seafront with a message for his brother Paulo (in the event that he was offshore in the ship's boat) to tell him to hide aboard the ship; obviously, he thought a trap was being set. But Paulo da Gama was nowhere to be found; the other boats could not be arranged either; and so the Portuguese spent the night at Pantalayini, in the house of a Muslim merchant.

Chickens and rice were fetched, and Gama and his party dined well. By now, his humour had improved, and he is reported to have even said that it was just as well they had not embarked in the dark; perhaps their local interlocutors meant well after all, despite the 'bad suspicions' the Portuguese had, as a result of their earlier dealings in Calicut. But the next morning, the pendulum swung the other way once more. Gama again demanded boats, and the others replied that he had best ask for the ships to come closer to the shore first, to make the passage easier. Gama refused, giving as a reason that if he told his brother to bring the ships closer, he would think that he was being held prisoner ashore, and hence simply sail away to Portugal. In fact, he obviously suspected a trap once more. A lengthy argument now began. Gama at first said that it was at the Calicut King's orders that he was returning to the ships; but if the Bale and others did not wish to let him go, he was happy to stay there, as a Christian among other Christians. The others then told him he could go, but on seeing the doors of the house shut, and a number of armed men there, the Portuguese did not dare to stir out. The anonymous account claims besides that their local interlocutors demanded that they hand over their 'sails and rudders' (*as velas e os governalhos*), apparently as a sign of good faith that they would not sail away without paying port-duties. Gama refused, and now asked that some of his men (who were hungry) be allowed to leave and find food. This request was denied.

At least three members of the group had gone astray by this time. However, one of them returned with the news that Nicolau Coelho was on the strand with the ships' boats, awaiting Gama and his party. Rather than take this chance, Gama sent a secret message to Coelho, asking him to leave quickly before he was trapped; the latter fled to the ships, after being chased briefly by some local boats. The nervousness of the Portuguese party mounted, as the strength of the guard around them was increased. Gama steadfastly refused to ask his ships to come closer to the shore; and the Bale's suspicions at his comportment also seem to have grown. At the core of this dispute were two factors: Portuguese suspicion of the Muslim influence in this 'Christian' court, and local fears that the Portuguese would sail away, without paying the usual dues for the use of the port.

The party thus spent the entire night in doubt, taking turns to keep awake, in case they were attacked. The next morning, of 2 June, the entire tension was quite simply deflated. Gama agreed to bring some goods on shore, pretending that it was the whole of his cargo, and sent a letter to his brother Paulo to this effect; and on being assured that the Portuguese would not sail off summarily, the party was allowed by the Bale and others to return on board ship, 'giving many thanks to Our Lord for having taken us from the clutches of such men, who were incapable of any reason, as if they were beasts'. Once on board ship, Gama did not unload the rest of the cargo, leaving only a portion ashore with two Portuguese to take care of it.

Five days later, Gama sent off a letter of complaint to the court, stating that he had been ill-treated on the way, and detained for a day and a half; further, he stated that he did not wish to unload the rest of his goods, as the Moors were damaging those that were ashore. He received a prompt response: the ruler said that those who had ill-treated him were 'bad Christians' who would be punished, sent seven or eight chosen merchants to inspect the goods, and sent out instructions that the Muslim merchants were to be kept away from the Portuguese. But the Portuguese were unhappy with the new merchants with whom they had to treat as well; and as for the Muslim traders, tensions with them mounted. The anonymous account assures us that when any Portuguese went ashore, the Muslim merchants would spit on the ground near them, and say 'Portugal, Portugal' in an insulting fashion; besides, we are told that 'from the beginning they sought means to capture us and kill us'.

Gama thus found it difficult to prosecute the commercial pursuits he had in mind at Pantalayini, and he hence asked that he be allowed to move the goods to Calicut proper. The King not only allowed this, but sent men to carry the goods physically (and free of charge) to Calicut, which earned him but little credit with the Portuguese. By now, the latter were so convinced that they faced a local conspiracy that the anonymous text insists that even this seemingly friendly act was 'with the intention of doing us some harm, on account of the incorrect information that he [the King] had of us, that we were thieves and went about stealing'. The goods were eventually moved to Calicut on 24 June, and through the month of July an intermittent trade was carried on, both on account of the Portuguese Crown, and to profit the crew.

This latter aspect merits some mention, and we shall have further occasion to return to it. Let us retain for the moment both the characteristics of this private trade, and its ostensible justification in the eyes of our anonymous author.

> And once the said merchandise was there [at Calicut], the Captain ordered that everybody should go to Calecut in this manner: that from each ship one of its men should go, and when they had returned, that others should set out, and in this manner they could go and see the city, and each one could buy what he wanted.

The trade goods that were taken were shirts and other cloth, and bits of tin, as well as chains; however, the prices of these in metal coin at first surprised the Portuguese, since they were far lower than in Portugal. In turn, the crewmen are reported to have purchased cloves, cinnamon, and even precious stones in Calicut, equally for very low prices.

These men were, it is reported, all rather well received by the 'Christian people' of the area, who took them home, and gave them food and a place to sleep. On the other hand, the local people also began to approach the ship to sell fish, and even brought along their young children. Indeed, so many people came to visit the ships that the Portuguese were somewhat irritated, especially when they had to make them leave at nightfall. The text refers to considerable poverty, a measure of this being that when Portuguese sailors were repairing sails, the local people would come and take biscuits from them! 'And this is because of the many people that there are in this land, and the foodstuffs are very few', concludes our author.

Gama now began to toy with the idea of leaving behind a factor (namely a certain Diogo Dias), and a scribe (Álvaro Braga), besides Fernão Martins, and another four men, as a sort of longer-term establishment. However, as July drew to a close, he also sent word to the King, stating that he would like to take back some local men with him to Portugal, as a form of counterbalance to the men that he intended to leave behind; he further sent a *serviço* at long last to the King, comprising some coral and other goods, as well as some metalwork. These goods were sent with Diogo Dias, who was given a rather cold reception at the court, having to wait four days for an audience. The King told him to give the goods to his *feitor*, as he had no desire to see them; he also ignored Gama's request for a *bahar* each of cloves and cinnamon to be sent to Dom Manuel at a price to be agreed upon. He finally reminded the emissary that he was owed 600 *ashrafīs* by way of duties.

There appear to have been some suspicions that Dias would make off promptly without paying. He was hence followed to the Portuguese trading-post, his goods sequestered, and orders sent out that no boats were to go out to trade with the Portuguese ships. Dias for his part sent out an Indian messenger-boy (*um moço negro*) to the ships by night on a fishing boat, advising Gama of what had come to pass. By now it was 13 August, and those on board had begun to feel nervous, seeing their departure delayed until the men who were ashore could be recovered. How could a Christian king be guilty of such atrocious 'dog-like' behaviour (*perraria*), the Portuguese asked themselves, and soon came up with an answer. To be sure it could not be his fault alone; he must have been instigated by the 'merchants of Meca', who had poisoned his mind, and told him that if the Portuguese began to frequent those parts, soon no ships from the Red Sea or Gujarat would put in at his port. A local Muslim and two 'Christians' equally warned the Portuguese that a plot was afoot to kill them, and Vasco da Gama in particular.

The Portuguese thus waited a day, but no boats approached them, save one with four boys carrying precious stones to sell. Eventually, on 19 August, about twenty-five men, including six of the city's notables, approached the vessels to parley; Gama did not hesitate, and seized eighteen of them, sending the rest back with a message that he would exchange his hostages for the men held ashore. There was no response, and so on 23 August the Portuguese vessels lifted anchor, saying that they were leaving for Portugal, 'and that we hoped to return very soon, and that they would then know if we were indeed robbers'.

Unfavourable winds prevented a prompt departure though. On 26 August, the vessels were still not far from Calicut, waiting for the wind to turn, when a boat approached them with word that Diogo Dias was now in the King's palace, in order to parley. Gama suspected this was a trap, to make his fleet stay on until local ships came from the Red Sea to attack him. He hence asked that his men who were ashore be brought back immediately, or at least that he be given letters written by them, to prove that they were alive. He also threatened to behead anyone he captured if this were not done. As it happened Dias was really in the palace at Calicut, and the next morning he appeared with a small flotilla of seven boats and one of his companions; an exchange was gingerly effected, in which six Calicut notables (*os mais honrados*) were released. Gama also handed over still another commemorative pillar (or *padrão*) to the men who had come from Calicut, instructing them to place it on land. He also agreed to release the remaining prisoners once the rest of the goods that were ashore were brought to the ships.

Once aboard, Dias reported the nature of his interview with the Samudri. It turned out that once Gama had taken his hostages, Dias was summoned to the palace, welcomed with some ceremony, and asked why the men from Calicut had been captured. Dias replied that it was by way of reprisal, since he himself was being detained on land. The King now claimed that he had no hand in what had transpired, and that it was all the fault of his factor (*feitor*) in Calicut, and said somewhat ominously: 'It was only a short while ago that I killed another factor, because he extracted bribes from some merchants who had come to this land.' Dias was instructed to go back to Gama and negotiate the release of the hostages, and further told that he could remain in Calicut and trade. And finally, a letter was dictated by the King, which Dias (no doubt after a process of multiple translation, from Malayalam, to Arabic, to Portuguese) eventually wrote in Portuguese on a palm-leaf (*olai*), which ran:

> Vasco da Gama, fidalgo of your household, came to my land, and I rejoiced at that. In my land, there is much cinnamon, and much cloves and ginger and pepper and many precious stones. And what I want from your land is gold and silver and coral and scarlet.

Further minor developments were to take place. On 28 August, early in the morning, one of the Tunisian Muslims (a certain 'Bontaibo' or 'Monçaide') whom the Portuguese *degredado* had first encountered on setting foot in Calicut, arrived on board ship, and asked that he be taken to Portugal, since his relations with the other merchants at Calicut had deteriorated to such a point that he feared for his life. A little later, seven boats full of people approached; they said they carried the goods the Portuguese had left on land, and asked to exchange these against the remaining hostages. But Gama flatly refused the exchange, saying that 'he did not want the goods, but rather to take the men to Portugal, and that they should wait patiently, for he hoped that he would return soon to Calecute, and that then they would know if indeed we were robbers as the Moors had told them'. The bargain struck by Dias at the palace was thus not kept, and the Portuguese fleet prepared to depart. Thus, the text tells us:

> One Wednesday, which was the twenty-ninth of the said month of August, seeing that we had found and discovered what we had come to seek out, both spices and precious stones, and that we could not manage to leave the land in peace and as friends of the people, the Captain-Major decided, on consulting the other captains, to leave . . .[63]

THE LONG RETURN

But the fleet's return voyage was not to be a simple one. The next day, the ships were still a mere league north of Calicut, awaiting favourable winds, when a fleet of seventy small boats full of men approached them; suspecting hostile intentions, the Portuguese fired their bombards; a minor skirmish began that lasted for an hour and a half, at the end of which the offshore winds began to blow taking their ships into the open seas. However, the winds soon ceased, and on 10 September the fleet was still only just to the north of Cannanur; here, Gama had a letter written in Arabic (by 'Bontaibo') and sent to the Calicut ruler with one of the hostages. Five days later, the fleet put in at some small islands off the south Kanara coast, where another *padrão* was placed on land; here they encountered some fishermen, and came to the conclusion that these too were 'Christians'.

On 19 September, still sailing very slowly up the Indian west coast, the fleet arrived at the Anjedive Islands, a small group of six islands not far south of Goa, to take on board drinking water and firewood. The Portuguese encountered a young man, who told them there was wild cinnamon to be found on the island; on being assured that he too was a 'Christian', the Portuguese went out with him to collect forest products, while other local people brought them chickens and milk. However, Gama soon grew suspicious of the situation, the more so when one of the mariners spotted eight ships at some distance from them, on the open sea. Giving chase to one of them, the Portuguese found it rapidly

[63] Velho, *Roteiro*, p. 75.

abandoned, but still containing foodstuffs and arms; their suspicions were confirmed the next day when seven men in a boat approached, and told them that these ships had in fact come from Calicut to attack them.

Prudently, the Portuguese decided to move on to another island, where they found a semi-abandoned place of worship ('a church of large dimensions which had been destroyed by the Moors'), where the island's inhabitants worshipped before three 'black stones which were in the middle of the body of the chapel'. The text gives us to understand that the islands were frequented by pirates or corsairs in ships, and the Portuguese fleet in fact had an encounter with one such group, in two large foist-like boats. Being forewarned by their local informants that these men 'went about armed, and entered ships in good faith, and once they were inside, if they found themselves powerfully placed, took over the ship', the Portuguese vessels fired on the boats as they approached, and forced the men aboard to flee, shouting 'Tambaram, Tambaram!'.[64]

After a few other desultory encounters, some days later, a well-dressed man aged about forty, who spoke the Venetian dialect fluently, is reported to have approached Gama's vessel, which was in the process of being cleaned and readied for the further voyage. He is reported to have embraced Gama and the other captains, declared that he himself was a Christian from the Levant who had come to India many years ago, and that he served a local potentate, 'who commanded forty thousand horsemen'. This overlord being a Moor, his elegant subordinate too had become one, 'though the desire that he felt in him was to be a Christian'. He went on to state that he had set out from his place of residence expressly to meet the Portuguese:

> For when he was at home, people had come and told him that in Calecute there were men whom no one understood, and that they went about fully dressed and that when he heard that, he had said that such men could only be Franks (*francos*), for it was thus that we were called in these parts. Then, he asked for permission [from his master] to be allowed to go and see us, and that if he were not permitted, he would die of chagrin, and that then his master had said that he could come, and that he should tell us that if we

[64] The later chronicles suggest that these corsairs were in fact commanded by a certain Timoji or Timmayya, who operated out of Honawar; he was later to be an ally of Afonso de Albuquerque. On this personage, see Geneviève Bouchon, 'Timoji, un corsaire indien au service de Portugal (1498–1512)', in *Ciclo de Conferências, Portugal e o Oriente* (Lisbon: Fundação Oriente-Edições Quetzal, 1994), pp. 7–25.

> needed anything from his land, he would give it to us, offering ships and supplies, and further, that if we should wish to reside in his lands, he would rejoice greatly.

Gama thanked him profusely, and a lengthy conversation followed, in which their interlocutor spoke at some length about all sorts of things. Paulo da Gama, less seduced by his talk than his brother, is reported however to have made enquiries with the local 'Christians', who assured him that the man was really the one behind the attack that had been mounted on them by corsairs, and that he had 'his ships with many people' waiting not far from there. The Portuguese thus seized him, carried him aboard their ships, and began to flog him, until he confessed to being a spy who wished to see how many men and what arms the Portuguese had; he also informed them that 'the entire land wished them ill', and that in various bays and creeks along the coast, armed men in boats were waiting for them. The fleet nevertheless remained in the place twelve days in all, and set sail from there at last on 5 October, after having burnt one of the ships that they had captured (Gama having refused to sell it back, even for a thousand *fanões*).

Later, once the ships were in full sail in the western Indian Ocean, the man (who was still on board) confessed that his intentions were rather complex. In fact his master had heard that the Portuguese were wandering about lost on the west coast, not knowing how to return to Portugal. He had hence sent his agent to see if they could be brought to his land, and be used 'to make war on other neighbouring kings'. It is of course of some significance that this talkative man, referred to here still as a Moor, was in fact a Jewish merchant, and that by the time of his return to Portugal (to be precise, in the Azores) he had been baptised and given the name Gaspar da Gama (for his godfather, Vasco da Gama!). The name 'Gaspar' was taken from that of one of the three Magi, the kings of the east who came to seek out the infant Jesus (even as this Gaspar sought out Vasco da Gama). It was this Gaspar, together with the Tunisian Muslim 'Bontaibo', and the Gujarati pilot from Malindi, who formed the main sources of information of the Portuguese concerning the Indian Ocean, until the arrival of Cabral's fleet there in 1500. Of the three, it is the highly articulate Gaspar who is likely to have been the major informant on commercial and political matters.[65]

[65] For further details, see Elias Lipiner, *Gaspar da Gama: Um converso na frota de Cabral* (Rio de Janeiro: Editora Nova Fronteira, 1986).

The return across the western Indian Ocean took the Portuguese nearly three months, in the course of which thirty men died of scurvy alone. A mere seven or eight men were left as active mariners on each ship, and even those were in indifferent health. The fleet had been reduced to such straits, writes our anonymous author, that had the voyage lasted a few more days, there would have been 'no one left to sail the ships'. Promises and pledges were made by the crew to all the saints, and the captains (the two Gamas and Coelho) even decided at one stage that if the winds did not turn favourable, they would all simply turn back to India. However, the wind did turn, and on 2 January 1499, the ships briefly sighted the African east coast near Muqdisho, but decided not to stay there, partly for fear of disease in the nearby islands, and partly on account of being informed that the city was 'of the Moors'. Sailing down the coast, the vessels were in a minor skirmish with eight boats just off Pate; and at last, on 7 January 1499, arrived before Malindi.

Though welcomed once more by the King, and offered fruits, eggs and chicken, it was too late for many of those on board who were in an advanced stage of sickness. Nevertheless, diplomatic formalities were gone through; another *padrão* was erected, a gift of ivory was taken for Dom Manuel, and a local Muslim was sent by the ruler of Malindi, to accompany the Portuguese back. On the 11th January, they left Malindi, passed Mombasa, and on the 16th burnt and abandoned the *São Rafael*, 'for it was an impossible thing to navigate three ships with as few people as we were'. The goods on board were, however, transferred onto the two other vessels. Finally, after another ten days of rest, the two surviving ships set sail, passed the island of Zanzibar ('which is populated by many Moors'), and in early February returned to the *ilhéu de S. Jorge*, in the vicinity of Mozambique.

From now on, the sailing was relatively smooth. On 20 March, the ships rounded the Cape of Good Hope, and despite the intensely cold winds (which the Portuguese felt all the more, in contrast to the lands they had left behind), made quick headway as far as the island of Santiago, which was attained in twenty-seven days. Here, the winds calmed once more, but with the aid of some land-breezes, the two ships had arrived as far as the *baixos do Rio Grande* on 25 April. And here, on the so-called Guinea coast, the anonymous text breaks off abruptly, either because the original text stopped as such, or because the copyist of the later text (the sole that we dispose of), had an incomplete version before him.

With the aid of the chronicles – Castanheda and Barros, in particular – and other contemporary documentation which we shall discuss below, we can broadly reconstruct the rest of the voyage. The first vessel to return to the banks of the Tagus, in early July 1499, was the *Bérrio*, commanded by Nicolau Coelho, and Castanheda claims that he returned post-haste (taking advantage of the fact that he had a faster and lighter vessel) to steal Vasco da Gama's thunder. The second vessel, the *São Gabriel*, came in to Lisbon in August, under the command of João de Sá, which suggests that Gama himself was in no hurry to return home. Where was he during this period?

Vasco da Gama, it turns out, had taken a caravel from Santiago to the island of Terceira in the Azores with his brother, and then tarried in the town of Angra to bury Paulo da Gama, who had sickened on the return voyage. It is thus more or less clear that – contrary to Castanheda's version – he allowed the captain of the *Bérrio*, Nicolau Coelho, to carry the news back to Dom Manuel, who then announced the discovery of the true India to his European neighbours in July 1499 (the letter to the Catholic monarchs of Spain being dated the 12th of that month).[66] Venice, we have seen from the Priuli diaries, had equally received the news through Alexandria, and found it hard to accept. On the 10th of the same month, a Florentine merchant wrote back to his native town with the same news.

> The Most Illustrious Lord Manuel, King of Portugal, sent three new ships to discover new lands, that is two vessels (*balonieri*) of 90 *tonelli* each and one of fifty tons (*tonelli*) and besides a navette of one hundred and ten tons filled with supplies, which carried 118 men in all; and they departed from the city of Lisbon on the 9th of July Anno 1497, and of this fleet there went as captain Vasco da Gama. On the 10th of July 1499 there returned to this city of Lisbon the vessel of 50 tons; the Captain Vasco da Gama remained in the islands of Cape Verde with one of the vessels of 90 tons in order to set on land there a brother of his, Paolo da Gama, who was extremely ill. And the other vessel of 90 tons they burnt

[66] Teixeira de Aragão, *Vasco da Gama e a Vidigueira*, docs. 8 and 9, pp. 217–20; for the context, also see A.A. Banha de Andrade, *Mundos Novos do Mundo: Panorama da difusão pela Europa de notícias dos Descobrimentos Geográficos Portugueses*, 2 vols. (Lisbon, 1972), Vol. I, pp. 206–24.

> because there were not enough people to man it and navigate, and they also burnt the navette . . .[67]

The last but one reference is to the *São Rafael*, abandoned and burnt near Mombasa in early 1499, for reasons that we have seen (and which accord with the version of the Italian letter-writer). The same writer goes on to describe how fifty-five of the men died on the return voyage of a sickness that attacked them first in the mouth, then caused bodily pains, obviously a reference to scurvy. The report goes on then to describe Malindi, a city populated by Moors, and waxes ecstatic about Calicut, 'which is larger than Lisbon, and populated by Christians'. A second letter to Florence, written a few days later, corrects a number of errors in the first, now noting for example that Paulo da Gama had been buried in the Azores and not the Cape Verdes as first stated. Also mentioned in this letter are two characters of some significance: the first the Jew whom Gama had captured in the Anjedive Islands, and the second the Gujarati Muslim pilot who had guided Gama from Malindi to Calicut, and who it is noted was at least partly Italian-speaking. Both these men were sent by Gama to Lisbon even before his own arrival in the Portuguese capital, and the Italian letter-writer had apparently had occasion to gather information from them.

A third letter of 1499, this one written by a certain Guido di Messer Tomaso Detti on 10 August, surveys the implications of the new discoveries, rejoicing that the Venetians will have to leave the Levantine trade 'and become fishermen'. Vasco da Gama, we may note, had *still* not returned from the Azores, and Signore Guido thus still found only one ship out of the original four at Lisbon. He concludes his letter as follows.

> As has been seen, this is an excellent finding (*un bel trovato*) and this King merits great commendation from all Christians, and certainly all the Kings and great and powerful lords (*signori potenti e massime*) who are sea-powers should always send out to find and give news of unknown things because it brings honour and fame,

[67] Biblioteca Riccardiana, Florence, Codex 1910, fl. 61r (a), 'Copia di una lettera avuta da Lisbona della nuova terra trovata colle speziarie l'anno 1499 a di 10 di luglio', in Radulet, *Vasco da Gama*, pp. 169–81, the author being Girolamo Sernigi. For an earlier translation, see Ravenstein, *A journal of the first voyage*, pp. 123–36. This letter is almost identical to a version attributed to Sernigi from the same month, and published as 'Navigazione di Vasco da Gama, Capitano dell'armata del re di Portogallo, fatta nell'anno 1497 oltra il Capo di Buona Speranza fino in Calicut, scritta per un gentiluomo fiorentino che si trovo al tornare delle detta armata in Lisbona', in Giovanni Battista Ramusio, *Navigazioni e viaggi*, ed. Marica Milanesi (Turin, 1978), Vol. I, pp. 607–17.

> reputation and riches, and in fine, because of it they are praised by all men. And to such men it is well that Lordship and State (*signoria e stato*) is given . . . And thus we may say: this King of Portugal should be praised by all men.[68]

Thus, even if the Portuguese did not leave Calicut with altogether pleasant thoughts in late August 1498, by certain standards, the expedition could be deemed a success, and it certainly strengthened the hand of Dom Manuel, among other reasons because Gama brought back word of the purported existence of many Christian kingdoms in Asia (including Calicut itself). With this new-found confidence, the next fleet, commanded by Pedro Álvares Cabral, was a fully fledged affair, far larger than Gama's rather paltry trio of vessels. It comprised thirteen ships, carried over a thousand men on board, and not only discovered Brazil en route to India, but also brought matters to a head where relations with Calicut were concerned.

The information brought back by the fleet was considerable, but also based on the persistent confusion that located a large Christian population in India. One part of the commercial information may already be found in the body of the anonymous text (fls. 64–6), and appears to have been derived from 'Bontaibo' and the Gujarati pilot; a far more elaborate, and fantastic, tableau is set out as a sort of appendix, and almost certainly derives from the fertile imagination of Gaspar da Gama (or Gaspar da Índia). The first set of informations centres on Calicut, described as the land from which there originates 'the spice that is eaten in the west, and in the Levant, and in Portugal, and equally so in all the provinces of the world'. The ginger and pepper of Calicut are singled out for notice, though the cinnamon is said to be inferior to that of 'an island called Cillam [Ceylon], which is eight days' journey from Calecut'. As for the cloves, they come from 'an island that is called Melequa', which lies fifty days' sailing from Jiddah (even with favourable winds). In Jiddah, the Grand Sultan (*o Gram Soldam*) collects his duties; the goods are then taken up the Red Sea (here, the *Mar Ruivo*, not the later *Mar Roxo* or *Mar Vermelho*) to a port near Mt Sinai.

> From here, the merchants load this spice on camels, rented out at four *cruzados* a camel, and they take it to Cairo in ten days, and

[68] Biblioteca Riccardiana, Florence, Codex 1910, fls. 68–70v, 'Copia d'una terza lettera di Lisbona di Guido di messer Tomaso Detti, di X d'agosto 1499 di questo medesimo, che viene a verificare tutto', in Radulet, *Vasco da Gama*, p. 196.

> there pay another tax. And on this route to Cairo, they are frequently attacked by robbers that there are in that land, who are Arabs, and others. From there, they are loaded once again on some ships that go on a river, which is called the Nile, which comes from the land of Prester John of the Lower Indies. And they sail for two days up this river until they reach a place called Roxete, and here they pay another tax. And again, they load up camels, and they take it in a day to a city which is called Alexandria, which is a seaport. To this port of Alexandria, there come the galleys from Venice and from Genoa, to seek out this spice, from which it is found that the Grand Sultan gets six hundred thousand *cruzados* in duties, from which he gives a hundred thousand every year to a King called Cidadym, so that he may make war on Prester John. And this name of Grand Sultan is purchased for money, and is not handed over from father to son.

Here then is a convenient – mildly inaccurate – potted account of the overland spice trade, of succession under the Mamluks, and of Middle Eastern politics, with the reference to 'Cidadym' (Sa^c^ad al-Din) undoubtedly being to the wars between the Salomonid rulers of Christian Ethiopia and the Sultans of Adal (later Harar) to their east.[69] Far more elaborate is Gaspar's account, which listed the Christian kings of India, and their armies that literally ran into tens of thousands of men, potential allies of D. Manuel in his anti-Islamic ventures. It would appear from his account that most of the kings of India were Christian, save perhaps those of Kayal and Bengal (where Moorish kings reputedly ruled over a largely Christian population). The translative rule in operation is quite simple: anyone who is not a Muslim may be termed a Christian! The list of Christian kingdoms thus appears as follows:

Kollam (Coleu): 10,000 men
Kayal (Cael, Muslim king, Christian subjects): 4,000 soldiers and 100 war-elephants
Cholamandalam (Chomandarla): 100,000 men
Ceylon (Ceilão): 4,000 men, and many war-elephants for sale
Sumatra (Camatarra): 4,000 soldiers, 1,000 horsemen, 300 war-elephants

[69] Triaud, 'L'expansion en Afrique', in Garcin *et al.*, *Etats, Sociétés et Cultures du Monde Musulman Médiéval*, pp. 420–3. One Sultan Sa^c^ad al-Din of Adal was killed in battle against Ethiopian Christians in 1415, and one of his successors, Shihab al-Din Ahmad Badlay, also died in similar circumstances in 1445; the text however surely refers to Muhammad ibn Azhar al-Din ibn ^c^Ali ibn Abu Bakr ibn Sa^c^ad al-Din, Sultan of Adal (r. 1487–1520).

Siam (Xarnauz, for Shahr-i nau): 20,000 soldiers, 4,000 horsemen, 400 war-elephants
Tenasserim (Tenaçar): 10,000 soldiers, 500 war-elephants
Bengal (Bengala): (Muslim King, and some Christian subjects): 20,000–25,000 soldiers, including 10,000 horsemen, and 400 war-elephants
Melaka (Malequa): 10,000 soldiers, including 200 horsemen
Pegu (Peguo): 20,000 soldiers, including 10,000 horsemen, and 400 war-elephants.
Conimata (?): 5,000–6,000 soldiers, and 1,000 war-elephants
Pater (?): 4,000 soldiers, and 100 war-elephants

The identification of these last two kingdoms proves difficult; they are identified as being 50 days' sail from Calicut, which suggests that they are somewhat to the east of Melaka (estimated at 40 days' journey from Calicut). Perhaps they represent kingdoms on the Malay peninsula or in Indo-China; Conimata might thus well be Cambodia. A further appendix, derived either from Gaspar or (more likely) from the hostages from Kerala, is a vocabulary of Malayalam terms ('This is the language of Calecut'). Gaspar, it is clear, took some pains to identify himself as a genuine 'native informant'. Indeed, it is obvious that the converted Jew cut a great figure in the Portuguese court, and within a few days of his arrival there in August 1499 was being seen by Dom Manuel as a major asset.[70] In a letter dated 28th August 1499, and addressed to Dom Jorge da Costa, the Portuguese Cardinal-Protector at Rome, Dom Manuel stated for example:

> And above all [the fleet] brought back another [man] who was a Jew, and has already turned Christian, a man of great discretion and ingenuity, born in Alexandria, a great merchant and jewel-trader, who had been trading in India for some thirty years and knows all that there is to be found in it in detail, and also all the lands around and their affairs, from Alexandria therewards, and from India into the interior (*sartaão*), and Tartary unto the Great Sea (*mar mayor*), which goes to show that that land was found on account of a great mystery of Our Lord, for His Holy Service and the well-being of Christendom, for He at once ordained that this man should be brought to us, so that we should be able to grasp what was around us to the extent possible, for had he not come, it

[70] He was also one of Amerigo Vespucci's major informants; cf. Luciano Formisano (ed.), *Letters from a New World: Amerigo Vespucci's Discovery of America*, tr. David Jacobson (New York: Marsilio, 1992), pp. 23–6, letter to Lorenzo di Pierfrancesco de' Medici, dated 4 June 1501.

> would have taken many years to know what had been found, in as great length, and as deeply, as we now know. May God be praised. This man knows how to speak Hebrew, Chaldean, Arabic, and German; he also speaks Italian mixed with Spanish so clearly that one can follow him like a Portuguese, and he can understand our people no less.[71]

We may note the characteristically mystical tone of the letter, in keeping with Dom Manuel's patented messianism: the discovery of India as a Divine Mystery, the Jew as its sign. From the other sixteenth-century chronicles (Barros and Castanheda in particular), we can put together some other elements of Gaspar's career. Born in Alexandria of parents who were Polish Jews, and who had fled there from pogroms in Poznan, Gaspar (sometimes known as Muhammad) had made his way to western India in the 1460s or 1470s. Here, he had taken service with Yusuf ᶜAdil Khan (known as 'Sabaio' to the Portuguese, on account of his *nisbat* of Sawaᶜi) in the 1490s, when the latter took control of Goa; at the same time, he had a Jewish wife and at least one son (later also converted, and christened Baltasar after another of the Magi), either at Goa or in Cochin.[72] Gaspar was to play a role of some significance, accompanying first Cabral's fleet, then that of Vasco da Gama on his second voyage, and then that of the viceroy Dom Francisco de Almeida in 1505. Three letters written by him to Dom Manuel survive, the last of these dated to late 1507 or early 1508; it is to be supposed that he was in India until at least 1512, after which he may have returned to Portugal.

DIVINISING THE VOYAGE

Our primary purpose in this chapter has been to subject to a close reading the only contemporary text that details Vasco da Gama's voyage to Calicut and back, between August 1497 and April 1499. We have had occasion to refer to two other texts intermittently, even though they date from the middle decades of the sixteenth century, Barros's *Décadas da Ásia*, and Castanheda's *História*. We have eschewed another text, Gaspar Correia's *Lendas da Índia*, almost wholly, though we shall return intermit-

[71] AN/TT, Colecção São Vicente, Vol. XIV, fl. 1, cited in Lipiner, *Gaspar da Gama*, p. 79–81, with a facsimile of the text. Another copy of the text may be found in AN/TT, Colecção São Vicente, Vol. III, fl. 513.

[72] Barros, *Da Ásia*, Década I, Book IV, Ch. 11. This is confirmed in most details by Damião de Góis. Gaspar Correia as usual differs from them, and claims that he was a Jew from Granada, which seems highly unlikely.

tently to discuss its significance in later chapters. Our purpose would however not be met unless we were to turn our attention, even if only briefly, to the opposite textual pole, as represented by Luís Vaz de Camões's *Lusíadas*.[73] Though ironically deriving, in part at least, from a reading of our anonymous author, and of Barros's and Castanheda's chronicles, Camões transforms Gama's voyage in tone and content, from the mundane to the divine.[74] Gama here jostles not with his own partly paranoid sensibility (on display through much of the text we have examined), but with the Gods, as he strikes one happy pose after another.

It would be presumptuous to claim to add a great deal to what is already a cottage industry at an advanced phase of proto-industrialisation: namely, the study of Camões in the Lusophone world. But it is nevertheless essential to our purpose to address Camões, for both within Portugal and the Lusophone world, and outside it, the *Lusíadas* has been one of the major vehicles that has propelled Gama's voyage and the voyager himself into the limelight; Sir Richard Fanshawe's mid-seventeenth-century translation into English, for example, enjoyed wide circulation.

Let us recall the structure of the text in its broad outlines. Comprising ten cantos, the *Lusíadas* tells of the exploits of the Portuguese nation, with a brief panorama of national history before the Age of the Discoveries, and a last section that goes quickly over both the regions of Asia 'discovered' by the Portuguese, and the leading actors in the saga of the establishment of the *Estado da Índia* to the mid-sixteenth century. We may remind ourselves that Camões in fact spent roughly fourteen years, from 1553 to 1567, in Portuguese Asia, as a soldier and a minor administrator (in which capacity he participated in trading voyages to South China, for example). Yet the core of the text comprises the epic struggle of Vasco da Gama to conquer the titanic forces of nature, in a struggle in which the pagan Greek gods are used as a metaphor; some of them aid the Portuguese, while others impair their progress. The key struggle is with the malevolent giant Adamastor, a figure whom Camões invented rather than drew upon, who is often seen as representing the Cape

[73] For an earlier essay on a similar theme, and highly critical of official Portuguese historiography, see Sílvia Maria Azevedo, 'O Gama da História e o Gama d'*Os Lusíadas*', *Revista Camoniana* (2nd Series) 1 (1978), 105–44.

[74] Cf. Luís de Albuquerque, 'Sur quelques textes que Camões consulta pour écrire *Os Lusíadas*', *Arquivos do Centro Cultural Português* 16 (1981), 35–50; also an earlier article by Albuquerque, 'A viagem de Vasco da Gama entre Moçambique e Melinde, segundo *Os Lusíadas* e segundo as crónicas', *Garcia da Orta* (1972), 11–35.

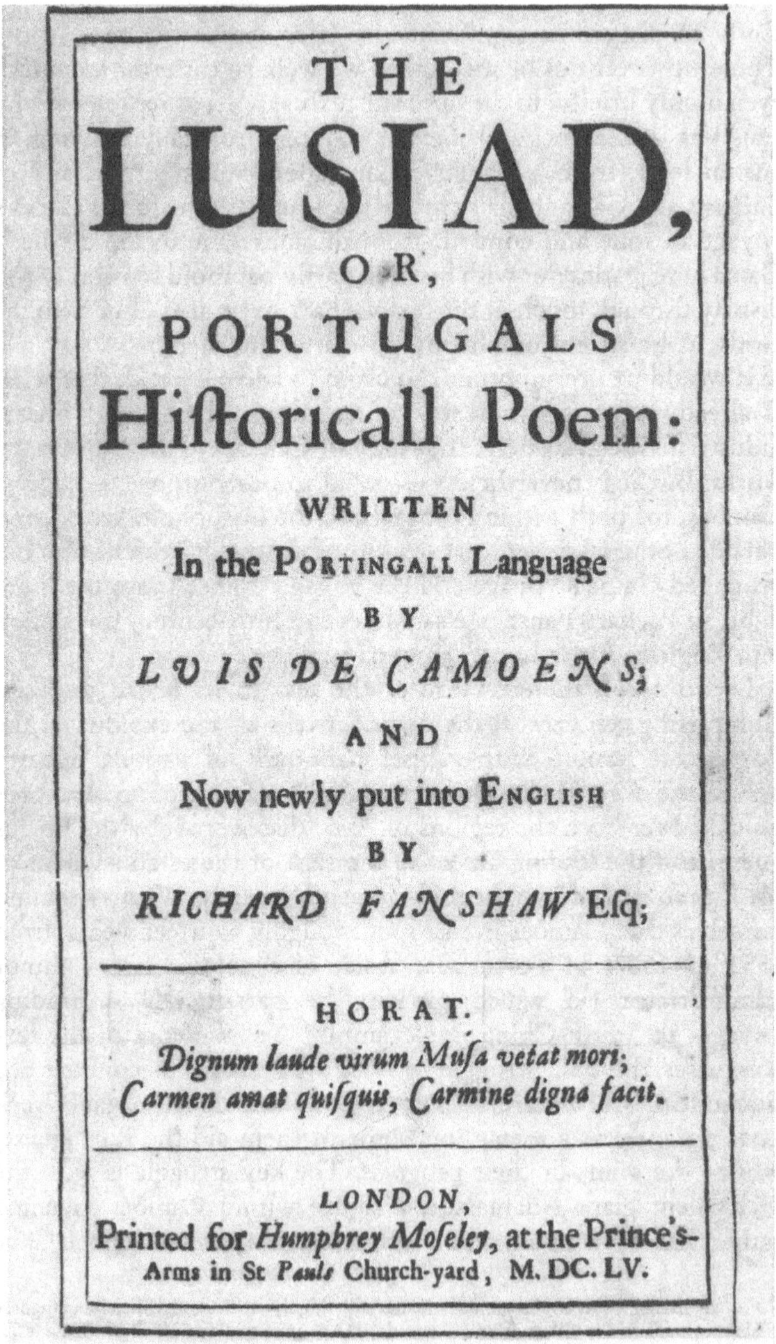

THE

LUSIAD,

OR,

PORTUGALS

Hiſtoricall Poem:

WRITTEN

In the PORTINGALL Language

BY

LUIS DE CAMOENS;

AND

Now newly put into ENGLISH

BY

RICHARD FANSHAW Eſq;

HORAT.

Dignum laude virum Muſa vetat mori;
Carmen amat quiſquis, Carmine digna facit.

LONDON,
Printed for *Humphrey Moſeley*, at the Prince's-
Arms in St *Pauls* Church-yard, M. DC. LV.

Plate 14 Title page of Sir Richard Fanshawe's English translation of *Os Lusíadas* (1655).

of Good Hope, that barrier between the Portuguese and the Indian Ocean. Significantly, the Goddess of the Sea, Tétis, rejects Adamastor and instead offers herself in marriage to Gama; the symbolism could hardly be clearer.

In the *Lusíadas*, the hero Gama is shown as clearly surpassing his 'precursors' from antiquity, thus 'stealing the fame of Aeneas' among a number of other accomplishments. A key aspect of the *Lusíadas* is the play between Christianity and Paganism, the latter represented by the Greeks, and to a lesser extent the Romans. Camões, writing in the high flush of the Counter-Reformation (his text was completed in 1572), could hardly be seen to use the pagan card overly seriously. Yet, it is this very play that permits the partial divinisation of Gama more or less explicitly in a pagan context, for what it is worth, as well as hints in the Christian context that are somewhat disturbing in their flavour. Gama is portrayed whenever he lands, be it in Malindi or Calicut, as surrounded by twelve companions, in an obvious comparison to Christ and the Apostles. There is also the literal fact of his setting out from Belém (Bethlehem), the port of departure on the banks of the Tagus, to reclaim the world for Christianity. One of the text's most famous verses, declaimed by Gama himself to the King of Malindi in Canto IV, is today to be found inscribed on the marble tomb that has been built for him in the Jeronimite Monastery at Belém, and claims:

> We set out thus from the Holy Temple,
> that on the shores of the sea rests,
> which has the name, by way of example,
> of where God to this world came in the flesh.[75]

Much has been made by authors of another aspect of the *Lusíadas*, namely its allegedly pessimistic vision of Portuguese expansion. Reference is usually made in this context to a particular episode in the poem, which occurs just as the fleet is about to set sail (Canto IV, 94–104). An old man 'of venerable aspect/ who remained on the strand, among the people' (and known in the literature as *o velho do Restelo*) berates the Portuguese for sailing off to discover the world, comparing their enterprise to that of

[75] Luís de Camões, *Os Lusíadas*, ed. Maria Letícia Dionísio (Lisbon: Publicações Europa-América, 1985), IV, 87: 'Partimo-nos assi do santo templo/ Que nas práias do mar está assentado/ Que o nome tem da terra, para exemplo/ Donde Deus foi em carne ao mundo dado' (the translation is mine). For a useful discussion of this theme, see António Cirurgião, 'A divinização do Gama de *Os Lusíadas*', *Arquivos do Centro Cultural Português* 26 (1989), 513–37.

Prometheus, who brought fire to the earth from heaven, a fire that was used to burn, to kill, and to dishonour. Do they not know that this discovery will lead to a dimunition in the nation's energies, and will eventually be their downfall?[76]

> To how many new disasters art thou determined
> to take these Kingdoms and this people?
> What perils, what deaths dost thou determine,
> under pretext of gaining some outstanding name?
> What promises of kingdoms and of mines
> of gold, wilt thou make them with such ease?
> What fame wilt thou promise? What histories?
> What triumphs? What prizes? What victories?[77]

Now, various interpretations have been given to this statement, depending on whether or not we assume that the old man represents an authorial voice. If so, it is possible that Camões meant (like his friend and contemporary Diogo do Couto) to make a statement about the moral decline of the Portuguese in the East by the early 1570s. Or alternatively, Camões may have quite simply been stating a fairly standard 'bucolic' theme of pastoral nostalgia, for the country's agrarian interior rather than its destiny overseas (thus, the evocation of 'another state, more than human/ of quiet, and of simple innocence/ a Golden Age', in Canto IV, 98), reflecting the views of other poets such as Brás da Costa and Francisco de Sá de Miranda.[78] This seems altogether less likely, as Camões had, we have noted, spent a good deal of time overseas himself, even if this was partly the result of the pressure of circumstances rather than choice. And finally, there is the possibility that the *velho do Restelo* does not represent an authorial voice at all, but a *recognition* on Camões's part that overseas expansion had been severely opposed in some quarters in Portugal in the late fifteenth and early sixteenth centuries.

It remains true that whatever the message concealed in these particular verses, the *Lusíadas* simply cannot be read as a critique of Portuguese expansion from an insider; one cannot baldly state, as one recent author does, that 'at the heart of the *Lusiads*, then, is a condemnation of the heroic ethic, a denial of the epic values

[76] See, for example, Gerald M. Moser, 'What did the Old Man of Restelo Mean?', *Luso-Brazilian Review* 17 (1980), 139–51.

[77] Camões, *Os Lusíadas*, IV, 97 (the translation is mine). The line that begins the Old Man of Restelo's tirade is 'O glória de mandar, o vã cobiça', which gave the title of Manoel de Oliveira's celebrated film of 1990, *Não, ou a vã glória de mandar*.

[78] For a discussion, see José Sebastião da Silva Dias, *Os descobrimentos e a problemática cultural do século XVI*, 3rd edn (Lisbon: Editorial Presença, 1988).

epitomized in da Gama'. To do so would be to set aside the thrusting weight of the text in its dominant (and most understood) mode, a most artificial procedure.[79]

It is another matter that Camões had to skip delicately over a number of issues in order to mythify Gama. Thus, as has been demonstrated by Roger Bismut, there is a tension between Gama's claims before the Samudri Raja (Canto VII) and Camões's own posterior knowledge of the history of Portuguese Asia.[80] Gama is made to state (as he indeed does in the anonymous text as well!) that the Moors who claim that the Portuguese will take over ocean trade by force are perfidious, and that he is there to affirm a sincere peace, since all accords made there are inviolable. Yet, the experience of Cabral's massive fleet was to show that the Moors were not altogether wrong: the Portuguese did indeed intend to impose their will on Calicut, and gain extensive trade privileges at the expense of their rivals. Camões's reading of this can only have been that such trifling matters were of little import, when compared to the fulfilment of Portugal's national destiny, and the consummation thereof, which was – of course – left to Vasco da Gama.

CONCLUSION

The mood in the Portuguese court in July–August 1499 can only have been triumphal from the viewpoint of Dom Manuel and his close ideologues, in particular Duarte Galvão. At a very limited financial and human cost, the predictable loss of the supply-vessel, the burning of the *São Rafael*, the loss of about half the crew, the process that had been more or less interrupted in 1488, with the return of Bartolomeu Dias, could now be resumed definitively. Against his detractors, Dom Manuel had some forceful arguments now: first, the fact that a large number of Christian kingdoms supposedly existed in India, as potential allies against the Mamluks; second, that India lay a mere month's voyage across from the east coast of Africa, if the winds were favourable; third, that no major standing maritime fleets existed in

[79] Robert Finlay, 'Portuguese and Chinese Maritime Imperialism: Camões's *Lusiads* and Luo Maodeng's *Voyage of the San Bao Eunuch*', *Comparative Studies in Society and History* 34, 2 (1992), 225–41, especially p. 239. Finlay appears to derive inspiration, in turn, from David Quint, 'Voices of Resistance: The Epic Curse and Camões's Adamastor', *Representations* 27 (1989), 118–41.

[80] Roger Bismut, '*Les Lusiades* de Camões, confession d'un poète' (Paris: Centre Culturel Portugais, 1974), pp. 198–200.

Plate 15 Tomb of the Portuguese national poet Camões, Mosteiro dos Jerónimos, Lisbon (c. 1880).

and around the coast of south-western India, so that the Portuguese could count on very limited naval opposition in the first instance. The triumphalism is particularly eloquently expressed in one of the Florentine letters of 1499, where we have noted the sarcastic prediction that the Venetians would have to become fishermen, since they have no future in trade. It is equally reflected in the ambitious new titulature of the Portuguese rulers: Dom Manuel took on the title of not only 'King of Portugal and of the Algarves on this side and beyond the sea in Africa, and Lord of Guinea' (which may have seemed cumbersome enough), but 'Lord of the Conquest, Navigation and Commerce, of Ethiopia, Arabia, Persia and India'.[81]

At the same time, the reader of the Florentine letters of 1499 realises rapidly that the 'native informants' whom the Portuguese brought back were apt to tell different stories to different

[81] It is sometimes mistakenly stated that this title was assumed after the return of Cabral's fleet in the second half of 1501; but it is clear that it was already in use by August 1499, as we see in Dom Manuel's letter to Dom Jorge da Costa, for example, or later in his letter of February 1501 to the Doge of Venice, for which see *Epistola de El-Rei D. Manoel ao Doge de Venezia, Agostino Barbadico 22 de Fevereiro de 1501* (Coimbra, 1907).

audiences. Thus, to Dom Manuel, Gaspar da Gama may have painted a rosy picture of an India filled with Christian kingdoms; to the Italians, he told a rather different story. We have seen how in his first letter of 10 July 1499, Girolamo Sernigi had accepted that the natives of Calicut were largely Christians. But in a later letter of the same year, he notes his conversations with Gaspar da Gama, who 'appeared to be a Sclavonian, and turns out to be a Jew, born at Alexandria, or in those parts'. From him, he had gathered that 'in those countries there are many gentiles, that is idolaters, and only a few Christians; that the supposed churches and belfries are in reality temples of idolaters, and that the pictures within them are those of idols and not of Saints'. He concludes pragmatically: 'To me, this seems more probable than saying that there are Christian but no divine administrations, no priests and no sacrificial mass. I do not understand that there are any Christians there to be taken into account, excepting those of Prester John, whose country is far from Calichut.'[82] Gaspar's information to Sernigi seems of a piece with the attitude of men like João de Sá, who are reported (by Castanheda at least, who may have spoken with him in the 1520s) to have had doubts about the 'Christianity' of the population they were dealing with. Nevertheless, Sernigi's conclusion is in marked contrast to that of Dom Manuel, even after the arrival in Lisbon of the *São Gabriel*. For in his letter to Dom Jorge da Costa (cited briefly above), Dom Manuel continues to assert that the *Samorim* and his people should be considered Christians, albeit deviant or heretic ones.[83]

This firm conviction of Dom Manuel and the court was also translated into action, for the instructions given to Pedro Álvares Cabral were based on the assumption that the ruler of Calicut was indeed Christian. Cabral, in 1500, thus carried a letter to the Samudri Raja of Calicut full of obscure eschatological references on the basis of the idea that most of Dom Manuel's millenarian ideas would find an echo in distant Kerala. A sample of this text, dated 1 March 1500, reads:

> Since the creation of the world, there have been in those regions over there, and those that are here, great powers and seignories of princes and kings, and of the Romans and other nations who possessed the greater part of the earth, of whom one reads that they had a great wish and desire to make this voyage and they

[82] Translated in Ravenstein, *Journal of the first voyage*, pp. 137–8.
[83] Letter published in J.M. da Silva Marques (ed.), *Descobrimentos Portugueses*, Vol. III (Lisbon, 1971), pp. 549–50.

> attempted it, but it did not please God to give them such an opportunity at that time, as we ourselves would not have been able to do, had we not had it from His Hand and consequent on His Will . . . For one should truly believe that God, Our Lord, has not permitted this feat of our navigation solely in order to be served in trade and temporal profits between you and us, but equally in the spiritual profit of souls and their salvation, which we ought to place higher. He considers Himself better served by the fact that the holy Christian faith is communicated and joined between you and us as it was for six hundred years after the coming of Jesus Christ, until the time that, for the sins of men, there arose some sects and contrary heresies as predicted, of which Christ had said that they would arise after him to test and prove the good, and to confound the evil of those who merited condemnation and perdition . . . and these sects occupy a great part of the Earth between your lands and ours.[84]

This last rather direct reference to Islam, like the rest of the text, could not have meant a great deal to the Samudri Raja, and was drafted by the greatest of the Manueline ideologues, Duarte Galvão.[85] It is of a piece with Galvão's grandiose formulations that are equally in evidence in his *Crónica de D. Afonso Henriques*, commissioned in 1503, and purporting to be a history of the founder-king of Portugal; though never published in the sixteenth century, there were several deluxe contemporary copies of this chronicle, and references in other texts which suggest that it circulated quite widely as a manuscript. The chronicle is in fact in large measure a mystical celebration of Portugal, seen in somewhat anthropomorphic terms. Thus, Galvão writes, 'God ordered and wished to constitute Portugal as a kingdom for a great mystery of his service and for the exaltation of the Holy Faith'. And earlier, in the prologue to the same text, addressing D. Manuel, he reflects on the 'great marvel and mystery of the discovery, or more truly the conquest of the Indies, never hoped for nor believed in among men, until it was seen to be done and accomplished by your order'. The comparisons with the ancients then follow: 'Neither that great king Alexander, conqueror of the world, nor the Carthaginians, lords of Africa and of a part of Europe, nor the Romans, who exceeded all the others in the

[84] Text cited in Castanheda, *História*, Livro I, pp. 78–80, and reproduced in Silva Marques, *Descobrimentos Portugueses*, Vol. II, pp. 568–70.

[85] For a detailed study of this and other texts, see Jean Aubin, 'Duarte Galvão', *Arquivos do Centro Cultural Português* 9 (1975), 43–85. Aubin points out that the usual versions of this text are defective (p. 66, n. 91), preferring that of Castanheda.

extent of their empire could attain this with all their effort, as one reads.' Further, for Galvão, Dom Manuel was a direct instrument in the hands of Divine Will in the matter: 'The King is not King of his own self, or for his own self . . . The King's heart is in the hand of God, and God inclines it where he will, as the Holy Scripture says.'[86] Armed with this type of conviction, then, the Portuguese monarchy set out to consolidate the gains made in the first voyage of Vasco da Gama.

[86] Quotations from Duarte Galvão, *Crónica de D. Afonso Henriques*, ed. Castro Guimarães (Cascais, 1918), pp. 2–5, 83–5, and also cited in Aubin, 'Duarte Galvão', pp. 64–6. A more recent edition of the *Crónica* also exists (Lisbon: Imprensa Nacional-Casa da Moeda, 1995).

CHAPTER FOUR

THE ATTEMPT TO CONSOLIDATE

When Vasco da Gama came back after the discovery of India, the Count [of Vimioso] asked him what goods were there to be brought back, and what goods they wanted from here in exchange for them. And on Vasco da Gama saying to him that what was brought back from there was pepper, cinnamon, ginger, amber, and musk, and that what they wanted from us was gold, silver, velvet, scarlets, the Count said to him:
– In this fashion, it is they who have discovered us . . .

Sixteenth-century Portuguese anecdote[1]

THE PROBLEM OF A REWARD

THE RETURN OF THE PORTUGUESE FLEET, THAT IS THE TWO vessels that remained of it (first the *Bérrio* commanded by Nicolau Coelho, and then the *São Gabriel*, commanded by João de Sá), was an occasion for the Crown to launch a major propaganda effort in Europe, in the form of letters to the Catholic Monarchs of Spain, and to the Papacy, some of which were widely circulated. This effort, repeated once more after Cabral's voyage, was intended to show that Portugal was at the cutting edge of the *Respublica Christiana*, and that Dom Manuel, though monarch of a relatively modest kingdom in Europe, nevertheless had at his disposal the possibility of constructing an immense empire. Besides, merchant intelligence networks, be they those of the Italians who were directly interested in the commerce in Asian spices, or of the south Germans (who had a slightly less direct but

[1] José Hermano Saraiva, ed., *Ditos Portugueses Dignos de Memória: História íntima do Século XVI* (Lisbon, n.d.), p. 113. For a discussion of the Count of Vimioso, see Jorge Borges de Macedo, 'Para o estudo da mentalidade portuguesa do século XVI – uma ideologia de cortesão: As sentenças de D. Francisco de Portugal', *ICALP – Revista* 7–8 (1987), 73–106.

nevertheless considerable involvement), rapidly spread the news of the Portuguese achievement. By the time of Vasco da Gama's eventual return to Portugal from the Azores, in late August or early September 1499 (the dates are unclear), the first basis of a personal legend had been laid.[2] But it was an as yet fragile basis, that had to be consolidated by a series of efforts, not least of all his own. Yet, his fame in at least European court-circles was assured, as we see from the letter sent to the Catholic Monarchs in July 1499, by Dom Manuel:

> Very high, most excellent Princes and most powerful lords. Your Highnesses know how we had sent out to discover (*a descobrir*), Vasco da Gama, *fidalgo* of our household, and with him his brother Paulo da Gama, with four ships through the ocean, who had been gone for about two years now; and as the principal purpose of this enterprise for our predecessors had always been the service of God and our profit, He in His Mercy decided to bring it about, according to a message that we have had from one of the captains who has now returned to us in this city, so that India and other neighbouring Kingdoms and Seignories have been found and discovered; and they entered and navigated her [India's] sea in which they found great cities and great edifices and rivers and great settlements, in which is conducted all the trade in spices and stones that passes in ships, which the same discoverers saw and found in great quantities, and of great size, to Meca, and from there to Cairo, from where it spreads out throughout the world.[3]

Once more the insistence on that indeterminate transitive verb, 'to discover', without a clear object; but the agent of the discovery, Vasco da Gama (and his brother Paulo) are identified clearly enough, at least in this version. Now, it is a commonplace that Gama marks the beginning of a new type in the history of Portuguese exploration; as opposed to men like Diogo Cão and Bartolomeu Dias who were professional mariners, Gama was already a petty nobleman when he left.[4] As the maverick

[2] One anonymous source even suggests that Gama returned only in October 1499; cf. Luís de Albuquerque (ed.), *Crónica do Descobrimento e primeiras conquistas da Índia pelos Portugueses* (Lisbon: Imprensa Nacional, 1986), p. 41.

[3] AN/TT, Colecção de São Vicente, Livro XIV, fl. 1, reproduced in Teixeira de Aragão, *Vasco da Gama e a Vidigueira*, doc. 8, pp. 217–18.

[4] Diogo Cão is a somewhat mysterious figure, between corsair and explorer, who had by 1484 attained the position of *cavaleiro* in the royal household. For the most recent discussion of his career, see Carmen M. Radulet, 'As viagens de Diogo Cão: Um problema ainda em aberto', *Revista da Universidade de Coimbra* 34 (1988), 105–19, as also Radulet, 'As viagens de descobrimento de Diogo Cão: Nova proposta de interpretação', *Mare Liberum* 1 (1990), 175–204.

Portuguese historian, António Sérgio, irreverently wrote in the first half of the century:

> Once Bartolomeu Dias had passed the southernmost point of the African continent, the problem of the route to India – the nautical and geographical problem – was a problem that had been solved. This was why Bartolomeu Dias was the last representative of the type of discoverer created by the Infante Dom Henrique: a man of modest condition, dedicated to affairs of the sea. From now on, it was no longer a question of discovery: it was necessary to organise the purchase and despatch of merchandise from India, and to negotiate diplomatically, to this end, with Oriental sovereigns. Hence, the king did not choose a seaman as the head of the expedition, but a noble, capable of the task for which he was sent: Vasco da Gama.[5]

We may debate the merits of Sérgio's dismissal of the 1497–8 voyage, which after all did not follow the African coast (unlike Dias) but instead made an innovative wide loop in the Atlantic; it is, of course, another matter whether this trajectory was due to Gama's initiative (which seems unlikely), a result of his instructions, or (as seems most likely) the result of his pilots' expertise. It is also amusing to note that Sérgio here in fact partly reproduced a rather inaccurate sneer that had been directed at Gama as early as July 1500 by none other than Amerigo Vespucci, who in a letter to his patron Lorenzo de' Medici had declared: 'Such a voyage as that I do not call discovery, but merely a going to discovered lands, since, as you will see by the map, they navigate continually in sight of land, and sail along the southern part of Africa, which is to proceed upon a way discussed by all the authorities in cosmography.'[6]

But there was a somewhat piquant consequence of Gama's initial, relatively high, social status compared to that of his predecessors. Whereas men like Cão and Dias could be contented with relatively minor rewards – such as a pension (*tença*), or admission into the ranks of the lower service-nobility – the stakes were already bound to be somewhat higher in the case of Gama,

[5] António Sérgio, *Breve interpretação da História de Portugal* (Lisbon: Sá da Costa, 1972), p. 59. It is interesting to note that Sérgio (technically António Sérgio de Sousa [1883–1969]), who made a major contribution to demystifying Portuguese expansion in this century, was nevertheless a great admirer of the Infante Dom Henrique, and more oddly still of D. Francisco de Almeida, and of Duarte Pacheco Pereira. For a rather kinder, and more sober, evaluation of Gama's role, see J.H. Parry, *The Discovery of the Sea* (Berkeley: University of California Press, 1981), pp. 164–83.

[6] Cf. Luciano Formisano (ed.), *Letters from a New World*, p. 17.

already a cavalier of the Order of Santiago, already a *fidalgo* of the royal household, and linked in one fashion or the other to powerful men in and around the court. The rewards came, however, in a trickle, following on from the first audience at the court, probably held on 18 September 1499.

On 24 December 1499, Gama received his first formal honour, namely a royal grant (*alvará*). Its purpose is noteworthy, and suggests that it may have been set down at the request of Gama himself, for it is a letter for the possession of the *vila* of Sines with all its revenues, as a heritable property. This was significant, for Sines was a sort of ancestral base, from which Estêvão da Gama, his father, had acted on behalf of the Order of Santiago. But the right over Sines was also hedged in with all sorts of hesitancies – since the town still pertained to the Order of Santiago, and the post of *alcaide-mor* therein to a certain Dom Luís de Noronha (half-brother of the Countess of Odemira, allied to the Braganças, and one of those who had been disgraced in the mid-1480s). The grant-letter thus ran:

> We, the King, make known to all those who see this grant, that we bearing in mind the merits of Vasco da Gama, *fidalgo* of our household, and of the many services that he has done for us in the discovery of the Indies, have decided to give, gift and grant him the Vila of Sines, by sworn right and as heritable property, with its revenues and taxes, save God's tithe on the sea and on the land, and with civil and criminal jurisdiction over it, and as it belongs to the Order of Santiago, before making out the grant-letter formally, we have first to render satisfaction for it to the said Order, once we have received dispensation from the Holy Father to be able to exchange it against another village in this kingdom pertaining to the Crown. And also, we have to render satisfaction to Dom Luís de Noronha, *alcaide-mor* of the said Vila for the said post of *alcaide*. However, it pleases us, and we promise him by this [order] that if the said Dom Luís does not wish to come to an agreement with us in order to leave the said *alcaidaria*, as soon as the said dispensation arrives to make the said exchange, we will order that the said Vasco da Gama be given his grant of the jurisdiction, lordship and revenues of the said Vila, in the form and manner in which we are accustomed to give similar Vilas to other persons, and as for the said castle, whenever we come to an agreement with the said Dom Luís about it, or if he leaves it for any other reason, we will give it to the said Vasco da Gama, as his right, in the same way as the said Vila.[7]

[7] Teixeira de Aragão, *Vasco da Gama e a Vidigueira*, doc. 10, pp. 220–1.

We have begun to enter into the first signs of a potential conflict between Gama, and the Order to which he and his father belonged. The conflict does not seem to have been fomented by Dom Manuel (who is unlikely to have been at the origin of the idea of granting Sines to Gama); however, the king did take advantage of it. Thus, a royal letter dated nearly two years later, in September 1501, notes that Gama has represented that he has still not been handed over Sines. The letter then continues to the effect that 'since, until the present, for certain reasons we have not been able to comply with and satisfy the obligation we have to give the said Dom Vasco as free and unencumbered the said Vila de Sines with its jurisdiction', Gama and his descendants should be given the sum of 1,000 *cruzados* a year, payable at the Casa da Mina in quarterly instalments with effect from 1 January 1501. A letter was hence issued to this effect to Fernão Lourenço, the royal factor in charge at that time of the Guinea and India trades.[8] Another grant-letter and receipt from later in the same year refer to the handing over of a quantity of wheat (15 *móios*, to be precise) from the Casa de Ceuta to Gama, in partial payment of the sums owed to him.[9] Gama thus found himself ranged with the monarch, in his struggle with the Order of Santiago, and more explicitly with its master, Dom Jorge. We may imagine that Dom Manuel pressurised the Order to hand over Sines, on the grounds that Gama's services needed to be recognised urgently; for its part, the Order is likely to have seen the affair as the thin edge of the wedge of royal infringement on its prerogatives. The affair thus dragged on for several years before being resolved in Gama's favour, but only temporarily, as we shall see below. In the interim, at least one minor outbreak of violence is reported in 1500 or early 1501, between some of the servants of Vasco da Gama, and those of D. Luís de Noronha, *comendador* of Sines. A royal letter of late March 1501, directed to a certain Álvaro Afonso, a servant (*criado*) of Vasco da Gama, and an unmarried resident of Sines, who had been on the voyage to India in 1497–9, notes that Afonso had been in a skirmish between his master's men and those of Dom Luís in Sines itself, in which he had wounded a black slave of the latter with his sword. However,

[8] Teixeira de Aragão, *Vasco da Gama e a Vidigueira*, doc. 13, pp. 222–4. For an English summary, see Ravenstein, *Journal of the first voyage*, p. 229, where the date of the document is given as 22 February 1501.

[9] AN/TT, CC, II–5–42, dated 19/28 November 1501, reproduced in Teixeira de Aragão, *Vasco da Gama e a Vidigueira*, pp. 51–52, and in Ravenstein, *Journal of the first voyage*, pp. 229–30.

since the slave had recovered from his wounds, and D. Luís had agreed to forgive Afonso, the king agreed to pardon him, on the condition that he paid a fine of 1400 *reais* for pious works.[10]

What was the grant of Sines in fact worth, in relative and absolute terms? From the papers of the Order of Santiago, we have at our disposal a 'List of commanders and cavaliers' (*Rol dos comendadores e cavaleiros*) from the very first years of the sixteenth century, which includes a number of illustrious names of the period, with their pensions (*tenças*). The largest of these, from what we can gather, was held by D. João Mascarenhas, with 800,000 *reis*, deriving from Mértola and Alcácer; next was D. João de Meneses, Conde da Tarouca, with 500,000 *reis* revenue from Sesimbra and the church of Santiago de Beja. There follow João de Sousa with 480,000 *reis* from Alvalade, Feira and Represa, D. Gonçalo Coutinho with 350,000 *reis* from Arruda and Rebaldeira, and then a number of grantees with 320,000 *reis*. Vasco da Gama (listed simply as *o allmirante*), has 80,000 *reis* from his old grants of Mouguelas and Chouparria. Sines is listed with D. Luís de Noronha, and its worth stated at 150,000 *reis*, but there is also a supplementary item for the *ribeira de Sines*, evaluated at 50,000 *reis* in revenues. Of particular significance is that both Sines and the *ribeira* are listed, like a number of other grants, as pertaining to the 'Master's Table' (*mesa mestral*); thus, in fact, D. Jorge probably had the discretionary right to dispose of them as he saw fit, unlike in the usual *comendas*.[11]

To be sure, Gama had more than just Sines by way of royal grants. We are aware that, despite a certain confusion in the minds of historians, Gama had in fact also been given the title of *Dom*, admitted into the Royal Council, as well as given a series of maritime honours, possibly in imitation of what the Catholic Monarchs had done for Columbus, all by January 1500. We owe this insight to a rather useful, but often neglected, study by Anselmo Braamcamp Freire, who showed, using a set of eight documents, that Gama was referred to repeatedly in documents as *Dom* by February 1500, and hence proposed that the major grant-letter from Dom Manuel to Gama, usually dated to 10 January 1502, in fact dated from two years earlier.[12] If this is

[10] AN/TT, Chancelaria D. Manuel, Livro 45, fl. 75v, reproduced in Brito Rebello, 'Navegadores e Exploradores Portuguezes', Doc. 17, pp. 63–4.

[11] AN/TT, B-51-135, 'Privilégios da Ordem de Santiago', fls. 193–201; and for a brief discussion of the text, Isabel Lago Barbosa, 'A Ordem de Santiago', pp. 150–3.

[12] Anselmo Braamcamp Freire, 'O Almirantado da Índia: Data da sua criação', *Archivo Historico Portuguez* 1 (1903), 25–32. The argument is complex, for the title of Admiral

correct, then Gama also received the title of Admiral of India, or Admiral of the Ocean-Sea, in January 1500; though it was not then that he was given the pre-emptive right to be the Captain-Major of any fleet being sent to India, whether for trade or for war, a right which only indirectly stemmed from his title of Admiral (*almirante*).

The grant-letter of January 1500 (accepting Braamcamp Freire's dating), is a truly elaborate document, more so than any that we have dealt with thus far concerning Gama. Here, we are presented with a potted history of the Portuguese discoveries, the role of Gama therein, and a series of grants not only to him but to his family (his surviving brothers, and sister). Besides the version that has come down to us in the late copy made in D. Manuel's reign (in the *Livro dos Místicos*), draft versions of the narrative part of the document (which precedes the grant) also exist, stating that it is in this way that all grants to Gama must be framed.[13] Let us pause a moment to examine the text (which has been published on a number of occasions, and partially translated by Ravenstein):

> Dom Manuel etc. [titles]. To those who see this letter of ours, make it known: that the discovery of the land of Guinea having been begun in the year 1433 by my uncle the Infante Dom Henrique with the intention and desire that through the coast of the land of Guinea, India was to be found and discovered, which until now had never been known through there, not only with the intention that great fame and profit might follow to these Kingdoms from the riches that there are therein, which were always possessed by the Moors, but so that the faith of Our Lord should be spread through more parts, and His Name known. And afterwards, the King Dom Afonso my uncle, and the King Dom João my cousin, wanting with the same desires to prosecute the said work, with a good number of deaths and expenses in their time, until the Rio do Infante was discovered in the year 1482, which is 1885 leagues from where they first began to discover.[14]

An astonishing error has crept into the text here, for the Rio do Infante was the last point reached in the voyage of Bartolomeu Dias in 1488 (and not Diogo Cão in 1482), and lies in south-

is used in several documents of 1500, but not in some others of 1501, where Gama is merely termed Dom and a member of the Royal Council.

[13] AN/TT, CC, III–1–9, 'Extracto de como se avião de fazer as mercês a Vasco da Gama em remuneração do descobrimento da Índia, e do q̃ na dita viagem passou e da gente q̃ lhe morreu'.

[14] AN/TT, Livro dos Místicos I, fl. 204, in Teixeira de Aragão, *Vasco da Gama e a Vidigueira*, Doc. 14, pp. 224–5.

eastern Africa. A Freudian slip, given D. Manuel's desire to minimise the role of D. João II? At any rate, three other aspects are worth noting in this introductory passage. First, the insistence on the Catholic faith, posed here in slightly different terms from what we shall find later in the same grant-letter; second, the genealogy that is clearly established from Dom Henrique through Dom Afonso V, to Dom João II; and third, the anachronistic insistence that even in the time of Dom Henrique, the discovery of India had been the real purpose of the whole enterprise. And finally, of minor interest is the fact that the name of the last spot (Rio do Infante) touched by Dias has been associated implicitly with the Infante Dom Henrique, rather than with Dias's companion João Infante, for whom it was actually named!

The text now continues, and we shall follow it literally only part of the way through:

> And we, with the same desire, wishing to accomplish the work that the said Infante, and Kings, our predecessors, had begun, trusting that Vasco da Gama, *fidalgo* of our household, was such that in pursuit of our service and in order to carry out our order, he would take upon his own person the entire peril, and risk his own life, we sent him with our *armada*, as its Captain-Major, sending along with him Paulo da Gama, his brother, and Nicolau Coelho, also a *fidalgo* of our household, in order to seek out the said India, in which voyage he served us in such a way that whereas in all those years since the said discovery (*descobrimento*) had commenced, and so many captains sent out, the said 1885 leagues had been discovered, he in this voyage alone discovered 1550 leagues.

Mention is then made of a great gold-mine (*mina d'ouro*) found on this voyage, a reference undoubtedly to the gold trade of East Africa, and also to the many great cities that had been found. The crowning achievement nevertheless lay in having 'found and discovered India, which by all of the writers who have described the world and all its provinces was posed as the richest of them all', besides of course being coveted by all the emperors and great kings of the world. Mention is made of the great costs and perils incurred by Gama in person, the death of his brother Paulo and of half the men on the fleet, and the fact that the voyage had endured as long as two years. Naturally, the results of this voyage would benefit not just Portugal alone, but all of Christendom, 'from the damage to the Infidels that is expected'; the defeat of the Mamluks and the imminent reconquest of the Holy Places is thus an explicit undercurrent here. Indeed, the text continues to labour under the impression that the bulk of the Indians are Christians. Thus, mention is made of 'the great service that we hope will follow to Our Lord, for it seems that one can easily bring all the people of India around to the true knowledge of His Holy Faith, on account of the great [knowledge] that some of them already have, in which they may be entirely confirmed'.

The second part that follows on this extended preamble details the actual grants to be made, as indeed 'every prince should do to those who serve them greatly and well'. These begin with the sum of 300,000 *reis* anually, in perpetuity to Gama and his descendants, divided as follows:

– A fifth part, or 60,000 *reis*, was to come from a tithe on fishing (*dízima do pescado*) newly imposed on Sines and Vilanova de Milfontes, which had until then been collected as a benefice by the *vedor da fazenda* Dom Martinho de Castelo-Branco. The risk of fluctuations in the collections was to be born by Gama and his family.
– Another 130,000 *reis* were to come from excise (*sisas*), levied on Sines; if the collections exceeded this amount, the Crown would get the difference, but if there was a shortfall, it would be made up from the excise at the neighbouring town of Santiago de Cacém.
– A further 40,000 *reis* were to be derived from the excise at Santiago de Cacém itself, to be paid in quarterly instalments.
– And finally, the remaining sum of 70,000 *reis* would come out of the royal timber trade at Lisbon (*a casa do paaço de madeira*).

The letter then goes on to name him Admiral of India, 'with all the honours, prerogatives, liberties, power, jurisdiction, revenues, quit-rents, and duties that by right should accompany the said Admiralty, and which the Admirals in these kingdoms have, as is contained in more detail in its manual (*regimento*)'.[15] Besides Gama and his descendants were given the right to send out 200

[15] Vasco da Gama's title here ran parallel to that of Admiral of Portugal, a post created in 1322 by D. Dinis for the Genoese mariner Manuel Pessagna (or Pessanha), which later passed on to his descendant by the female line, Lopo Vaz de Azevedo, and eventually in the seventeenth century to D. Luís de Portugal, Conde de Vimioso. For Gama's *regimento*, see Vicente Almeida d'Eça, 'Almirante da Índia', *Revista Portugueza Colonial e Marítima*, (special commemorative number, 20 May 1898), 26–31, and more recently Carmen M. Radulet and António Vasconcelos de Saldanha (eds.), *O Regimento do Almirantado da Índia* (Lisbon: Inapa, 1989). A dispute subsequently arose over jurisdictions; cf. AN/TT, Chancelaria D. João III, Livro 5, fl. 28, reproduced in Brito Rebello, 'Navegadores e Exploradores Portuguezes', Doc. 37, pp. 154–8, and the discussion in Chapter 5 below. Finally, also see António Vasconcelos de Saldanha, 'O Almirante de Portugal: Estatuto quatrocentista e quinhetista de um cargo medieval',

cruzados in cash every year to India, to be employed in the purchase of goods, which would pay no duties to the Crown, and only a five per cent tax (*vintena*) to the Order of Christ.[16] The Crown's captains and factors in India were to cooperate in this affair. And lastly, there was the grant of the title of Dom, to Gama himself, to his brother Aires, and to his sister Teresa, as well as to such of their direct descendants who maintained the name 'Gama'.

Let us return to the first part of this grant. Its significance is the renewed concentration of fiscal resources in Gama's hands in the region of Sines, that is in the town itself, and in Vilanova de Milfontes and Santiago de Cacém, both in its neighbourhood. This, taken together with the first grant of December 1499, constituted a substantial accumulation in perpetuity in the hands of a single family, in a region which had traditionally been dominated by the Order of Santiago. It compared very well indeed with what was given to Nicolau Coelho, which is to say a *tença* of 50,000 *reis* a year.[17] And yet, it fell short of a real title, and certainly did not place Gama in the ranks of the upper echelons of the nobility. Another route existed to consolidation, and this was marriage.

In the years 1500–1, Vasco da Gama moved to utilise this means as well to consolidate his fortune and lineage. The bride was Dona Catarina de Ataíde, daughter of the late Álvaro de Ataíde, *alcaide-mor* of Vila de Alvor and of Maria da Silva (in whose house Dom João II had died in 1495).[18] A more important branch of the Ataídes was to be linked later again with the Gamas; Vasco da Gama's grandson, Dom Vasco Luís da Gama, third Count of Vidigueira, married Dona Ana de Ataíde, daughter of the Count of Castanheira – the adviser and close confidant of the

Revista da Universidade de Coimbra 34 (1988), 137–56, especially pp. 146–7, for a comparison.

[16] This privilege was eventually withdrawn by D. João III; cf. Jorge Manuel Flores, 'Os Portugueses e o Mar de Ceilão, 1498–1543: Trato, Diplomacia e Guerra', Master's thesis (Universidade Nova de Lisboa, 1991), p. 199.

[17] Still further down the pecking order, the pilot Pêro Escolar (or Escobar) received a mere 4,000 *reis* (later revised to 6,000); cf. Teixeira de Aragão, *Vasco da Gama e a Vidigueira*, pp. 50, 221; the documents are in AN/TT, Livro 2° dos Místicos, fl. 245v, and AN/TT, Chancelaria D. Manuel, Livro 13, fl. 7v. For further documents, showing receipts of the years 1506 onwards, see Brito Rebello, 'Navegadores e Exploradores Portuguezes', Docs. 27 to 32.

[18] Cf. Felgueiras Gayo, *Nobiliário de Famílias de Portugal*, vol. XIV (Braga, 1939), pp. 75–6; Affonso de Dornellas, 'Bases genealógicas dos Ataídes', in his *História e Genealogia*, Vol. I (Lisbon, 1913), pp. 107–42, is less than wholly reliable in this matter. Álvaro de Ataíde had already died in 1497 or 1498.

King Dom João III. The significance of Vasco da Gama's marriage is quite different: Álvaro de Ataíde, his father-in-law, was a maternal uncle of Dom Francisco de Almeida, and the marriage thus reaffirmed Gama's ties to that family. We do not know when precisely the marriage took place, but it was certainly prior to 5 October 1501, for a royal grant of that date confirms a pension of 50,000 *reis* to Dona Catarina de Ataíde, 'wife of Dom Vasco da Gama, of our council'.[19] Dona Catarina's brothers, Álvaro de Ataíde and Tristão de Ataíde, were to play a role later in Gama's career, as was another future brother-in-law (later to be the husband of Teresa da Gama), Lopo Mendes de Vasconcelos. But the alliance had a larger significance, for we know that the Almeidas contributed to facilitate Dona Catarina's dowry, suggesting that they too saw its advantages.[20] At the same time, the years 1499–1501 witnessed a delicate realignment of forces, as the Almeidas, the Ataídes and the Gamas seem to have sought collectively to distance themselves in these years from the waning star of Dom Jorge, and the Order of Santiago. It was in the aftermath of this (and before his departure for India as viceroy in 1505) that Dom Francisco de Almeida, for example, shifted his allegiance to the Order of Christ.[21]

THE IN-BETWEEN VOYAGES: CABRAL AND JOÃO DA NOVA

In these years, two fleets were sent out to India, one a huge affair under the command of Pedro Álvares Cabral, the other a modest and somewhat mysterious one (believed to have been exploratory in nature), led by João da Nova. Now Cabral's fleet, later to become famous as the one that discovered Brazil, comprised

[19] Cf. AN/TT, Chancelaria D. Manuel, Livro 38, fl. 86v, also reproduced in AN/TT, Chancelaria D. João III, Livro 3, fl. 166, published in Brito Rebello, 'Navegadores e Exploradores Portuguezes', Doc. 71, pp. 473–5; this document is also cited in Braamcamp Freire, 'O Almirantado da Índia'. We may note that another document which refers to Vasco da Gama, dated 28 May 1501, and written to D. Manuel by the *desembargador* Pêro Jorge at Évora, almost certainly pertains to another Vasco da Gama, not the object of our attention here; for this letter, see AN/TT, CC, I-3–60.

[20] The transaction for D. Catarina's dowry was between her brother, Nuno Fernandes de Ataíde, who had succeeded his father as *alcaide* of Alvor in 1498, and D. Pedro da Silva, son of the Conde de Abrantes (and hence brother of D. Francisco de Almeida).

[21] AN/TT, Gavetas, VII/10–14, Papal Bull permitting D. Francisco de Almeida to pass from the Order of Santiago to the Order of Christ, January 1505. Note that Almeida was named Captain-Major of India only on 27 February 1505; cf. AN/TT, Gavetas, XIV/3–14, António da Silva Rego (ed.), *As Gavetas da Torre do Tombo*, 12 vols. (Lisbon: CEHU, 1960–77), Vol. III, pp. 599–602.

thirteen ships, and carried about 1500 men; the second-in-command to Cabral was probably a certain Sancho de Tovar (in fact a very well-born Castilian noble, exiled in Portugal). The expedition was a disaster from a number of points of view. Of the thirteen ships, one was lost off the Cape Verde Islands, and still another (a supply-ship) returned directly from Brazil. Four more were lost in a great storm in the south Atlantic in May 1500, and still another (commanded by Diogo Dias) strayed off to Madagascar, and then to the Red Sea. Thus, only six of the original thirteen vessels made it to the Malabar coast; and on the return voyage, still another (*El-Rei*, commanded by Sancho de Tovar) had to be abandoned near Malindi.[22] There are several facts worthy of note concerning the organisation of this fleet. First, it involved a far more extensive and explicit private participation than the first voyage, and one of the ships that successfully made the round trip was the *Anunciada*, belonging to D. Álvaro de Portugal and three other Lisbon-based partners (including the Florentine firm of Marchionni). Second, several of those who took part are familiar names; most obvious is Nicolau Coelho, who had just returned from India, but Bartolomeu Dias too took part (and commanded one of the four ships wrecked in the Atlantic); also on board one or the other vessel were men like João de Sá from the first voyage.[23]

Cabral's fleet was intended to execute D. Manuel's plans, establish a factory definitively in Calicut (the factor named to the post being one Aires Correia, with two scriveners called Martinho Neto and Afonso Furtado), consolidate relations between that kingdom and Portugal, in general promote relations with other 'Christian kingdoms' in India (including Cannanur), and at the same time do all that was possible to impede the navigation of Muslims to the Red Sea.[24] We have already taken note of the altogether curious letter given to Cabral by D. Manuel, and we also know that the instructions (*regimento*) that were prepared for his use were made ready with the close consultation of Vasco da

[22] See the exhaustive, and at times exhausting, study of Moacir Soares Pereira, 'Capitães, Naus e Caravelas da Armada de Cabral', *Revista da Universidade de Coimbra* 27 (1979), 31–134.

[23] On Dias's participation, see Albuquerque, *et al.*, *Bartolomeu Dias: Corpo Documental*, Docs. 10 and 11.

[24] AN/TT, CC, II–3–9, *regimento* for the scriveners Martinho Neto and Afonso Furtado, dated Lisbon, 4 February 1500; cf. Isaías da Rosa Pereira, 'Documentos inéditos sobre Gonçalo Gil Barbosa, Pêro Vaz de Caminha, Martinho Neto e Afonso Furtado, escrivães da despesa e receita do feitor Aires Correia (1500)', in *Congresso Internacional, Bartolomeu Dias e a sua época: Actas*, Vol. II (Oporto, 1989), pp. 505–13.

Plate 16 Mosteiro dos Jerónimos, a major Manueline monument from the sixteenth century in Lisbon, exterior.

Gama.[25] Some authors go even further, and it has even been claimed that Gama was offered the command of the expedition in 1500, but turned it down in order to be able to consolidate his position with respect to Sines, and to be able to marry. This argument, based on the slenderest thread of evidence (the fact that Gama's name appears on the margin of a draft document in which Cabral is named Captain-Major of the fleet), need not detain us; all that this shows is that there were some in the court who may have proposed that Gama should return to India.[26] Nevertheless, we are aware that we know all too little concerning the circumstances surrounding the nomination of Cabral, or even the man himself. Briefly, what has come down to us is that Pedro Álvares de Gouveia (c. 1467–1520), later known as Cabral, was born in Belmonte into the lower nobility, and was made *moço fidalgo* under D. João II by 1484; he was also a cavalier of the Order of Christ, certainly by the early 1500s. We also know that he married D. Isabel de Castro, a niece of Afonso de Albuquerque, and that the latter interceded for him at the court in 1514, at a time when Cabral was still in disgrace, albeit to no real avail.[27] This was a state in which he passed most of his life after his return from India and Brazil in 1501; in 1509, he retired to Santarém, where he is buried in somewhat obscure surroundings (though the Brazilian government has helped to dignify the church and gravestone!).[28]

It seems reasonable to assume that Cabral was initially supported by D. Manuel, in view of the fact that the expedition of 1500 was one in which the monarch massively invested in terms of prestige, and also given the fact that the monarch's power was

[25] Cf. the instructions given to Cabral, in Ramos Coelho (ed.), *Alguns Documentos da Torre do Tombo acerca das navegações e conquistas Portuguesas* (Lisbon, 1892), pp. 101–2, supplemented by the important study of Alexandre Lobato, 'Dois novos fragmentos do regimento de Cabral para a viagem da Índia em 1500', *Studia* 25 (1968), 31–50, showing the conflict between different groups of D. Manuel's advisers on the strategy to be followed. Also see W. Brooks-Greenlee, *The voyage of Pedro Álvares Cabral to Brazil and India* (London: The Hakluyt Society, 1937).

[26] Cf. Francisco Leite de Faria, 'Pensou-se em Vasco da Gama para comandar a armada que descobriu o Brasil', *Revista da Universidade de Coimbra* 26 (1978), 145–85. Faria's argument is based on the close reading of a single document, AN/TT, CVR, No. 178.

[27] AN/TT, CC, I-17-1, letter from Afonso de Albuquerque at Calicut to D. Manuel, dated 2 December 1514, in R.A. de Bulhão Pato (ed.), *Cartas de Afonso de Albuquerque*, Vol. I (Lisbon: Academia Real das Sciências, 1884), pp. 353–5.

[28] For materials towards a biography of Cabral, see L.M. Vaz de Sampaio, *Subsídios para uma biografia de Pedro Álvares Cabral* (Coimbra, 1971); for earlier studies with collected documents, see Jaime Cortesão, *A expedição de Pedro Álvares Cabral* (Lisbon, 1922), and the altogether unfortunately titled work of Abel Fontoura da Costa, *Os sete únicos documentos de 1500 conservados em Lisboa referentes à viagem de Pedro Álvares Cabral* (Lisbon, 1940).

quite considerable at this moment in his reign. Yet, we cannot go much beyond this, lacking as we do a Namierite study (or even an adequate prosopography of the nobility) of the court of Dom Manuel. Also, in the final analysis, Cabral proved unequal to D. Manuel's expectations, and we must not forget that the discovery of Brazil (though it may be seen as a capital event in retrospect) was hardly of great significance at the time. To lose half of the fleet, to break off commercial relations with Calicut (and have the factor Aires Correia killed in the process), was bad enough; the definitive news that Christians were relatively scarce in the Indian context only made matters worse. A sense of this can be gathered from the dispatch sent in June 1501 by the Cremonese merchant, Giovanni di Francesco Affaitati, to his Venetian correspondents, wherein he reports the return of the first of the vessels, the hostilities at Calicut, and the decision made by Cabral (reputedly on the advice of Gaspar da Gama, whom he took with him) to move on to Cochin in search of a cargo.[29]

It may be worthwhile here to recapitulate the principal steps of Cabral's voyage. First of all, let us note that his fleet left Lisbon in early March 1500, in what was to become the established pattern (rather than much later in August, as with Vasco da Gama in 1497). After the accidental landfall in Brazil, and the storm in the South Atlantic that we have already mentioned, the first places in the Indian Ocean that the fleet probably touched on were the road at Sofala (where the Portuguese were particularly interested in the gold trade), and then Mozambique (where the fleet remained about six days). Thereafter, they put in at Kilwa (for the King of which they carried letters in Arabic and Portuguese), and then Malindi. In Kilwa, where the fleet arrived in late July, relations with the *de facto* ruler Amir Ibrahim appear to have been rather uneven, prefiguring later tensions of a more serious sort; in Malindi, on the other hand, the fleet was very well received, and the ruler was also handed over a letter and various gifts from D. Manuel, in thanks for his treatment of Gama on his first voyage. Here, as on the first voyage, the Portuguese took on board a pilot to guide them as far as Calicut, and left behind two of the twenty *degredados* (convict-exiles, facing commuted capital punishments) whom they had on board, to explore inland.[30] After just over a

[29] Nicolò Barozzi (ed.), *I Diarii di Marino Sanuto*, Vol. IV (Venice, 1880), pp. 66–9, 99–102.

[30] It is generally thought that these two men were Luís de Moura, and João Machado, the latter a rather celebrated character in the early history of Portuguese Asia. For a study of

month's sailing, the fleet reached Calicut in mid-September, six months after its departure from Lisbon, and at a rather sensible time (unlike Gama, whose men had waded through the waterlogged streets of Calicut in the monsoon).[31]

It is of some importance to note that some crucial aspects of Calicut politics had changed in the interim. The old Samudri (the eighty-fourth of his line, according to tradition) was dead, and his younger and more vigorous successor appears to have been eager to develop his port further, to the detriment of rivals such as Cochin, which was notionally subordinate to him but which in fact sought an increasingly autonomous role.[32] This change of ruler is lent great significance by at least one sixteenth-century Portuguese chronicle, which states in no uncertain terms that it was 'certain, that if this King [the old Samudri] had lived, we would always have had peace and friendship with Calicut, but shortly thereafter [1499], this one died and another reigned in his place, from whom our people received much harm, though he paid for it later, as we shall see'.[33] The Portuguese, for their part, arrived at first determined to prosecute a particular form of hostile trade with hostages, in continuation of the tenor of Vasco da Gama's voyage. Thus, Afonso Furtado, who was first sent ashore, demanded a number of notables from the Calicut court as hostages before Cabral himself would come on land. This request was granted, and the Samudri Raja now came as far as the strand to receive the Portuguese Captain-Major. The Samudri Raja impressed them with the richness of his personal finery and trappings; the Portuguese for their part were rather better prepared with gifts (including gold and silver objects) than on the earlier occasion. The letters from Dom Manuel, in Portuguese and Arabic, were handed over; we can only wonder what the Samudri and his court made of them. After this Cabral was given permis-

his career, see Maria Augusta Lima Cruz, 'As andanças de um degredado em terras perdidas – João Machado', *Mare Liberum* 5 (1993), 39–47.

[31] The principal sources for this voyage are the chronicles of Barros, Castanheda and Góis; Gaspar Correia's *Lendas* differs from them in some measure, as does the anonymous *Crónica do Descobrimento e primeiras conquistas da Índia*, ed. Albuquerque, pp. 44–9. More complex is the controversial letter of D. Manuel to the ruler of Castile (Rome: Johan de Besicken, 1505), reprinted in Teixeira de Aragão, *Vasco da Gama e a Vidigueira*, pp. 232–50, often alleged to be apocryphal.

[32] The best chronological account we have to date of these rulers is K.V. Krishna Ayyar, *The Zamorins of Calicut (From the earliest times down to A.D. 1806)*, (Calicut: Norman Printing Bureau, 1938), pp. 80–90, but it is nevertheless unsatisfactory. Royal successions are more or less clearly established there, but not shifts in policy and political factions.

[33] *Crónica do Descobrimento*, ed. Albuquerque, p. 35.

sion to leave, and the exchange of hostages was accomplished, though not without a good deal of panic and confusion. The Portuguese factor Aires Correia now went on land with his trade goods, while the nephews of a Gujarati merchant based at Calicut were retained on board the fleet as hostages for him. The Samudri appears to have given the Portuguese a written grant (on either a golden or silver plate, in different versions) to guarantee them security of trade. At the end of two months, once the Portuguese had been given a large house near the seafront as a factory, they released their hostages; meanwhile, the Samudri Raja offered them all assistance in the matter of finding cargoes for the vessels.

In contrast to the earlier Samudri, who had preferred to have the Portuguese come to his palace, the new ruler of Calicut appears to have been in the habit of personally visiting the seafront and keeping a rather close eye on what took place on the maritime margins of his domains. Thus, he is alleged to have sent a message to Cabral that a ship was being loaded up in Cochin by Muslim merchants (in some versions, Cherian and Mammali Marikkar), in which one or more war-elephants on board belonged by right to him. He hence asked the Portuguese to seize the vessel, on its way to Cambay in Gujarat, on his behalf. Cabral, eager to ingratiate himself, sent a caravel armed with artillery under the command of Pêro de Ataíde, and a combat ensued. The Samudri himself is reported to have turned up on the seafront (*na práia*) to witness the combat, but to have left in disgust, saying that the single caravel was unequal to the task. However, after a chase as far as Cannanur, the ship was captured, and brought into Calicut some days later, 'so that the King was amazed and filled with pleasure'.[34]

On the other hand, the resentment of both Middle Eastern and local merchants against the Portuguese seems to have increased from this moment on (even though some leading Mappilas of Calicut such as Koya Pakki supported them), and matters came to a head on 16 December 1500. The proximate reason for the conflict was the seizure by the Portuguese of a Muslim-owned ship that was about to leave the port for Jiddah; the Portuguese interpreted this departure as an abrogation of their understanding with the Samudri Raja that they would be given priority in

[34] *Crónica do Descobrimento*, ed. Albuquerque, pp. 54–6. For a succinct, and useful, summary of this conflict and other aspects of the expedition, see Geneviève Bouchon, *Mamale de Cananor: Un adversaire de l'Inde Portugaise (1507–1528)* (Geneva/Paris: Librairie Droz, 1975), pp. 53–9.

finding a spice cargo. When Cabral, on the express advice of Aires Correia, seized the ship, a group of Muslim merchants in the port reacted violently, and attacked the factory; after a combat lasting three hours, Aires Correia and the others tried to flee to the boats, but were killed. Portuguese losses were numbered at fifty-four men, and goods of considerable value.

Cabral waited a day for a message from the Samudri; when none came, he captured ten (or in some versions, fifteen) large ships that were in the port, and the next day bombarded first Calicut and then Pantalayini, causing a good deal of damage to these two unfortified towns.[35] The fleet then went on to Cochin, reputedly – we have seen – on the expert opinion of Gaspar da Gama who suggested that this was the best option in the circumstances, but still tarrying en route however to seize two more vessels from Calicut, which were summarily burnt. By 24 December, when the fleet arrived in Cochin, the Portuguese had caused several times the damage in terms of human lives and goods than they themselves had incurred in Calicut; the bombardment alone is said to have resulted in four to five hundred deaths on land. In Cochin, the Portuguese were rather well received, despite the earlier episode of the Mappila vessel that they had seized; the Cochin ruler, Unni Goda Varma, appears to have grasped directly the political advantages that would accrue to him by diverting the Portuguese away from Calicut. The fleet remained in Cochin, making up that part of the cargo that could not be loaded in Calicut, until mid-January 1501; they then departed in some haste, fearing an attack from Calicut, and in the process took back to Lisbon two of the Cochin Raja's servitors (Idikkela Menon and Parangoda Menon), as well as the celebrated Joseph of Cranganor, who became an important informant of the Portuguese Crown while in Europe.[36] In turn, from the Portuguese fleet, some persons were left behind at Cochin, in what would turn out to be the beginnings of a permanent factory; these included the celebrated Gonçalo Gil Barbosa, whom we shall

[35] For an account of this strife, see 'Navigazion del capitan Pedro Alvares scritta per un piloto portoghese . . .', in Giovanni Battista Ramusio, *Navigazioni e viaggi*, ed. Marica Milanesi, Vol. I (Turin: Giulio Einaudi, 1978), pp. 647–9.

[36] Cf. Geneviève Bouchon, 'Les Musulmans du Kerala', pp. 13–14. On Joseph and his account, also see Georg Schurhammer, 'The Malabar church and Rome before the coming of the Portuguese: Joseph the Indian's testimony', in Schurhammer, *Orientalia*, ed. László Szilas (Lisbon: Centro de Estudos Históricos Ultramarinos, 1963), pp. 351–63. Joseph had in 1490 already been, as one of three representatives, to the Catholicos of the East Syrian church at Gazarta Bet Zabdai in Mesopotamia, to request bishops for the Indian church.

have occasion to mention below. The fleet then went on to pay a visit at Cannanur in mid-January, where the Portuguese appear to have been well received by the Kolattiri Raja, and where a cargo of (as it happens, rather poor quality) cinnamon was taken on board.[37] Thereafter, following a brief halt at Malindi, the fleet turned back to Portugal, and having lost one more vessel en route (that of Sancho de Tovar, as noted above) arrived in either June or July 1501.[38] The cargo that was brought back has been estimated very roughly at 4,000 *cantari* of mixed goods (but largely comprising pepper, one would imagine).[39]

By the time these vessels returned, however, another fleet of four vessels had already been sent out: this was the *armada* of João da Nova, with extensive private participation on the part of the Florentine Bartolomeo Marchionni (who may have sent Girolamo Sernigi as his captain and factor on one of the ships), and of D. Álvaro de Portugal (who once more sent out a ship, under the command of Diogo de Barbosa); D. Álvaro's ship was in fact fitted out in partnership with the Marchionni, in keeping with the arrangement he had followed in Cabral's fleet.[40] Accounts differ considerably on the activities of Nova's ships, and in general the expedition is one of the most obscure of the early period, on which no contemporary documents have been found to date.[41] Nova himself was from Galicia, and had apparently been in North Africa for a time, before taking office as a petty official – *alcaide pequeno* – in Lisbon. His major asset, from all accounts, was his close association with Tristão da Cunha, a powerful *fidalgo* of Dom Manuel's court, whom we shall have further occasion to notice. It appears that the fleet left the Tagus on 9 or 10 March 1501, and once having rounded the Cape, made

[37] Bouchon, *Mamale de Cananor*, pp. 56–7.

[38] Letter from Affaitati at Lisbon to Domenico Pisani, dated 26 June 1501, in Barozzi (ed.), *I Diarii di Marino Sanuto*, Vol. IV, pp. 66–9; also see ibid., pp. 99–102, letter from Pisani.

[39] The figure in *cantari* is naturally very approximate; for a discussion, see Geneviève Bouchon, 'Glimpses of the beginnings of the Carreira da Índia (1500–1518)', in Teotónio R. De Souza (ed.), *Indo-Portuguese History: Old Issues, New Questions* (New Delhi: Concept Publishing Co. 1984), pp. 40–55.

[40] The speculation that Sernigi was on the fleet is due to Carmen Radulet, who suggests that he was also known by the name Fernando Vinete; cf. C.M. Radulet, 'Girolamo Sernigi e a importância económica do Oriente', *Revista da Universidade de Coimbra* 32 (1985), 67–77, especially pp. 74–6.

[41] The best study to date of this voyage remains that of Geneviève Bouchon, 'A propos de l'Inscription de Colombo (1501): Quelques observations sur le premier voyage de João da Nova dans l'Océan Indien', *Revista da Universidade de Coimbra* 28 (1980), 233–70, reproduced in Bouchon, *L'Asie du Sud à l'époque des Grandes Découvertes* (London: Variorum, 1987). The account that follows here diverges from it only in minor details.

for Kilwa. Nova had apparently been instructed to leave two ships there, in order to obtain gold, but finally decided against this idea, since the trade-goods he carried were deemed worthless in east Africa. After briefly touching Mozambique, Mombasa and Malindi, the fleet seems to have left the African east coast in August.[42] We next encounter trace of the fleet in November 1501, having by then made landfall on the Indian coast at Cannanur, where they were welcomed by the Kolattiri and offered a cargo; but Nova instead decided to head for Cochin, where Cabral had left a factory. In one version, this fleet had already captured two ships headed from Calicut to the Red Sea laden with spices, and from them learnt of the conflict between Cabral and the Calicut merchants; these two ships were then burnt, though their cargos were unloaded and their crews taken off. In another version, Nova had already had news of this affair at São Brás, where a letter had been left for him by Pêro de Ataíde of Cabral's fleet. En route to Cochin, Nova passed Calicut, where he again attacked and sank three vessels in the port; he is also reported to have captured another ship (reputedly 'belonging' to the Samudri) with jewels and navigational instruments on board, as well as a pilot and navigational charts. The pilot had the good fortune of escaping with his life, and being taken to Lisbon; the rest of the crew was not so fortunate, and was burnt with the vessel in sight of Calicut, once the valuables had been unloaded. The Portuguese fleet then put in at Cochin, where it was found that the factory was in poor shape, as the Portuguese goods did not find a market. In December 1501, eventually reacting to the Portuguese raids on shipping that had lasted intermittently for several weeks, the Samudri is reported to have had a fleet of small vessels made ready at Pantalayini, but these were unable to match the Portuguese for firepower in an engagement fought north of Calicut. Then, after a relatively uneventful return visit to Cannanur to set up a permanent factory and make up the cargo, these ships left rather hastily in late February (for fear of missing the monsoon), leaving behind D. Álvaro's factor Paio Rodrigues and three or four others at Cannanur. Belying Venetian hopes, they eventually returned to Lisbon in September 1502, carrying (and here various Italian

[42] Bouchon (ibid., pp. 254–5, passim), has suggested that in September–October 1501, or early 1502, the fleet of João da Nova may have touched Colombo and left an inscription there. This hypothesis has been questioned by Jorge Manuel Flores, 'Os Portugueses e o Mar de Ceilão, 1498–1543: Trato, Diplomacia e Guerra', Master's thesis (Universidade Nova de Lisboa, 1991), pp. 114–18. In general, there are large gaps in our knowledge of the activities of the fleet, and the reconstruction summarised here is tentative.

reports diverge) between 950–1000 *cantari* of pepper, 450–550 *cantari* of cinnamon, 30–50 *cantari* of ginger, and some lac as their principal cargo.[43] We may also imagine that they carried the accumulated loot of several months of successful raiding, though they are reported to have returned with partly empty bottoms.

Retrospectively seen, from the vantage point of, say, Zain al-Din Ma^cbari's Arabic chronicle of the late 1570s, the *Tuhfat al-mujâhidîn*, the actions of Cabral and Nova prefigured a situation of more or less continuous conflict off the coast of Kerala through the sixteenth century. But the conflict shifted in terms of its main centres and Asian protagonists as the sixteenth century wore on; as has been noted often enough in the recent literature, the Mappilas were not the principal targets of early Portuguese aggression, which was instead directed at the villainous 'Moors from Mecca', who were seen as responsible for the problems faced by Cabral at Calicut in 1500.

Besides commodities, it is worth noting that each such early voyage also brought back precious information, both commercial and political. In the case of Nova's fleet, the Italian letters of the period mention that he brought back from Kerala a certain sixty- or seventy-year-old Venetian called Benvenuto d'Abano, who had been in India for some twenty-five years, and who was in a situation of abject poverty; also on the same fleet was a Valencian (called Antão Lopes), who had been in India for fifteen years.[44] Benvenuto d'Abano claimed to have voyaged extensively in Persia, and to have visited Hurmuz, Gujarat and even Melaka; he also brought back an Indian wife and two children, ostensibly to 'make them Christians'.[45]

THE RETURN OF AN ADMIRAL

Let us consider, then, how matters stood in early 1502, a time when – we might remind ourselves – João da Nova's fleet had not

[43] Letters from Lunardo Nardi and Giovanni di Francesco Affaitati at Lisbon, dated 20 and 26 September 1502, in Barozzi (ed.), *I Diarii di Marino Sanuto*, Vol. IV, pp. 545–7, 664. Also see Bouchon, *Mamale de Cananor*, pp. 58–61, for a discussion.

[44] This Lopes was eventually sent back to India in 1503, as Portuguese ambassador to Ethiopia; however, he died *en route* at Jiddah; cf. Jean Aubin, 'L'ambassade du Prêtre Jean à D. Manuel', *Mare Luso-Indicum* 3 (1976), 4.

[45] Letter from Bartolomeo Marchionni at Lisbon, dated 20 September 1502, in *I Diarii di Marino Sanuto*, Vol. IV, pp. 544–5. Benvenuto d'Abano, here named 'Bonajuto d'Albano', is described as lame; on arriving at Lisbon, he was almost immediately taken to see D. Manuel, at that time in Sintra. He apparently had relatives in the Campo di San Bartolomeo, and it is believed his real name was Benvenuto del Pan; cf. Bouchon, *Mamale de Cannanor*, p. 31.

yet returned to Portugal. Vasco da Gama was now Admiral of the Indies, a nobleman with the title of *Dom* and the right of entry into the royal council, and a not inconsiderable pension. His ambitions in respect of Sines (or for that matter a territorial base of any sort) remained frustrated; Dom Manuel, for his part, was frustrated too by the lack of real success of the first major fleet he had sent into the Indian Ocean, a fleet which – moreover – confirmed that the notion of a large number of Christians in India, who would serve as allies against the Mamluks, was a chimera. The king's position and credibility, always fragile throughout his quarter-century-long reign, had suffered considerable damage, despite diplomatic efforts to put a bold face on affairs. It would even appear that the majority of the Royal Council manifested its opinion against continuing the expeditions to the Indian Ocean.[46] An excellent window onto the political situation is afforded by two Venetian embassies to Lisbon in these very years.

These embassies are a curious testimony, as Donald Weinstein has remarked, to how 'Portuguese and Venetian destinies became intertwined at one of the most dramatic moments in the history of each state'.[47] The Venetians, even after the return of Gama's fleet, were unable to believe that the Portuguese had sailed around Africa; we have already noted Girolamo Priuli's confused account of August 1499 (based on merchant letters from Alexandria), in which it is reported that Columbus had led three Portuguese ships to Calicut and Aden, that two of them were wrecked and the third was too damaged to return. The Venetian Republic in 1499–1500 was too concerned with another pressing problem, in which they actually solicited Portuguese aid: this was to mount a joint Christian force to resist the Ottoman naval forces of Sultan Bayezid II (r. 1481–1512), which were raiding the Adriatic and Aegean possessions of Venice. To this end, the Venetian ambassador to Spain was sent to D. Manuel's court in January 1501 (while Cabral was still off the coast of Kerala), to solicit Portuguese naval help. This envoy, Domenico Pisani, was, at one and the same time, to propose such an alliance in the interests of the Catholic faith, to condole with the King for the death of his only

[46] João Paulo Costa and Luís Filipe Thomaz, 'D. Manuel I', in Albuquerque and Domingues (eds.), *Dicionário de história dos Descobrimentos Portugueses*, Vol. II, p. 675.

[47] Donald Weinstein, *Ambassador from Venice: Pietro Pasqualigo in Lisbon, 1501* (Minneapolis: University of Minnesota Press, 1960), p. 19.

son Dom Miguel (on 19 May 1500), and to congratulate him for his recent, second, marriage with Dona Maria, sister of his late first wife, and daughter of the Catholic Monarchs of Spain. Ferdinand and Isabel, for their part, were happy enough at any project that drew D. Manuel away from his stated objective (for which he had received a Papal dispensation in 1496) of attacking the Wattasid ruler of Fez, Muhammad al-Shaikh (r. 1471–1504).

Pisani's letters to Venice suggest that he was well received, but that D. Manuel complained to him (characteristically, in a private chamber, far from the ears of prying courtiers) of the lack of enthusiasm shown for his own North African project by other rulers in Christendom. However, he stated his own willingness to aid in the war against the Ottomans, even offering intermittently to lead the naval expedition personally (save that he had no successor other than the erratic Duke of Bragança)! Eventually, a fleet of thirty-five caravels, under the command of D. João de Meneses, Conde da Tarouca, was made ready, and Pisani declared himself content; he also noted in his letters to Venice that D. Manuel had announced the sending of thirteen ships to Calicut for spices, which were expected to return shortly, and that four other vessels (the fleet of João da Nova), were also ready to depart in the same direction. These letters arrived in Venice at much the same time as an official missive from D. Manuel to the Doge, where he proudly announced the same.[48]

The Signoria followed up this embassy almost immediately by another one, that of Pietro Pasqualigo, who was elected in April 1501 to the task. Pasqualigo arrived in Lisbon shortly before the return of the vessels of Cabral's fleet, in late June 1501, and was able at once to report on the significance of this event to Venice. This time, at least some Venetians, and notably Girolamo Priuli himself, took the news with a far greater degree of seriousness than in 1499; the commercial threat, he at once decided, was 'more important to the Venetian State than the Turkish War or any other war that might take place'. Now that the King of Portugal was in a position to bring spices directly to Lisbon, the Hungarians, Germans, Flemish and French all would abandon Venice, since the spices that arrived there came through the Mamluk Sultanate, paying 'the most burdensome duties', to which were added the levies in Venice itself. Priuli hence foresaw a Venice in great commercial and financial difficulty (although not

[48] Weinstein, *Ambassador from Venice*, pp. 20–2.

altogether in ruins), whose merchants would be 'like a baby without its milk and nourishment'.[49] But not everyone saw matters in such dark hues. Public opinion in Venice soon divided, with some 'optimists' noting that Cabral had lost a large number of vessels, that the hostility the Portuguese had faced on this expedition from the natives of the Indian Ocean might daunt them, and that the quantities of spices too were not as large as might have been expected from the outlay of resources.

But most importantly, this divided Venetian opinion came to reflect the divisions in Portugal itself, a fact of which the Venetian diarists soon caught wind. Priuli himself notes that it was said that many in Portugal thought that too many lives had been lost, and that the venture was financially not a paying one; this surely reflects not 'public opinion' in some general sense, but divisions in the Portuguese court and élite. Pasqualigo, for his own part, seems to have encouraged these doubts and divisions, allegedly even going so far as to approach the envoys brought back by Cabral from the rulers of Cannanur and Cochin, to tell them that Portugal was a poor and marginal country, and his own *patria*, Venice, the veritable centre of Europe! At the same time, in his official oration before the ruler, pronounced at Lisbon on 20 August 1501, Pasqualigo made it a point to play to the monarch's secret weaknesses and universalist pretensions, assuring him that his 'high character' had already been 'known and reported far and wide throughout the world', giving rise to 'great hope even beyond the city of Venice, in all the peoples, realms, and nations of Europe'. Referring *sotto voce*, it would seem, to D. Manuel's Joachite leanings, and his inherently messianistic trajectory, Pasqualigo assured him that 'everyone has been convinced by those assiduously circulated reports of your virtues that, as the Christian religion teaches, the good and great God will soon give them a king who in his virtue, wisdom and felicity will not only protect the weary and tottering Christian Commonwealth but even extend it far and wide'. Moreover, D. Manuel had justified these universal hopes in the most unheard-of fashion, by sending out to discover lands 'entirely unknown to Ptolemy and to Strabo'. Then, in a remarkable passage which takes a leaf very blatantly out of D. Manuel's own royal rhetoric, Pasqualigo goes on to state:

[49] *I Diarii di Girolamo Priuli*, Vol. II, pp. 156–7, as cited in Weinstein, *Ambassador from Venice*, pp. 29–30.

> That which neither the Carthaginians of old achieved, nor the Romans who held the power after the overthrow of Carthage, nor Alexander, that great world explorer, nor all of Greece in the days when she flourished, nor the Egyptian and Assyrian kings, your excellence and good fortune have achieved. At your command, the whole coastline of outer Libya, from the Atlantic Ocean as far as the Barbarian Gulf (*Sinus Barbaricus*) which is joined to the Red Sea, has been navigated. Peoples, islands, and shores unknown until now have either surrendered to your military might or, overawed by it, have voluntarily begged for your friendship. The greatest kings and unconquered nations of the past used to boast justifiably that they had extended their power to the ocean, but you, invincible King, are entitled to take pride in having advanced your power to the lower hemisphere and to the Antipodes. What is greatest and most memorable of all, you have brought together under your command peoples whom nature divides, and with your commerce you have joined two different worlds.[50]

This mildly astonishing text reflects Pasqualigo's considerable skills in catching a precise flavour of the thought of men like Duarte Galvão and D. Manuel himself; the metaphor of the lion and the lamb ('whom nature divides') being joined together, and the genealogy reaching to the Ancients are perfectly exploited by this Venetian ambassador.

D. Manuel's dreams of declaring himself Universal Emperor (or at least Emperor of the East), and his project to attack North Africa and thus mount a two-pronged attack on Jerusalem (of which the campaign against Fez was intended as a mere foretaste), were obviously not quite as well guarded as has been thought. None of this made his plans in the Indian Ocean more palatable to his opponents; and in view of D. Manuel's own delicate position at this point with respect to the opposition to his policies, it is no surprise that the sending out of the next fleet to India was a most tortuous affair, hedged in by highly complex negotiations. Besides the mixed fortunes of Cabral's fleet, there was the mismanaged naval sortie against the Moroccan fortress of Mers el-Kebir in late July 1501, as part of the activities of the fleet sent out under D. João de Meneses ostensibly in support of Venice and against the Ottomans. Besides, the main fleet sent out into the eastern Mediterranean, comprising thirty ships and about three thousand men, tied up valuable resources until the spring of 1502,

[50] Weinstein, *Ambassador from Venice*, pp. 45–6 (English translation), p. 37 (Latin text). Weinstein himself (pp. 55–6) makes a Joachite connection, but fails perhaps to realise its political significance.

and achieved nothing of any military note. Further, we may legitimately wonder if the expeditions of Gaspar Corte-Real in 1500–1 into the North Atlantic, which resulted in sightings of Greenland, Newfoundland, Nova Scotia and the Chesapeake, did not further confuse the situation, and divide the opinions.

But even if this were not the case, the situation in 1502 had evolved considerably from that in 1497. The question now was less one of *whether* the route from the south Atlantic to the Indian Ocean should be defined and exploited, than – if it were exploited – in what manner, and by whom. Some real pressure to abandon the project altogether seems to have been forthcoming, but there is also evidence of a tussle over who would control the next expedition, and hence its outcome. To what extent should it be used to prosecute the king's own major obsession (of which Pasqualigo caught wind), to blockade the Red Sea, asphyxiate the trading economy of the Mamluk Sultanate, and thus prepare for the *coup de grâce* to be delivered at Jerusalem? D. Manuel himself appears hesitant, often the case with him at such moments. Thus, the Portuguese chroniclers of the sixteenth century all appear to agree that the command of the fleet of 1502 had initially been given to Pedro Álvares Cabral, despite his lack of success in his earlier voyage to Kerala. Then, for reasons that have been much discussed and disputed, Cabral eventually did not command this fleet, and it fell to Vasco da Gama to do so. What were the reasons for this sudden change of command?

To arrive at a resolution to this puzzle, we have to scrape away at the carapace of the chroniclers' reticence. Now, some of the chroniclers have suggested that deteriorating relations with Calicut, and the violent conflict there at the time of Cabral's first visit, led Dom Manuel to send back Vasco da Gama, well known for his strict and uncompromising temperament. But there is also another factor mentioned at times, including by the chronicler Damião de Góis, which may be of greater significance.[51] This is the fact that Cabral could not get along with Vicente Sodré, one of the captains in the fleet, who had a more or less free hand with a sub-fleet of five vessels under his command, with which he was to remain in the Indian Ocean, even after the return of the

[51] It is of course true that Góis's account of the voyage is largely derivative, and draws heavily on Barros and Castanheda; cf. Jean Aubin, 'Como trabalha Damião de Góis, narrador da segunda viagem de Vasco da Gama', in Helder Macedo (ed.), *Studies in Portuguese Literature and History in Honour of Luís de Sousa Rebelo* (London: Tamesis Books, 1992), pp. 103–13.

remaining spice-ships. Now Sodré, according to João de Barros, was given instructions separate from those for Cabral, which required him to 'guard the mouth of the Strait of the Red Sea, to ensure that there neither entered nor left by it the *naos* of the Moors of Meca, for it was they who had the greatest hatred for us, and who most impeded our entry into India, as they had in their hands the control of the spices which came to these parts of Europe by way of Cairo and Alexandria'.[52] Cabral, for his part, was to concentrate on the Malabar coast, where he was formally to set up factories at Cochin and Cannanur (ports from which he had brought back emissaries on the return from his last voyage). Cabral was also to take a much larger fleet than Sodré, of fifteen ships, but the latter – as we have stressed – was partly independent of the former on account of his separate instructions (*regimento*) from the ruler Dom Manuel. It would even appear that the division of spoils from the captures (*presas*) made by Sodré in the Indian Ocean was initially such as to exclude the Captain-Major and crews of the main fleet; only on the eve of the departure, and obviously bearing in mind that it was now Gama and not Cabral who commanded the main fleet, was this division altered, to allow Gama a share.[53]

The quarrel between Cabral and Sodré, which Barros claims resulted from Cabral's hypersensitivity in matters of protocol ('*elle era homem de muitos primores acerca de pontos de honra*'), is far from devoid of larger significance, once we realise who Vicente Sodré in fact was. In point of fact, Sodré was none other than Vasco da Gama's maternal uncle, and the quarrel was convenient enough reason for the nephew to step in and relieve Cabral of the charge! This was facilitated by a supplement to the earlier royal grant of 1500, dated 2 October 1501, which now explicitly gave Gama the right to assume the Captain-Majorship of India-bound fleets whenever he wished.[54] Gama thus quickly transformed the fleet of 1502 into a family affair. Besides Vicente Sodré, there was among the captains another of Gama's maternal uncles, Brás Sodré, and the three left together (as part of a fleet of fifteen vessels) in early February. Vicente Sodré's extraordinary position was made clear from the start; he was named to succeed as Captain-Major, in the event some mishap befell Gama; also the

[52] Barros, *Da Ásia*, Década I/2, pp. 21–2; Góis, *Crónica*, part i, ch. lxviii.
[53] AN/TT, CC, I–4–40, letter from D. Manuel to Gama, dated 6 February 1502.
[54] AN/TT, Cartas Missivas, Maço 4, No.36, and CC, I–4–42, reproduced in Teixeira de Aragão, *Vasco da Gama e a Vidigueira*, Doc. 12, pp. 221–2, but wrongly dated to 1500.

captains of the ships were given a general order that if any of them parted company with Gama's ship, they were to follow Sodré, accepting his orders 'as you would do and act if the said Admiral demanded and ordered it'.[55] Later, in early April, another five ships left for India, to make up the total of twenty that had always been the target; this fleet was commanded by Vasco da Gama's first cousin, Estêvão da Gama, on the powerful *nau*, *Flor de la Mar*. Still another captain was an Álvaro de Ataíde, Gama's brother-in-law by his recent marriage; also on the fleet as captain was still another future brother-in-law, Lopo Mendes de Vasconcelos. The Gama–Sodré combine thus dominated the fleet entirely.

It seems very likely that the setting aside of Cabral, and the instatement of Vasco da Gama with such extensive powers reflected an elaborate set of manoeuvres in the court. The fact that Vicente Sodré was given such autonomy at the outset further suggests that Vasco da Gama and his supporters had already had a say in the initial distribution of functions, which probably took place in autumn 1501.[56] One may imagine that, thereafter, the incompetence of Cabral and his leadership was stressed in the court, and contrasted with Gama's own firm qualities; we should not forget that after his fall from grace at this time, Cabral was never reinstated in royal favour. Let us recall too that, at the same time, D. Manuel was in a somewhat curious position with Gama over Sines, which he had promised him but had been unable to deliver. The Admiral's absence for a year or more in India was thus not wholly inconvenient from the king's point of view; moreover, it was a way of weaning Gama even further away from his initial anchorage in court politics, as D. Jorge's agent. It is possible that Vicente Sodré, who had been intermittently associated with D. Manuel from before his accession, and who was also attached to him as a *fidalgo* of his household, and Cavalier of the Order of Christ, played the role of intermediary between Gama and the monarch.[57]

[55] AN/TT, CVR, No. 50, n.d., reproduced in Brito Rebello, 'Navegadores e Exploradores Portuguezes', Docs. 46 and 47.

[56] In early October 1501, D. Manuel by a special dispensation declared Fernando, the illegitimate son of Vicente Sodré and Isabel Fernandes, to be his legitimate descendant; this was at Sodré's request, and obviously done in view of his imminent departure for India; AN/TT, Chancelaria D. Manuel, Livro 17, fl. 89. The same appears to have been done for Sodré's other illegitimate son, João.

[57] Letter from D. Manuel, Duke of Beja, to the officials of Funchal, 8 July 1494, concerning the despatch of Vicente Sodré to improve the fortifications in Madeira, published in Silva Marques (ed.), *Descobrimentos Portugueses*, Vol. III, Doc. 297, p. 457;

As with Gama's first voyage, we are somewhat hampered by lack of access to the precise instructions (*regimento*) he carried. Indeed, the few contemporary documents available to us immediately prior to his voyage are largely receipts for supplies laden on board ships of this fleet, such as ships' biscuits for the *Flor de la Mar*, and supplies for Gama's own ship, *São Jerónimo*.[58] The longest of these is a detailed accounting of the wine and crockery (*vinho e louça*) for the vessels, from which we are able to confirm the names of a number of ships available to us from the chronicles, and also to note the clear demarcation between the ships of Sodré's sub-fleet and the others. Four vessels, *Esmeralda*, *São Pedro*, *Santa Marta* and *São Paulo*, are explicitly described as ships 'that are now going to India, to work as an *armada* there' (*para la amdar d'armada*); the first of these was Sodré's ship. Seven other vessels also find mention, without this qualifying phrase; three of them are mentioned with the names of their captains, Francisco da Cunha of the *Leitoa*, Rui de Castanheda of the *São Pantaleão de Batecabelo*, and Pêro Afonso de Aguiar, also confusingly enough of another *nau* called *São Pantaleão* (*do Porto*).[59] Besides, we are aware that another ship, *Leonarda*, was captained by a rather substantial nobleman, D. Luís Coutinho.

Once the change in the command had been effected in late 1501 or very early the next year, a solemn ceremony was held in Lisbon at the Cathedral Church in Lisbon on 30 January 1502. Here, in the presence of a number of dignitaries and ambassadors, Gama was publicly given the title of Admiral (the source of some confusion as to the date when he was actually *granted* the title), and handed the King's standard to carry on his voyage. A symbolic blow for equality of footing with Spain was struck; and the comparison with the other Admiral, Columbus (however inappropriate it might have been), could not have been far from the minds of those present. One of the eyewitnesses, Alberto Cantino, representative in Lisbon of the Duke of Ferrara (and

AN/TT, Gavetas, XV/9–3, draft of a royal letter dated 15 June 1500, concerning the despatch of Vicente Sodré on a mission to Safi, in ibid., Doc. 367, p. 611.

[58] AN/TT, CC, II–5–79, *conhecimento* dated 11 January 1502 for the *São Jerónimo*; AN/TT, CC, II–5–113, *mandado* dated 22 February 1502, for the *Flor de la Mar*. Other similar documents are AN/TT, CC, II–5–101, and II–5–105, dated January and February 1502 respectively, and the documents cited in note 59 below.

[59] AN/TT, CC, II–5–107, a volume of 31 folios, with both copies and original receipts, the latter signed at Belém, dating between 26 January, and 9 March 1502. Fragments of this document can be found in Brito Rebello, 'Navegadores e Exploradores Portuguezes', Docs. 48, 51, 53, 54, 55, 56, 57, 59, 60; for related documents, also see AN/TT, CC, I–3–79, I-3–80, and II–5–105, in ibid., Docs. 49, 50 and 52.

Plate 17 Palácio Nacional da Pena, Sintra, nineteenth-century stained-glass window from Nuremberg showing D. Manuel and Vasco da Gama.

later famous as the man who smuggled the Cantino planisphere out of Portugal), thus wrote to his master:

> As a good envoy in these parts, I wished to communicate to Your Excellency the aftermath of the little news that I had written in my other letter, of the present [month]. This Most Serene King of Portugal today, which is to say the thirtieth of the present [month], in the Great Church invested one Don Vascho as Admiral of India. And ten days from now, he will set sail for those parts with twelve ships, well-outfitted and armed, in order to guard and defend the mouth of the Red Sea, so that the ships (*zerme*) of the Soldano cannot come from Mecca to Calicut in order to lade spices. Even though the same King does not wish to allow any other powers or ships (*domini et navichi*) into those seas, still he has already told the Venetian ambassador that if this year his affairs do not turn out well, as it is believed, he will abandon the enterprise entirely, because the last year he lost eighty thousand ducats in it. And further, in the end of March, another twelve [ships] with their goods will be ready to follow those which are leaving now.

The precarious nature of the affair (and D. Manuel's own incertitude as well as the opposition in his court) is brought out here, as is the underlying expectation of the Venetians that the whole business would turn out badly for Portugal. The letter continues:

> It would not do if I were to neglect to mention the ceremonies that attended the handing over of the standard to the above-mentioned Admiral. First, everyone attended a sumptuous Mass, and when it was over, the above-mentioned Don Vascho, dressed in a crimson satin cape in the French style, lined with ermine, with cap and doublet matching the cape, adorned with a gold chain, approached the King, who was attended by the whole court, and a person came forward and recited an oration, praising the excellence and virtue of the King, and went so far as to make him superior in every way to the glory of Alexander the Great. And then, he turned to the Admiral, with many words in his praise and in praise of his late predecessors, showing how by his industry and vivacity he had discovered all this [*sic*] part of India, [and] when the oration was over, there appeared a herald with a book in his hand, and made the above-mentioned Don Vascho swear perpetual fidelity to the King and his descendants, [and] when this had been done, he knelt before the King, and the King taking a ring from his hand, gave it to him. Then, [the King] unsheathed a sword, while he was still kneeling, and placed it in his right hand, and the standard in

> the left. When this was done, he rose to his feet, kissed the King's hand, and the same was done by all the cavaliers and seigneurs, rejoicing in all the dignity and honour that would be added to his Crown, and the same was done by the Venetian ambassador. And thus it ended, with the most splendid music.[60]

It was only now, with these visible symbols of his status, that Dom Vasco da Gama, Admiral of the Indies, made his way back there in February 1502. The supplementary fleet, led by his cousin Estêvão da Gama, would only set sail several weeks later, in April, perhaps after the situation of D. João de Meneses's fleet had become clearer.

However, even before he sailed, it is reported that Gama had to face a minor challenge, apparently provoked by none other than Pasqualigo. The Venetian ambassador, as we have noted, had apparently managed to contact the 'ambassadors' brought back by Cabral from Cochin and Cannanur, to convince them that Portugal was in fact a poor and rather marginal country, whose voyages to the East actually had to be financed by Venice. Far better for the Kerala rulers then to have direct dealings with the Signoria than with the Portuguese. The Portuguese chronicles report that Gama eventually managed, while *en route* to India, to convince these ambassadors that this was simply not true: in any event, one of the men from Cochin died off the Cape of Good Hope, and the other off East Africa, so that the Venetian intrigue turned out to be fruitless; only the representative of the Kolattiri returned safely to Cannanur.[61]

THE SECOND VOYAGE

In the Indian Ocean for a second time, Dom Vasco – as he now was – comported himself with vigour – indeed even brutality. This second voyage has thus left historiographical ripples that at times even overshadow those of the first voyage of 1497–9. The sources that have been used most conventionally by historians to define the outlines of this voyage are the sixteenth-century

[60] Cf. Henry Harrisse, *Document inédit concernant Vasco de Gama: Relation adressée à Hercule d'Este, Duc de Ferrare par son Ambassadeur à la Cour de Portugal* (Paris, 1889).

[61] *Crónica do descobrimento*, ed. Albuquerque, pp. 69–70; also the discussion in Weinstein, *Ambassador from Venice*, pp. 31–32 (citing João de Barros). For a draft of a formula letter from D. Manuel to the rulers of Cochin and Cannanur, asking them to cooperate with D. Vasco da Gama, see AN/TT, CVR, No. 71 (n.d.), published in A. da Silva Rego (ed.), *Documentação para a história das missões do padroado português do Oriente*, Vol. I (Lisbon: Agência Geral das Colónias, 1947), pp. 22–4.

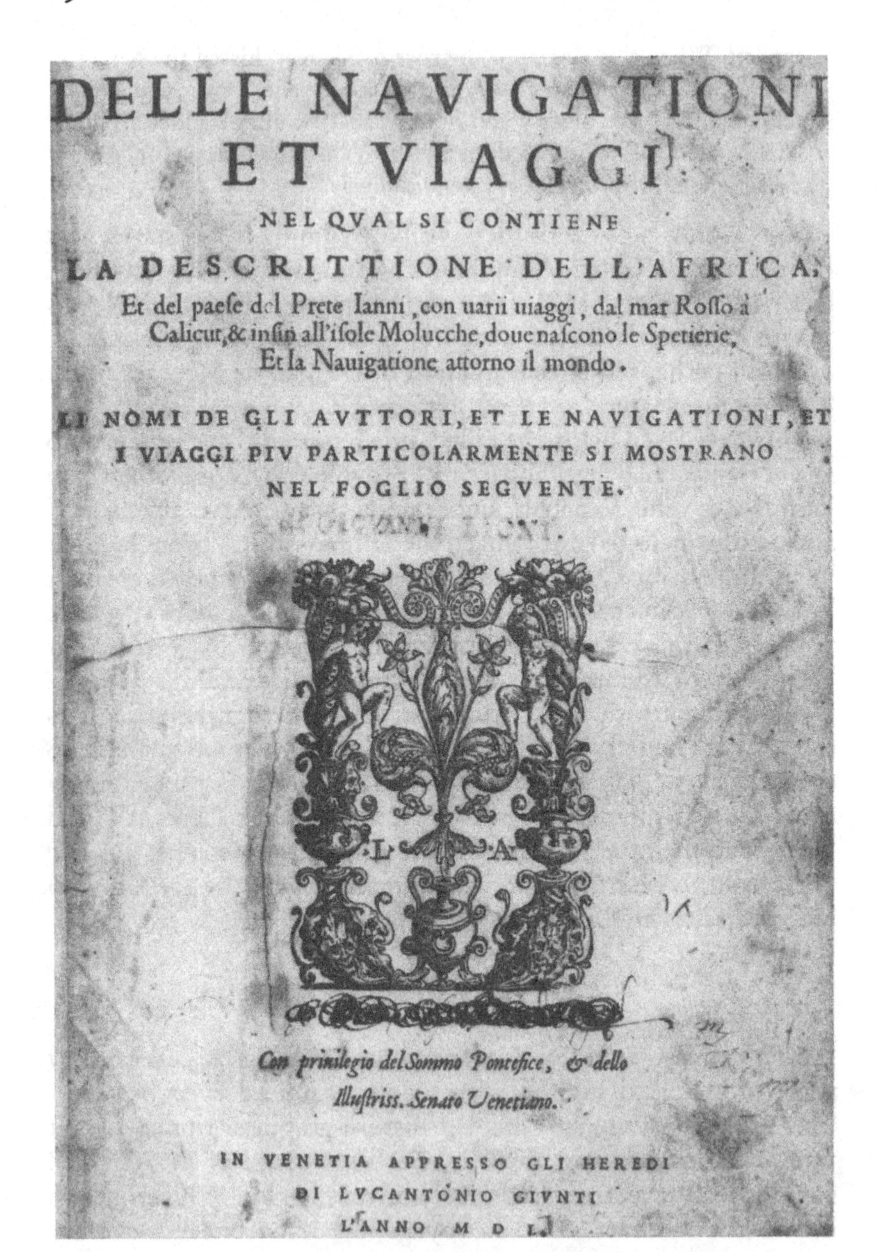

DELLE NAVIGATIONI ET VIAGGI

NEL QVAL SI CONTIENE

LA DESCRITTIONE DELL'AFRICA,

Et del paese del Prete Ianni, con uarii uiaggi, dal mar Rosso à Calicut, & insin all'isole Molucche, doue nascono le Spetierie, Et la Nauigatione attorno il mondo.

LI NOMI DE GLI AVTTORI, ET LE NAVIGATIONI, ET I VIAGGI PIV PARTICOLARMENTE SI MOSTRANO NEL FOGLIO SEGVENTE.

Con priuilegio del Sommo Pontefice, & dello Illustriss. Senato Venetiano.

IN VENETIA APPRESSO GLI HEREDI DI LVCANTONIO GIVNTI L'ANNO M D L.

Plate 18 Title page of Giovanni Battista Ramusio, *Navigationi et Viaggi* (1550), containing the text of Tomé Lopes (1503).

Portuguese chronicles, amongst which Gaspar Correia's is a particularly colourful account. But, once more, the chronicles can be checked against other more immediate accounts, written by persons on board the fleet, of which we have an unusually large number in the case of this voyage. These include the relation of the Italian Matteo da Bergamo, who sailed on a ship fitted out by the Italian Giovanni Buonagrazia, as well as the particularly valuable account of Tomé Lopes, scrivener on board one of the vessels which left in early April, and which was commanded by Rui Mendes de Brito.[62] A brief, but useful, anonymous account in Portuguese may also be found to supplement these, written by a member of the fleet that left Lisbon in early October.[63] And finally, there is the oft-cited text authored by an anonymous Fleming, also on board the fleet of Gama himself, on the ship *Leitoa Nova*.[64] This multiplicity of perspectives is particularly valuable, leaving us with considerably more play than in the case of the first voyage. Added to this is the fact that some documents from the fleet's sojourn in India have actually survived, which (even if not of capital importance) add certain useful details to the analysis.

Let us begin then with what may be the most celebrated of these itineraries, that of the anonymous Fleming, an account that begins confusingly enough.

[62] 'Navigazione verso le Indie Orientali di Tomé Lopez', in Ramusio, *Navigazioni e viaggi*, I, pp. 683–738, an account translated from Portuguese into Italian, of which the original is untraced; also Prospero Peragallo, 'Viaggio di Matteo da Bergamo in India sulla flotta di Vasco da Gama, 1502–1503', *Bollettino della Società Geografica Italiana* (series IV) 3 (1902), reprinted from *Studi bibliografici e biografici sulla storia della geografia in Italia* (Rome, 1875), pp. 92–129. For an overview of accounts concerning this voyage, see Banha de Andrade, *Mundos Novos do Mundo*, Vol. I, pp. 297–321. For a complete French translation of the accounts relating to this voyage, see *Voyages de Vasco de Gama: Relations des expéditions de 1497–1499 & 1502–1503*, trans. Paul Teyssier and Paul Valentin (Paris: Editions Chandeigne, 1995), pp. 201–355.

[63] Osterreichische Nationalbibliothek (National Library of Austria), Vienna, Codex 6948, first published by Christine von Rohr, *Neue Quellen zur Zweiten Indienfahrt Vasco da Gamas* (Leipzig, 1939), and recently re-edited with corrections by Leonor Freire Costa (with an introduction by João Rocha Pinto), 'Relação Anónima da Segunda Viagem de Vasco da Gama à Índia', in *Cidadania e História: Em homenagem a Jaime Cortesão* (Lisbon: Livraria Sá da Costa, 1986), pp. 141–99. There is still another brief and anonymous account, this one in German. For the text, and an English translation, see Miloslav Krása, Josef Polisensky and Peter Ratkos (eds.), *European Expansion (1494–1519): The Voyages of Discovery in the Bratislava Manuscript Lyc. 515/8 (Codex Bratislavensis)* (Prague: Charles University, 1986).

[64] The Flemish text, first published in Antwerp possibly in 1504, is reproduced in Teixeira de Aragão, *Vasco da Gama e a Vidigueira*, pp. 79–86; a more modern edition exists by Jean Denucé, *Calcoen: Recit flamand du second voyage de Vasco de Gama vers l'Inde en 1502–1503* (Antwerp, 1939). A summary of this text was already available in Venice by early 1504, and is reproduced in *I Diarii di Marino Sanuto*, Vol. V, ed. Federico Stefani (Venice, 1881), pp. 1064–5.

> This is the voyage that a man himself wrote, who sailed with 70 ships from the river of Lisbon in Portugal to Calcoen [Calicut] in India, and it happened in the year 1501. And they sailed along the coast of Barbary and came before a city called Meskebijl [Mers el-Kebir] and were driven off from there with great losses and affronted, and we lost many Christians there, may their souls rest with God. This battle took place on St Jacob's day in the abovementioned year. That stronghold lay one mile from a city called Oeraen, which is frequented by many bad Christians, merchants from Venice and Genoa, and they sell the Turks armour, muskets, and other supplies to fight against these Christians, and they have their entrepôt (*haer stapel*) there. I was six months on this coast of Barbary, and suffered great hardships in those Straits.

Obviously, our Fleming had participated in D. João de Meneses's expedition of 1501, to which reference has been made above. Returning to Lisbon, he had time to enlist for a second time, this time in the real India-bound fleet, departing on 10 February 1502 from the Tagus. Passing the Canaries, the fleet appears next to have made landfall in the Cape Verdes by late February or early March, a land that is described as being inhabited by blacks, 'who have no shame, for they do not wear clothes, and the women embrace their men like monkeys, they know neither good nor evil'. The usual astronomical phenomena are reported on crossing the Equator, the disquieting fact for men who had sailed in northern waters that the Great Bear and the North Star were no longer visible. Through April and May, reports our anonymous Fleming, there was nothing in the skies by way of a sign (*hemel teiken*) to guide them, and they depended entirely on the compass and charts. Unusual flying fish (as large as mackerels, herrings or sardines) were noticed, as well as black and white gulls that pounced on them even as they flew. In late April, contrary winds blew the vessels far west, so that it took them almost all of May to return to their rightful course; on 22 May, in the midst of a suitably dramatic winter thunderstorm and tempest, with rain and hail, the Cape was traversed. The fleet now headed north-east, and by 10 June the mariners once more sighted the North Star with relief.

The other anonymous account (in Portuguese) provides a slightly different chronology. The arrival in the Cape Verdes is dated to 28 February, Mass was heard there the next day, and the fleet is said to have departed from the archipelago on 6 March. The ship on which the anonymous Portuguese was had its share of misadventures. A mast partly broke on 11 April, and on the 28th the ship on which the text's author voyaged rammed the *Batecabelo* by mistake; the two could be separated only after confusion that endured a half-hour, but fortunately for them no permanent damage was suffered. This account is reticent about the entire Atlantic passage, and passes very quickly to the end of May, when the ship was already off the east coast of Africa.

In view of this, it may seem that by the time of this voyage – the fourth by the Cape route, we should recall – the entire proceeding had become somewhat banalised. The Flemish account helps us to catch some of the wonder that still inhered in this voyage from the western European perspective, and the author's emphasis on how everything depended on charts and compass is not to be missed. We are also aware that in these years, others who attempted to reach the Indian Ocean by the same route failed for lack of adequate information and skill. Thus, we have the case of the celebrated Binot Paulmier de Gonneville, a French navigator from Honfleur, who set out in the vessel *L'Espoir*, to make his way to the Indian Ocean in late June 1503.[65] Gonneville, who had been trading to Lisbon in the years preceding his voyage, did take some precautions; he was accompanied by two Portuguese, Bastião Moura and Diogo Couto (or Coutinho), who had been on one of the earlier Portuguese fleets to Asia. Passing the Canaries in mid-July, Gonneville reached the Cape Verdes at the end of that month, crossed the Equator in mid-September, and very rapidly lost all sense of direction. By early November, he believed he was at the Cape or very near it, and eventually, on 5 January 1504, made landfall in what he seems to have believed was an island of the Indian Ocean (but which was in fact near São Francisco do Sul in Brazil). After six months, the intrepid Frenchman, carrying back with him two 'Indians' named Essomeric and Namoa set sail for Honfleur, and with much difficulty reached Faial in the Azores in early March 1505. Then blown off course, towards Ireland, his ship was attacked by an English pirate Edward Blunt, forcing Gonneville to abandon *L'Espoir*. He eventually returned home in May 1505, convinced that he had reached the Indies; in the mid-seventeenth century,

[65] Binot Paulmier de Gonneville, *Campagne du Navire 'l'Espoir' de Honfleur (1503–1505): Relation Authentique du voyage du capitaine de Gonneville ès Nouvelles Terres des Indes*, ed. M. D'Avezac (Paris, 1869; Geneva: Slatkine Reprints, 1971); also the entry for 'Binot Paulmier de Gonneville', in *Nouvelle Biographie Générale* (Paris: Didot Frèves, 1865), Vol. XXXIX, pp. 408–9.

Etienne de Flacourt used his account to argue that the French had already been in Madagascar in the early sixteenth century, and thus had pre-emptive rights over that island, with respect to all other European powers![66] Clearly even geographical confusion has its political uses.

Returning to our anonymous Portuguese account, it is reported that once off the African east coast, by 3 June, the captains consulted; the bulk of the fleet, under Vasco da Gama, made for the first major port-of-call, Sofala, but some ships were sent on ahead to Mozambique. Other sources indicate that this was the sub-fleet commanded by Vicente Sodré, whose task it was to construct a caravel at Mozambique with wood brought from Portugal, the command of which was eventually given to João Serrão. A week later, on 10 June, the fleet arrived in Sofala, and anchored some distance off-shore; the next day, boats were sent ashore with parties of well-armed Portuguese, who were instructed though to keep their arms concealed, 'lest it seem that they went to make war'.

We are on familiar ground now, as Gama resumes his old habits. On 11 June, a small fleet of boats with armed men was made ready and approached the shore; a boat (*allmadia*) from the strand approached them, and Gama's chosen representative, a certain Rodrigo Reinel (in fact, attached to Vicente Sodré's ship as factor), began to parley, stressing Gama's new-found importance as Admiral of Portugal. Word was sent to the palace, and a positive response soon came back. Presents of figs, coconuts and sugarcane were brought for Gama, and he at once demanded two hostages, before sending his own men ashore. Two of the more respectable Muslims came aboard Gama's boat, and he for his part sent Rodrigo Reinel and Manuel Serrão to meet the King. Somewhat cordial relations having been established, Gama had soundings taken of the port (it was, after all, his first visit there), to see if his ships could enter; on the 13th, he entered the port with three vessels. On 15 June, the Portuguese began to buy gold, a transaction that went on for twelve days, and in the course of which 2,500 *misqâls* were reputedly obtained.[67] Then, on 26 June, Gama left for Mozambique, but while departing Sofala, his ship acciden-

[66] Etienne de Flacourt, *Histoire de la grande isle Madagascar, composée par le Sieur de Flacourt* (Paris: J. Henault, 1658; 2nd edn, Paris: G. Clouzier, 1661). The author was named commandant of Fort Dauphin in Madagascar (île Bourbon) in 1649.

[67] Estimates of the amount of gold vary wildly; cf. Freire Costa, 'Relação Anónima da Segunda Viagem', p. 180, fn. 11.

tally damaged that of João da Fonseca, which hence had to be abandoned and burned.

The Flemish account, in contrast, places the arrival at Sofala at 14 June, a minor discrepancy. On the other hand, it is noted that the Portuguese wished to buy and sell (presumably demanding gold in exchange for their goods), but the locals would have none of it. A confused passage follows concerning the land of Prester John (*Paep Jans lant*), which is described as being land-locked, with its only access to the sea being through the river of Sofala. The problem however was that the ruler of Sofala was perpetually at war with Prester John, and all this could be gathered by some of the Prester's folk who were held in Sofala as slaves! The great interest of contacting the Prester was emphasised, for his land 'overflows in silver, gold, and precious stones and riches'.

Both the anonymous accounts state that after leaving Sofala, the next port-of-call was Mozambique island, where it was possible (according to the Fleming) to obtain unminted gold and silver. In Mozambique, two ships that had been separated from the ships of the main fleet at the Cape, rejoined them; also the rendezvous with Vicente Sodré was kept. It was hence possible to make a suitably impressive entry into Kilwa, where the earlier Portuguese fleets had been given a rather ambiguous reception. The very large Portuguese fleet thus arrived before the port on 12 July, and Gama ordered a round of artillery to be fired, to impress on those ashore the fact that he would use force if necessary. It would appear that the *de facto* ruler in Kilwa at this time was a certain Amir Ibrahim, though the titular Sultan was Fuzail bin Sulaiman (to be set aside in 1505 by the Portuguese, and replaced by one Muhammad Rukn al-Din). Shortly after the Portuguese arrival, one of the convict-exiles (probably a certain Luís de Moura) whom Cabral had left behind in East Africa, arrived on board Gama's ship, with a letter for him from João da Nova, stating that no good was to be expected from dealings there, and also giving him news of goings-on in Calicut and Cannanur.

Messages sent ashore by Gama appear to have had no immediate response from Amir Ibrahim. After a consultation with the captains of his fleet, Gama decided to approach the shore and the palace in smaller vessels (*bateis*), and threaten bombardment if the ruler were unwilling to come to terms. Faced with this threat, Amir Ibrahim capitulated, and himself came to meet Gama in his boat. He was told in no uncertain terms that the Portuguese wanted not only gold by way of trade, but a tribute (*trabuto*) of ten pearls a year for the Queen, and 1,500 *misqâls* of gold. This sum was eventually paid in two instalments, and by 27 July, Gama was ready to leave the port. The anonymous Flemish account adds a further gloss, stating that the ruler of Kilwa was henceforth obliged to fly a flag as a sign of subordination ('*heeft hij van den selven coninc een banuier in een teyken dat hen kent voer sijn heer*'). Gama issued a proclamation of the new status of Kilwa, in a letter (unusually signed by him, one of the rare documents that we have of his authorship). Dated 20 July, once the negotiations for tribute were complete, the text runs as follows:

> The admiral Dom Vasco etc., makes it known to all the captains of any ships of the King, my lord, which come to this port of Quilloa that I arrived there on the 12th of this month of July of [1]502, and wished to see the King in order to make peace and friendship with him, and he did not wish to see me, but instead behaved very discourteously on account of which I armed myself with all the men I had, determined to destroy him, and I went in my boats before his house, and placed the prow on dry land, and had him sent for much more discourteously than he had behaved with me, and he agreed to do so and came, and I made peace and friendship with him on the condition that he should pay a tribute (*trabuto e páreas*) to the King, my lord, of one thousand five hundred *miticaes* of gold every year, which one thousand five hundred *miticaes* he at once paid me in this present [month] in which we are and he made himself a vassal of His Highness, on account of which I order you on his [D. Manuel's] part and thereafter on mine, that you keep the said peace as long as they keep it, as it is reasonable that one should do with the vassals of the said Seigneur, and I notify everyone in general of this, and to those who come to these parts while I am here, I order you not to delay here, but at once to go on to Melinde, and if you do not find me there, go on to Anjadiva, and if you do not find me there, go on to Cananor, and travel by day and night, taking rest so that you do not pass me, and if you do not find me there, in the same way go on to Calecut, and if you do not find me there either, go on to Cochim, and if before you enter this port, this letter is handed to you outside, do not enter it, because this port is difficult to exit from, but instead go on ahead, and follow eveything that has been said above. Written before Quiloa, 20 July 1502.
> The admiral Dom Vasco.
> And hand this over to the Moors as soon as you have read it, for the others who may come.[68]

[68] Biblioteca Nacional de Lisboa, Reservados, Mss. 244, No. 2, text in Brito Rebello,

The letter has two points of major interest in it. First, the arrogance with which Gama describes how he forced Amir Ibrahim to submit, treating him 'much more discourteously than he had behaved with me', taking an evident pleasure in his humiliation. Second, Gama's own planned itinerary, as stated here, that runs to first the Anjedive Islands, and then Cannanur, Calicut and Cochin in that order. In the event, the letter did not turn out to be of much use, for even as Gama's fleet began to leave the port (taking advantage of the high tide), there arrived the rest of the fleet, which had left Portugal in May, under Estêvão da Gama in the *Flor de la Mar*.[69] The whole fleet now being together, they decided to put in at Malindi, with whose ruler Gama had had relatively good relations on his last voyage, but adverse weather prevented this. Instead, Gama decided to cross the western Indian Ocean, and make directly for the Kanara coast.

This crossing is described in placid terms by the anonymous Portuguese text: leaving East Africa on 29 July, the fleet is supposed to have sighted a mountainous land in the broad vicinity of the Anjedives a mere two weeks later, and to have reached the island of Anjedive proper on 20 August. This placidity, and the general tenor of the account, leads one to suspect that it may not have been the first time that the author of the anonymous Portuguese account found himself in the Indian Ocean. The anonymous Fleming, on the other hand, recounts a far more fanciful version of the crossing; from Malindi, the fleet is said already to have gone as far north as Cape Santa Maria, between Arabia and the land of Prester John, by the end of July. He further claims to have seen a large city called 'Combaen' (in the vicinity of Chaldea and Babylonia) in passing, and states that at one point they were not far from Mecca, 'where lies Machomet, the heathens' devil (*die heydens duvel*)'. Mention is also made of a large urban centre called 'Oan', whose ruler commanded several thousand horse, and 700 war-elephants; it is claimed that the Portuguese captured 400 vessels of that region, killing the crews, and burning the vessels themselves. It is possible that this sojourn was not entirely a product of our Fleming's imagination, for

'Navegadores e Exploradores Portuguezes', Doc. 73, pp. 513–14; also reproduced with a photograph of the document, in Luís Keil, *As Assinaturas de Vasco da Gama: Uma falsa assinatura do navegador português, Críticas, comentários e documentos* (Lisbon, 1934), pp. 12–13.

[69] A few of the ships of the second fleet, including that on which Tomé Lopes was, were delayed at the Cape though, and joined the rest of the fleet at the Anjedive Islands on 21 August.

Gaspar Correia too claims for example that Gama, en route to Cannanur, put in at and raided the Kanara ports of Honawar (perhaps the mysterious 'Oan'?) and Bhatkal, from which he demanded and received tribute.[70] If this were indeed the case, the great king with horses and elephants might well be the ruler of Vijayanagara.

In any event, it is certain that by early September, Gama and his fleet appeared before Cannanur, having determined that they would await ships from the Red Sea at their normal point of landfall – Mount Eli. One of the ships of the fleet, which had lost track of the others, had apparently already put in briefly at Cannanur before making the rendezvous on the open seas with the others. What did Gama seek? It is clear enough that he had come prepared for hostilities with Calicut, and the letter from João da Nova that he had found at Kilwa had only emphasised the gravity of the situation. While we do not know what precisely his *regimento* asked him to do, the blockade of the pepper traffic between Kerala and the Red Sea was obviously high on the list of priorities. At the same time, if we accept Correia's version of his dealings on the Kanara coast, it appears that Gama was prepared to use violence not only against well-defined opponents, but on a rather larger scale. The anonymous Fleming explains the motives simply enough:

> On the sixth day of September we came from there [Anjedive] to a kingdom named Cannaer which lies by a mountain called Montebyl and there we awaited the ships from Mecha and these are the ships which bring over the spices to our lands and which we wished to prevent for thus the King of Portugal alone would bring the spices there, but we could not do this.

One failure is reported in the anonymous Portuguese account: the ship commanded by Fernão Lourenço attacked a four-masted ship (of between 300 and 400 *toneis*), but failed to take it, since 'it was very large and carried many people'. The ship of Rui Mendes de Brito for its part captured a *zambuco*, which was headed to Cannanur; the men and goods on board were handed over to the charge of the ambassador from Cannanur who was on board the fleet. The fleet continued to cruise off Mt Eli through September, waiting to make a major capture.

Gama's patience eventually, and inevitably, bore fruit, as the ships from the Red Sea began to return. The process is described

[70] For Gama at Honawar and Bhatkal, see Correia, *Lendas da Índia*, Vol. I, pp. 289–90.

in all its detail by Tomé Lopes, on board the ship of Rui Mendes de Brito. As he recounts it, on 29 September, the *São Gabriel* sighted a ship that was returning to Calicut from Mecca (that is the Red Sea), carrying 240 men, besides women and children of both sexes, 'who had gone from Calicut in pilgrimage to Mecca, and were coming back'. The Portuguese vessel fired warning shots, and to their surprise the other ship did not offer resistance, despite the fact that it carried firearms and artillery. It seemed to those on board the Asian ship, writes Lopes, that they had enough riches on board to buy off their lives, for there were ten or twelve of the richest Muslim merchants of Calicut, headed by Jauhar al-Faqih ('Ioar Afanquy'), who was said to be 'the factor in the city of the Sultan of Mecca', and who owned a fleet of four or five ships including that one.

When Gama met him, al-Faqih is reported to have offered the Admiral generous terms: some money, and to arrange a lading of spices for the entire fleet, if only the Portuguese would let the ship go. Gama refused this offer, and was then made a second one: that al-Faqih himself would give as a ransom, in exchange for himself, one of his wives, and a nephew, a full cargo of spices for four of the largest ships on the Portuguese fleet. To arrange this, he would send one of his nephews ashore, but remain hostage on board the Admiral's vessel. If the nephew did not keep his word within fifteen or twenty days, the Portuguese could do whatever they wanted with him, and he would have all the precious stuffs on board handed over to them, and even then arrange for peace and friendship between them and the Samudri. Gama once more refused, and instead ordered al-Faqih to tell the Muslim merchants on board his vessel to hand over all the precious goods they had. To this, the chief merchant is reported by Lopes to have responded: 'When I commanded this ship, they did as I commanded; now that you command it, you tell them'! A relatively modest sum of money collected from the merchants was now handed over, and Gama also had five or six *bateis* loaded up with the food-stuff and supplies on board the ship. Five days had elapsed since the ship's capture, and it was now 3 October, a day Lopes which assures us 'I will remember in all the days of my life'.

Having disarmed the ship, Gama now proceeded to embark most of the men who were with him on board the boats, and ordered his bombardiers to set fire to the ship in several places. They did so, and those on board the ship now realised at last that they had no hope of escape. Taking the few bombards they had, and stones in their hands, both men and women on board attacked the departing Portuguese boats; and meanwhile, they also managed to put the fire out. The Portuguese hence returned to set the ship alight again, and were met by a mixture of stones and pleas. As Lopes describes the matter evocatively, the women on board waved their gold and silver jewellery and precious stones, crying out to the Admiral that they were willing to give him all that for their lives; some of the women picked up their infants and pointed at them, 'making signs with their hands, so far as we could make out, that we should have pity on their innocence'. All of this was watched by Gama himself, from a vantage point through a spy-hole (*balestriera*), and it is clear that Tomé Lopes for his part could not understand the Admiral's intransigence. For, he claims, 'there is no doubt that with that [wealth], it would have been possible to ransom all the Christians who were imprisoned in the kingdom of Fez, and there would would still have remained great riches for the King, our lord'.

In desperation, those on the Calicut ship now decided to board one of the Portuguese vessels and rammed it; by this tactic, the other Portuguese vessels could not fire, for fear of hitting their own sister-ship. A furious hand-to-hand combat ensued, in which Tomé Lopes himself took part, and which endured until the late evening, with the Portuguese being given no quarter at any stage. He himself marvels at the ferocity of his opponents, who fought to the limit, as if 'they did not feel their wounds'. Eventually, other Portuguese vessels came to the rescue, and the Calicut ship was prised apart from the other; the Portuguese now attempted to bombard it for several days and nights, but because of the rough seas, could not get a clear field of fire. In the end, reports Lopes, one of the Muslims on board threw himself into the sea, swam across, and offered to tell the Portuguese how to blow up the ship, if his own life were spared. The treasure on board the ship already having been thrown into the sea by those on board, Gama now brought the combat to an end. 'And thus, after all those combats, the Admiral had the said ship burnt with the men who were on it, very cruelly and without any pity', reports Lopes.

This combat, set out in such graphic detail by Tomé Lopes, is described in rather laconic terms by our Flemish author as follows.

> But at the same time, we took a ship from Mecha in which there were three hundred and eighty men and many women and

> children. And we took from it some twelve thousand ducats and another ten thousand in goods and we burnt that ship and all those people with gunpowder (*pulver*) on the first [*sic*] day of October.

The anonymous Portuguese text adds a few details and provides a rather different set of numbers, noting like Tomé Lopes that the capture was made by the ship *São Gabriel*, that the Asian ship carried two hundred souls, that goods worth 26,000 *cruzados* were taken off the vessel, and that everyone on board was killed save seventeen children who were converted to Christianity (and made apprentice friars of Nossa Senhora de Belém), and a hunchback who was the pilot.

Gama's action on the occasion is often cited as a particularly conspicuous act of early Portuguese violence in Asia, and has troubled both contemporary and modern-day commentators, starting with Tomé Lopes.[71] Lopes implicitly criticises Gama's action for two reasons: first, because it would have been far better to sack the ship than destroy it, and second, because he was impressed by the courage and tenacity of the Muslims on board, and hence felt that they deserved consideration on that account rather than for purely humanistic reasons. From the chronicles, we can put together some additional details, even if the chroniclers like Barros have obviously closely read Lopes's account. It is claimed by them that the ship in question was the *Mîrî*, purportedly belonging to the Mamluk Sultan Qansawh al-Ghawri, a ship that carried many *hâjîs* on board besides gold and goods; this connection with the Mamluks is, however, less than certain.[72] Barros himself, though bellicose in the matter, is at a loss to explain Gama's attitude wholly, beyond pointing out that the Admiral was aggrieved at the accidental death of one of his entourage, who was crushed between Gama's vessel and the *Mîrî*. Gama's own reasoning appears to have been of an eye for an eye and a tooth for a tooth; the killing of the Portuguese at Calicut from the time of Cabral's voyage was mentioned by him as a cause for his actions in a letter to the Samudri from Cannanur,

[71] For a discussion, see Aubin, 'Como trabalha Damião de Góis', pp. 110–12.

[72] Also see Barros, *Da Ásia*, Década I/2, pp. 34–8. It is curious that the major Egyptian chronicle of the epoch carries no echo of this incident; cf. Ibn Iyas al-Hanafi al-Misri, *Badâ'i' al-Zuhûr fî Waqâ'i' al-Duhûr*, translated by Gaston Wiet as *Journal d'un bourgeois du Caire: Chronique d'Ibn Iyâs*, 2 Vols. (Paris: Armand Colin, 1955–60), particularly Vol. I (1500–1516). It has been suggested to me by Geneviève Bouchon that the *Mîrî* may in fact have been a Gujarati vessel, which the Portuguese misidentified (in contrast, see Bouchon, *Mamale de Cananor*, p. 62). Tomé Lopes, for his part, states in one place that the vessel was 'una gran nave di Calicut'; cf. Ramusio, *Navigazioni e viaggi*, Vol. I, pp. 701–2.

sent through the hunchbacked pilot of the *Mîrî.* In Barros's inimitable style:

> He let him [the Samudri] know that of the two hundred and sixty men who came in her [the *Mîrî*], he only granted that one [the pilot] his life, as well as twenty or so children. The men were killed on account of the forty or so Portuguese, who had been killed in Calecut; and the children baptised on account of a [Portuguese] boy, whom the Moors had taken to Meca to make a Moor. That this was a demonstration of the manner that the Portuguese had in amending the damage that they had received, and the rest would be in the city of Calecut itself, where he hoped to be very soon.[73]

In contrast, a modern-day local historian of Vidigueira expresses considerable unease at this taint on the career of one of his heroes. Thus, he writes, 'these actions blacken, without doubt, the memory of Vasco da Gama, although in their appreciation we should not neglect to take into account all the attenuating factors that present themselves', namely the attacks on Cabral's fleet, and the deep divide between Christians and Moors.[74] Or again, and somewhat more oddly, the French occult practitioner Dr Francis Lefebure, who in the 1950s made a visit to Portugal in the conviction that he was the reincarnation of Vasco da Gama, claimed to be haunted by the voices of the dead in the *Mîrî* (or the *Mérri*, as he called it).[75] Besides, the best known version in English of this incident comes to us through Gaspar Correia, whose *Lendas da Índia* contains a particularly vivid account of Gama's violent actions, and which was selectively translated as early as 1859 (within a few decades of its long-delayed publication in Portuguese).[76] In turn, in the twentieth century, this brought forth retrospective condemnations not only from nationalist Indian historians such as Sardar K.M. Panikkar, but a work with a curious imaginative turn, recalling in some respects our earlier discussion of the legend of Ibn Majid.

The reference here is to a work published in the 1930s (thus, under colonial rule) by the Bengali short-story writer and nove-

[73] Barros, *Da Ásia*, Década I/2, p. 43.
[74] Palma Caetano, *Vidigueira e o seu concelho*, p. 273.
[75] Cf. Docteur Francis Lefebure, *Expériences initiatiques*, 3 vols. (Paris: Omnium Littéraire, 1954), Vol. III, pp. 111–83.
[76] Henry E.J. Stanley, *The Three Voyages of Vasco da Gama and His Viceroyalty from the Lendas da Índia of Gaspar Correa* (London: The Hakluyt Society, 1859). For a discussion of Correia's account in relation to the other sources, see Anthony Disney, 'Vasco da Gama's reputation for violence: The alleged atrocities at Calicut in 1502', *Indica* 32, 2 (1995), 11–28.

list, Saradindu Bandyopadhyay, who later emerged as a spokesman of a form of militant Hindu (thus anti-Islamic) conception of medieval Indian history, in which the kingdom of Vijayanagara appeared as the last bastion of Hindu resistance to Islam. In his well-known story *Rakta sandhya* ('Bloody twilight') however, the targets are not the Islamic invaders of pristine Hindu India. Rather, the story opens with a motiveless murder in the lanes of twentieth-century Calcutta to which the author-narrator is witness: a Muslim butcher brutally stabs to death a passing Anglo-Indian whom he has never before met. The author goes to prison to meet the assailant, to question him on his seemingly inexplicable action. The butcher explains in the form of a rather complicated flashback. The story begins centuries ago, when he was himself a Muslim merchant at Calicut. Then, one day, there arrived three strange ships on their shores, commanded by a certain Vasco da Gama. The newcomers comported themselves strangely, not respecting the code of conduct and honour of local merchants. Various misunderstandings and altercations followed, in which the Portuguese demonstrated their chicanery, until at last the Muslim merchant was forced to fight a duel with swords with Gama. The Portuguese captain was badly bested, but his Muslim opponent spared his life in the best traditions of Indian hospitality, telling him that he had best leave the place. After this, some years passed. Then the Muslim merchant decided to make the *hajj* with his family. Returning from the Middle East, his vessel was approached in the middle of the sea by a huge and unfamiliar fleet: it was none other than Vasco da Gama, who had returned from Portugal to avenge his earlier humiliation. Since, after the earlier duel, Gama's life had been spared, our Muslim merchant naturally expected the Portuguese to reciprocate in kind. He put this proposition to Gama when his ship was captured, but the latter refused. The rules by which the game was played in the Orient and the Occident were different, it would seem. In the former, one behaved chivalrously with defeated foes; in the latter, when one had an enemy by the throat, one gave no quarter even if he were unarmed. The Muslim merchant's ship was thus mercilessly sent to the bottom by Vasco da Gama, and he and his family with it. Hearing their cries, as the sun set over the western Indian Ocean (in the bloody twilight of the title), the Muslim merchant swore that one day he too would be avenged. And sure enough, centuries later, he came to be reincarnated as a Muslim butcher in Calcutta, and Vasco da Gama as an Anglo-

Indian in the same city! The Muslim remained a Muslim, and the Christian a Christian (and a contemptible half-caste at that). Four and a half centuries after the sinking of the *Mîrî*, the Muslim merchant of Calicut thus had his revenge; by this fictional device, Saradindu condensed the opposition between honourable East and evil West, drawing an unbroken line between the Portuguese of the early sixteenth century, and the British Empire of the mid-twentieth century.[77]

ON THE MALABAR COAST, 1502–1503

The first port of call in Kerala that Gama's plans envisaged (as we have seen from his letter in Kilwa to the other captains) was Cannanur, where the Kolattiri had quite consistently been willing to deal with the Portuguese, and where a Portuguese factor, Paio Rodrigues (in fact a servitor of D. Álvaro de Portugal), had been left behind by João da Nova's fleet. Besides, Gama had brought back the ambassador sent by Cannanur to Portugal, and the potential to firm up relations existed. But the Admiral on this occasion was even more suspicious of native intentions than on his earlier visit. On entering Cannanur on 18 October, he secured an audience with the Kolattiri within a day, but refused to set foot on land. A complex negotiation hence ensued, which Tomé Lopes describes with some bemusement, and which is confirmed by other accounts as well. It would appear that the Kolattiri had a sort of improvised wooden pier constructed, extending out over the sea. Gama approached it in a caravel with its poop covered in green and crimson velvet, accompanied by the most important men on the fleet, and by boats with flags, trumpets, drums and so on. There was music and dancing, but also the firing of guns by way of salutation. Gama himself was dressed in all his finery, with a silk dress, and gold on his person; he wished his high status to be amply visible. The pier itself obviously had two entrances, one from the land and the other from the sea; both were festooned with painted cloths. The Kolattiri entered first on his palanquin (*amdor*), with about four hundred retainers, armed with swords,

[77] I am grateful to Amiya Kumar Bagchi for drawing my attention to this story, and to Kunal Chakrabarti for a detailed discussion of its contents. The story was first published serially in a Bengali magazine in 1930, then in a book of short stories entitled *Chua Chandan* in 1936. For the most easily accessible version of the text, see Saradindu Bandyopadhyay, 'Rakta Sandhya', in *Saradindu Amnibasa*, Vol. VI (Calcutta: Ananda Publishers, 1976). Thanks are due to Uma Das Gupta and Kaushik Basu for bibliographic help.

and some with bows and arrows, but they were dressed rather minimally in comparison to Gama's finery. Since it was a hot day (and the Cannanur ruler was getting on in years), they stopped to take rest, and then the Kolattiri with thirty retainers approached Gama's vessel, still expecting the Admiral to disembark. He was disappointed; Gama resolutely refused, saying that he had orders to the contrary from D. Manuel. The interview thus had to be conducted with Gama on his poop, and the Kolattiri on the pier, fanned by some retainers, and attended by the others. Gama handed over his gifts, which were of silver, and some of them gilded. In return, the Kolattiri gave some precious stones (but not with his own hand!) to Gama and the other Portuguese notables, indicating that they were a modest gift of no real worth. These courtesies coming to an end, Gama entered the issue of the price at which spices could be purchased, but the Kolattiri replied that this was not his affair, and that anyway the spices were not yet available. He promised to send the spice-merchants the next day; these merchants were, inevitably, Muslims (probably Mappilas).

The audience was thus wound up, and the next day, negotiations with the merchants began. Their prices were, however, much higher than on the earlier occasions, and the Portuguese began to suspect that they did not want to sell them anything in the first place. Gama entered into a fury, and sent a message to the Kolattiri, saying that it was apparent that he did not really value peace with the Portuguese. For why else had he sent *Muslim* merchants to deal with them, 'who as was well known had an old hatred against the Christians'! Further, he assured him that he would send back the bales of spices already on board the *São Jerónimo*, with a few rounds of shot to accompany them. In the midst of this tantrum, Paio Rodrigues arrived, and Gama ordered him not to return to Cannanur, since he was breaking off relations. But Rodrigues flatly refused, saying that he was answerable effectively to D. Álvaro de Portugal, rather than to the Admiral. Gama then sent a message through him to the Kolattiri, saying that he would seek his cargo elsewhere; he also warned him to ensure that the Christian Portuguese who were on land were safe, 'and he swore and promised that if any ill or dishonour befell them, that his Kafirs (*ciafferi*) would pay for it'. With this thunderous warning, on 22 October Gama and his fleet departed, and a little further down the coast captured a *zambuco* with men and supplies, and went on to bombard a small port where some ships were dry-docked, believing it to be within the territories

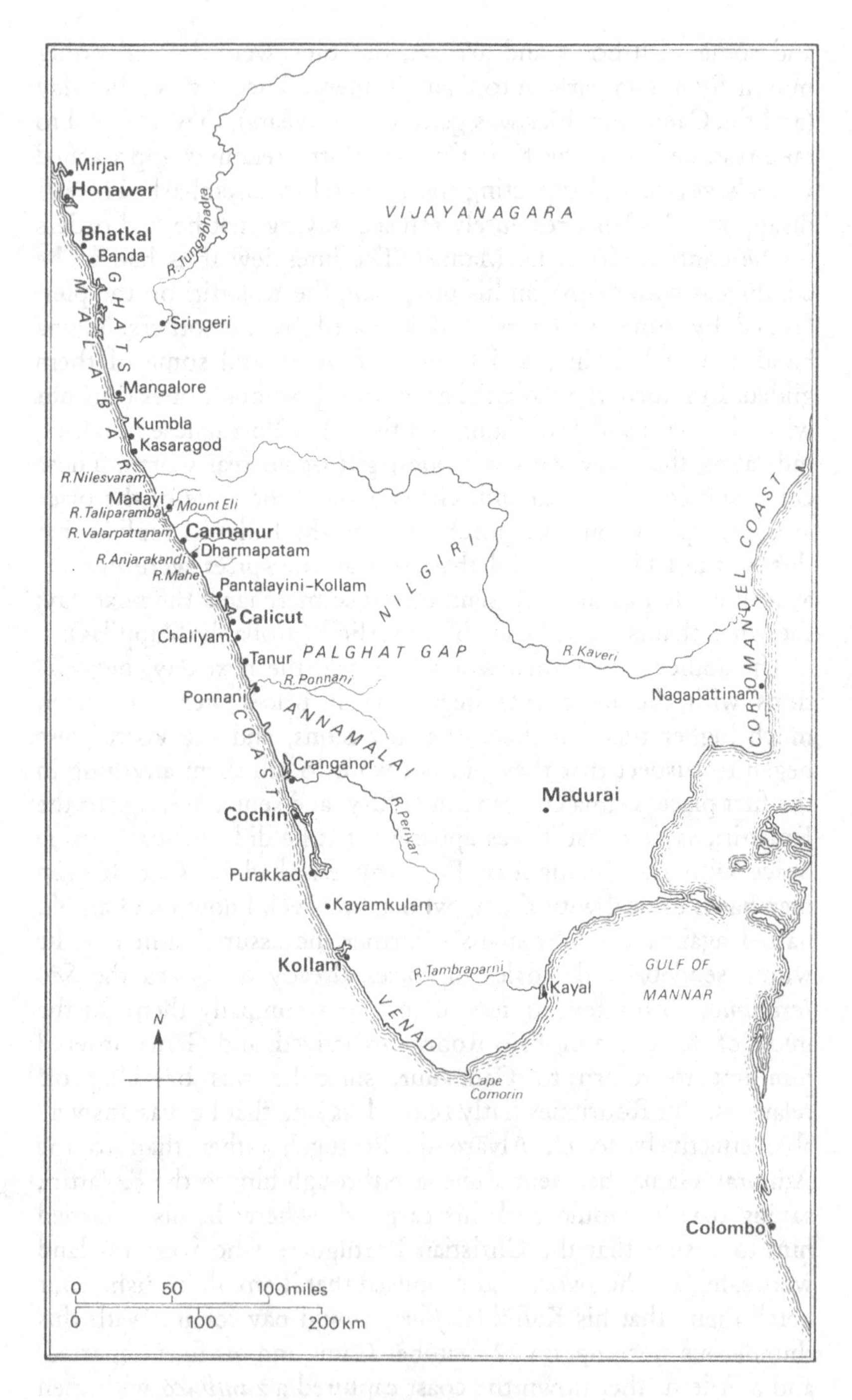

Map IV The coast of Kerala in the sixteenth century

under Calicut control. The local chieftain at once protested, declaring that he was subordinate to the Kolattiri, an enemy of Calicut, and a friend and relation of the Raja of Cochin. He was at peace with the Portuguese, and could not understand why he was being bombarded. Gama was simply out of his depth now, and his fury increased that night when a messenger arrived with two letters, one from Paio Rodrigues and the other from the Kolattiri (who Gama suspected had been tutored by Rodrigues). The Kolattiri's letter declared in effect that he was a loyal friend of the King of Portugal, and that even if Gama made war on him, he would continue to be one. Rather than take Gama's actions as those of the Portuguese Crown, he would complain against him to D. Manuel! This is, in fact, fairly clear evidence of complicity between the Kolattiri and Rodrigues, for such a letter could only have been written in the awareness that all was not well between the Crown and Gama. Outmanoeuvred, Gama made a reluctant peace with the local chieftain, and sailed on.

But Gama was not done with Calicut yet. While in Cannanur, he had received a letter from the Samudri, offering to make peace, and at the same time received a letter from Gonçalo Gil Barbosa (who was at Cochin) to say that the Samudri had heard of the powerful Portuguese fleet that was gathered in the Anjedives, and hence wished to make peace for tactical reasons. *En route* to Calicut, Gama's fleet spotted a large ship offshore and captured it; but since it belonged to 'Iuneos' (possibly a distortion of *vanias*, rather than Jainas – the usual interpretation), who were 'people who traded marvellously in India, and control a great part of the spices', it was decided after a consultation with the other captains to let the ship go. The Samudri's terms at this stage were simple enough: as against the goods from the Portuguese factory, there were the contents of the ship attacked by Cabral; the two sets of losses could be set off against each other, and whoever remained in debit would pay. As for the Portuguese who had died in the attack on the factory, the Portuguese had killed as many if not more in the ship from Mecca (the *Mîrî*), and the other ships they had burnt. To add to the Portuguese list, Gama had just hanged from the mast two Muslims found in a *zambuco* off Pantalayini, who were implicated as having been involved in the attack on Aires Correia's factory; evidence against them was taken from the children who had been taken off the *Mîrî* and converted. The Samudri's terms were repeated by a representative, sent by him to the Portuguese fleet on 29 October, while it was anchored off

Calicut. But Gama wanted far more, indeed nothing less than the expulsion of the entire Middle Eastern Muslim settlement at Calicut (*tutti i Mori della Mecca*), 'for since the beginning of the world, the Moors have been the enemies of the Christians, and the Christians of the Moors, and they have always been at war with each other, and on that account no agreement that could be made [between them] would be firm'. The uncompromising nature of his stand, and the reasonable tone adopted by the Samudri, is brought out by none other than João de Barros himself, who reproduces in his chronicle the following paraphrase of the letter to Gama from the Samudri:

> As for paying for the goods that the Portuguese lost in the uprising (*alvoroço*) of the people of Calecut, on account of the affronts committed by the Portuguese themselves, that he the Captain-Major should content himself with the taking of the ship from Meca, which amounted to more in terms of goods and people killed than ten times what Pedralvares had lost. That if the one side and the other had to add up the losses, the damages and the deaths, that he the Çamorij was the more offended of the two: and as he did not ask recompense for these things, though his people had requested with great clamour that he make amends for the wrongs that had been done by the Portuguese; he ignored this clamour from a desire for peace, and friendship with the King of Portugal. That he the Admiral should not repeat past affairs, and should be content by putting in at his city of Calecut, where he would find the spices that he needed. And as for what he said, that he should expel from his Kingdom all the Moors of Cairo, and of Meca, to this he did not respond, as it was impossible to root out more than four thousand households of theirs, who lived in that city not as foreigners, but natives, from whom his Kingdom received much profit.[78]

Gama, for his part, is reported by Barros to have declared this reply to be an insult (*afronta*). He detained the messengers, and when the Samudri sent another message, saying that he should not hold them hostage, Gama became even more 'indignant' (*indignado* in Portuguese, *con molta furia* in Italian) – a term frequently used by the chroniclers in connection with him. He declared his intention to begin the bombardment of Calicut the next day, and sent back the Samudri's emissary with orders that

[78] Barros, *Da Ásia*, Década I/2, p. 49. Compare Lopes, 'Navigazione verso le Indie Orientali', in Ramusio, *Navigazioni e viaggi*, Vol. I, p. 711, with a brief paraphrase of the same letter.

he 'should not return with any other message save with the price of the goods that had been taken from the Portuguese; and after this was handed over, then the matter of peace, and the spice trade could be dealt with'. Tomé Lopes for his part reports that Gama declared that he, as a vassal of D. Manuel, was a better man than the Samudri, 'and that a palm-tree would make a King as good as him', adding further disparaging remarks about his betel-chewing. If no favourable response arrived the next day, he stated, the Portuguese would deal with Calicut summarily.

The Samudri, having already been bombarded two years before by Cabral, had arranged for a limited stockade of palm-trees to be put up along the seafront, both to prevent a landing and to hold off artillery fire. There were also a few artillery pieces, of no great significance, in the defences. Gama had sent an ultimatum till midday; after this he began capturing passing small vessels, and had each ship of his fleet hang some 'Moors' on the mast, from a stock of earlier captives whom he distributed amongst them for this purpose, making a total of thirty-four victims according to Lopes. A crowd had gathered on the seashore to watch, and cannon-shot was fired into it by the *São Jerónimo* and by a caravel. The seafront quickly cleared, and a serious bombardment now commenced, which lasted to nightfall. Lopes describes its effects on the city clearly enough, as people fled to avoid the cannon-fire; since the houses were not of stone and lime, the damage was extensive but difficult to evaluate. Barros captures the essence of the next episode:

> [There was] a continuous storm (*torvão*) and a rain of iron and stone projectiles, which caused very great destruction, in which many people died as well. When night fell, to speed things up (*por espedida*), and for greater terror, he [Gama] had the heads, hands and feet of the hanged men cut off, and put on a boat with a letter, in which he said that if those men, though they were not the same who had been responsible for the death of the Portuguese, and only on account of being relatives of the residents, had received that punishment, the authors of that treachery could await a manner of death that was even more cruel.[79]

The version of the letter reported by Lopes is far less dignified, and states that the bodies were sent to the King, and that if he wanted their friendship he had to pay back not only the goods

[79] Barros, *Da Ásia*, Década I/2, pp. 52–3. These events occurred on 1 November, if one follows the account of Tomé Lopes, 'Navigazione', pp. 714–15.

that had been 'plundered in this port under your protection', but for the powder and shot expended in the bombardment. This boat was sent ashore with a certain André Dias, with the letter in Malayalam, and the rest of the bodies were thrown into the sea with the tide, so that they could be washed up on the beach. This matter cast a pall on the city, and the next morning no one appeared on the seafront. Things were so deserted, writes Barros, that Gama could well have sacked the city. But he desisted:

> For since the deaths of these people had been worked more for the King in terror to desist from hearing the counsels of the Moors than as a vengeance for the past, he did not wish to do as much damage as he could have, to give the King time to repent, and not to become indignant on account of the so great loss it would be were his city to be wholly destroyed.

It would seem then that Gama still held out hope of a complete capitulation, as well as the expulsion of the Muslims, and this was why the vessel captured some days before (of the 'Iuneos') was not looted. Nevertheless, on the next day, 2 November, the bombardment commenced once more, and since the houses near the seafront (which belonged to the poorer classes) had been largely demolished, this gave the Portuguese a clear field of fire to the houses of the notables (*case de Signori e gran maestri*), situated further to the interior of the city. Over four hundred cannonballs had been fired by midday by Tomé Lopes's reckoning. In the afternoon, a small fleet of boats approached the Portuguese *armada* from the shore; they were beaten off handily.

Thus it was, on 3 November, that Gama left with some satisfaction for Cochin, leaving behind Vicente Sodré with six large ships and a caravel, to keep up the blockade of Calicut and prevent ships (and supplies) from entering the port. He arrived in Cochin, where he was met by Gonçalo Gil Barbosa and Lourenço Moreno, on 7 November 1502.[80] But here too matters seemed to give him little reason for rejoicing. The Muslim merchants of Cochin had been advised by the letters of their counterparts at Calicut of the Portuguese doings, and there was hence some hostility in the air. Tensions were diffused at a meeting with the Cochin ruler, Unni Goda Varma, that was held roughly a week after Gama's arrival (after one attempted meeting that had to be

[80] Moreno and Gonçalo Gil Barbosa's nephew, Duarte Barbosa (already with him in these years), would eventually form part of a group that opposed the policies and actions of Afonso de Albuquerque; cf. Inácio Guerreiro and Vítor Luís Gaspar Rodrigues, 'O "grupo de Cochim" e a oposição a Afonso de Albuquerque', *Studia* 51 (1992), 119–44.

abandoned on account of rain). Presents were exchanged, there was the usual fanfare, and cannon-shot. At much the same time, emissaries arrived from Cannanur to reopen negotiations concerning trade, and offered to supply spices at the prices agreed on in Cochin. Gama now reconsidered his earlier decision, and sent off two ships to Cannanur to lade spices.

In Cochin, Gama did show a certain anxiety to maintain good relations with the ruler, realising probably that the crucial commercial stakes lay in that port. It would also seem that, unlike his unhappy dealings with Paio Rodrigues, he maintained relatively cordial relations with Gonçalo Gil Barbosa and Lourenço Moreno, who would appear to have been his local advisers. Thus, within about ten days of his arrival, a minor incident took place that permitted him to demonstrate his good will: three Muslims came and sold a cow to the Portuguese of the ship *Julioa*, to be slaughtered for meat. When the Cochin ruler heard of this he was incensed and complained to Gama, who – the next time the men came, to sell another cow – handed them over to the local authorities, who promptly impaled them as a rather extreme punishment. Within a few days of this incident, Vicente Sodré arrived from Calicut, and Gama took the opportunity to have another interview with the Cochin Raja. Sodré too was sent off to Cannanur, to lade spices on three vessels, including the *São Gabriel*. Gama's patience is equally in evidence in mid-December, when the commercial relations with the spice merchants of Cochin temporarily broke down, added to which was the fact that Gonçalo Gil Barbosa advised Gama of rumours that Calicut, Cannanur and Cochin were all preparing a joint front against the Portuguese. The Portuguese factors and scriveners were hastily called back to the ships, but an understanding was once more reached with the Mappila merchants within four days.[81] The pepper was to be paid for in a mix of three-quarters silver and one-quarter copper, at 12 gold ducats a *cantara*; cloves and benzoin would be exchanged against textiles; cinnamon, incense and alum would be bartered for brazil-wood.

Between November 1502 and January 1503, Gama used Cochin as a central point at which information was collected, and from which feelers were sent out. For instance, in the second half of November, he was paid a visit by some men, who stated that they were representatives of the Christians of Mangalor and

[81] Freire Costa, 'Relação anónima da Segunda Viagem', pp. 190–1.

several other places, a community allegedly numbering some 30,000 persons. They spoke amongst other things of the tomb of the Apostle St Thomas, and of their pilgrimages there; it must have been possible for the Portuguese of Gama's fleet to draw the connection between these Christians, and those described by Joseph of Cranganor.[82] Indeed, it would appear that these Christians too were in fact from Cranganor, and they seem either to have offered to submit themselves to D. Manuel (and thus to his representative, Gama), or at least to have proposed an alliance based on a common faith, giving Gama a red staff with silver ends, and three silver bells on it, as a ceremonial offering. This Christian network was also extended as far as the southern Kerala ports of Kollam and Kayamkulam, from which certain Syrian Christians arrived to see Gama at Cochin in mid-December. At their urging, Gama sent two ships there to lade spices, and then in early January, a third, the *Leitoa Nova*, on which our anonymous Flemish author voyaged. A letter of December 1504, written by a certain Matias, a Syrian Christian from Kayamkulam, to D. Manuel, claimed credit for this: 'Of the ships in which the Admiral came as Captain-Major, I arranged the lading for two.'[83]

The perspective of the Syrian Christians on early Portuguese activities in Kerala is an interesting one; they clearly support their co-religionists, rather than the local rulers, or the Muslim merchants. A letter sent by the four Syrian bishops of Kerala, Mar Jaballaha, Mar Denha, Mar Jacob and Mar Thomas, to the Catholicos, Mar Elias, and written in 1504, provides a rather confused chronology of Portuguese activity, but its tone is clear enough. Thus, it is noted that 'powerful ships have been sent to these countries of India by the King of the Christians, who are our brethren the Franks', by way of east Africa (Habesh), in order to buy pepper and other goods. A version of the Cabral fleet's experiences at Calicut in 1500 is next presented:

> In Calicut there are living many Ismaelites [Muslims], who, moved by their inveterate hatred against these Christians, began to

[82] For a discussion of these early relations between the St Thomas Christians and the Portuguese, see Luís Filipe F.R. Thomaz, 'A "Carta que mandaram os Padres da Índia, da China e da Magna China" – um relato siriaco da chegada dos Portugueses ao Malabar', *Revista da Universidade de Coimbra* 36 (1991), 119–81, especially pp. 131–2; also João Paulo Oliveira e Costa, 'Os Portugueses e a cristandade siro-malabar (1498–1530)', *Studia* 52 (1994), 121–78, especially pp. 126–36.

[83] AN/TT, CVR, No. 2, published in Silva Rego (ed.), *Documentação para a história das missões do padroado português do Oriente*, Vol. I, pp. 25–8. It is often wrongly asserted that Matias resided at Kollam.

> calumniate them to the pagan king saying: 'Those people came from the West and they were very well pleased with the town and the country. Therefore they will now return to their king as soon as possible and will come again bringing with their ships huge armies against you and they will press on you and take your country from you.' The pagan king believed the words of the Ismaelites and followed their advice and went out like a madman and killed all the said Franks, whom they found in the town, 70 men and 5 worthy priests, who accompanied them, for they are not wont to travel or to go any place without priests. The other ones who were in the ships weighed anchor and sailed away with great sadness and bitter tears and came to our neighbouring Christians to the town of Cochi. This town too had a pagan king, who, when he saw them in heavy distress and great grief, received them hospitably and consoled them, and swore never to abandon them until death.[84]

The account now jumps anachronistically to the events of late 1503 (the voyage of the Albuquerques to India), but then returns briefly to the subject of Gama's exploits of 1502–3, celebrating the fact that he and his men 'captured his [the Samudri's] ships and broke them to pieces and killed the Ismaelite sailors, whom they found in them, about 100, and devastated the town [Calicut] with their guns'. Mention is also made of Gama's dealings with the Kolattiri at Cannanur, and the fact that Portuguese are now settled there. The letter then concludes: 'The country of these Franks is called Portkal, one of the countries of the Franks, and their king is called Emmanuel. We beseech Emmanuel, that he may conserve him.'

While these relations with Kollam, Kayamkulam, Cranganor and Cannanur were in the process of being established, Calicut remained a problem. In early January 1503, a Brahmin ambassador came to Gama from Calicut, with three other persons (one the Brahmin's son, the other two probably Nayars), carrying still another letter from the Samudri, offering to restitute the goods lost in the affair of 1500, half in cash and the rest in spices, and inviting Gama back to Calicut. The fact that a Brahmin was sent for this task appears to have impressed the Portuguese. Tomé Lopes for example reports that the Brahmins were the sacerdotes of the Indians, and held in very high regard indeed: 'Even when

[84] Georg Schurhammer, 'Three letters of Mar Jacob Bishop of Malabar, 1503–1550', in Schurhammer, *Orientalia*, pp. 335–6; also see Thomaz, 'A Carta que mandaram os Padres'. The original Syriac text is in the Vatican Library, Codex Vatic. Fondo Siriaco 204, fls. 154v–60.

Plate 19 Seventeenth-century façade of a Syrian Christian church in Kerala, reflecting Portuguese influence.

they are at war with each other, no one can dare touch them [the Brahmins] or anyone who is in their company, because otherwise he would be deemed as damned and excommunicated, and can in no way be given absolution, and they are men who in these countries have much credit.' Some other aspects of this Brahmin's comportment appear rather strange, for he is reported by Lopes to have told Gama that he wished to go with him to Portugal, and that he had brought along funds for the journey, namely jewels worth 3,000 *cruzados*. Gama even acceded to his request to lade some spices on board on his own account, allowing him to purchase and place on board 20 *bahars* of cinnamon. In this atmosphere of growing bonhomie, Gama decided to go personally to Calicut on 5 January, and embarked on the ship of his cousin Estêvão da Gama, the *Flor de la Mar*. He also sent a caravel to Cannanur, advising his uncle Vicente Sodré to join him before Calicut.

On Gama's departure from Cochin, the Mappila merchants of the town, growing convinced that he would buy spices at Calicut, ceased the lading of the ships. D. Luís Coutinho, whom Gama had left in charge at Cochin, attempted to persuade them that they should continue as before, but to no avail; he then ordered the ship of Rui Mendes de Brito (on which Tomé Lopes was) to carry letters to Gama at Calicut, to apprise him of the situation. Not altogether unexpectedly, the invitation in Calicut had turned out to be a trap. On arriving before the port, some of the Calicut envoys were sent back on land; according to Lopes, it was the smooth-talking Brahmin himself who left the ship. However, he did not return by evening as he had promised, and instead another envoy appeared, assuring Gama that the amount that the Portuguese were owed in cash and spices was ready on land; all that he had to do was send a 'gentleman' (*gentilhuomo*) to take charge of it. Gama, greatly annoyed by now, flatly refused, saying that he had no intention of sending any part of his company ashore. The envoy returned on land, promising however to come back the next morning. However, that very night, the lookout on the *Flor de la Mar* spotted some vessels (which appeared to be fishing-boats) rapidly approaching their ship. Gama, asleep in his quarters, was woken up, and he hastily dressed and came on deck, thinking that the envoy had returned. On a closer look, it turned out that a fleet of seventy or eighty small vessels (*zambuchi*) were on their way, and before they could defend themselves, the Portuguese found themselves under close attack. So close were the Calicut vessels, that the ship's cannon were of no use; after a heated combat at close quarters lasting several hours into the next day, the Portuguese raised anchor and began to try to head for the open sea. However, it was discovered that the enemy had secretly placed a second anchor on the vessel, which hence remained immobile. From these straits, the *Flor de la Mar* was saved by the fortuitous arrival of Gama's uncle, Vicente Sodré, from Cannanur; on his approach, the attacking fleet dispersed and returned to land.

Gama had obviously made a rather grave error of judgement, a fact that obviously did not improve his humour. In revenge, he hanged the envoys whom he still held (including, in some versions, the Brahmin himself) on the masts of the two caravels, and made it a point to sail up and down the seafront of Calicut with them visible, finally placing the bodies in a boat with a letter in Malayalam for the Samudri, in what was becoming a habit for him. The letter stated, bluntly enough (in Tomé Lopes's version):

> O vile man, you had me called for, and I came in answer to your call. You did as much as you could, and if you could have done more, you would have done more. The punishment will be as you merit, and when I come back here, you will pay your dues, and not in money.

Gama and Sodré now went on to Cannanur, where they remained for some days, eventually returning to Cochin, to supervise the final lading of the vessels, and to enable the dispersed fleet to gather together. Gama also formally prosecuted negotiations with Unni Goda Varma for the setting up of a Portuguese factory in Cochin, in meetings held on 6 and 8 February, in the presence of all the captains of the fleet. It was agreed that one Diogo Fernandes Correia would be left as factor, with powers over not only the Portuguese but over any Indian Christians (including converts), and that he would have the right to punish any Christians who embraced Islam.[85] Letters were written on behalf of the Cochin ruler to D. Manuel, for Gama to carry back, and some envoys (*arefens*: technically hostages) were also taken on board.[86] On 10 February, Gama set sail for Cannanur, apprehending through local informants meanwhile that another fleet was being prepared in Calicut and Pantalayini, to attack him. These fleets, it would appear, almost always carried firearms, but relatively small in calibre (*bombarde molto corte e cattive*), rather than the large cannon that some of the more substantial Portuguese vessels had on board. Their tactics were thus inevitably to approach the Portuguese vessels in swarms, hoping by proximity

[85] Freire Costa, 'Relação anónima da Segunda Viagem', pp. 193–4.

[86] While we do not have the letters written by the Cochin ruler, we do have a draft of D. Manuel's reply, sent with Lopo Soares de Albergaria in 1504; cf. AN/TT, CVR, No. 59, published in Geneviève Bouchon, 'L'Inventaire de la cargaison rapportée de l'Inde en 1505', *Mare Luso-Indicum* 3 (1976), 126–7, reproduced in Bouchon, *L'Asie du Sud à l'époque des Grandes Découvertes.*

to avoid the cannon fire, and at the same time to board the Portuguese ships.

The final engagement off Calicut followed this general pattern, with some variations, for unlike most later Mappila fleets, there were thirty-two large vessels (*navi grosse*) each carrying four to five hundred men, besides the small ones. The Calicut war-fleet approached the Portuguese, sounding their drums (*loro naccaroni*) in a rhythmic fashion; Gama's fleet managed to hold them off with their cannon, realising that at close quarters they could not manoeuvre, the more so since they were heavily laden, and their bottoms had not been cleaned. In due course, unable to withstand the fire, the larger Mappila ships turned tail and fled towards Calicut, with the Portuguese giving chase half-heartedly. When they caught up with the vessels, the men on board threw themselves off and swam ashore, but on one vessel the Portuguese found a boy (*uno ragazzino nascosto*), whom Gama ordered hanged at once. However, he then changed his mind, and the boy, on recovering his senses recounted to his captors how the fleet had been readied. It turned out that the Samudri had made a great effort, getting together a force of seven thousand men who had taken vows of ritual suicide in combat, and forcing the Mappilas of Calicut and Pantalayini with threats to provide the ships and sailors, 'with blows and bastinadoes'; moreover, he had loaded 'all the artillery that there was in Calicut on board the fleet'. The reluctance of the Mappilas to participate in the engagement is understandable, for Gama had made it clear that his enemies were the 'Moors from Mecca', whom he wished to have expelled from the city. Yet, inevitably, the Mappilas, as the community best able to mount a fleet, found themselves caught in the crossfire, between their political overlord, the Samudri, and the Portuguese. Lopes himself reports that there were a large number of Indian casualties in the engagements, and even gloats somewhat: 'The night brought a great fury of wind in the sea, which threw all the dead on land, and they had the time to count them.' Two ships were burnt in (or near) the river at Calicut, with the Samudri watching from one of his palaces that was located at a height, and from which he had a clear view of the naval engagement.

We are now in the last leg of the stay in Kerala, as the fleet put in at Cannanur on 15 February. It is reported by Lopes that many of the principal Mappila merchants of Calicut had fled to Cannanur, having decided that it was pointless to continue in their former residence in the face of Portuguese attacks. Gama, for his part, finally decided that here, as in Cochin, he would set up a permanent factory, and the factor chosen for the purpose was Gonçalo Gil Barbosa, who had been relieved of his duties in Cochin. He and another twenty or so Portuguese were eventually left at Cannanur, where their presence is reported in the letter of the Syrian Christian bishops of 1504 as follows:

> The leader of the said Franks [Gama] came to another town called Cananor in the same country of Malabar with another infidel king and besought him: 'Give us a place in your town, where we can buy and sell when we return for traffic year by year.' And he gave them a place and a spacious house and received them gladly and treated them very well. And the Christian leader gave him vestments of gold and *ptutka* [brocade?] that is scarlet garments. Soon he bought 14,000 *tagar* pepper and sailed away to his country. Of his people there are about 20 living in the town of Cananor. When we started from the town of Hormizda [Hurmuz] and came to this town of the Indians, Cananor, we made them understand that we were Christians and indicated them our condition. We were received by them with the greatest joy and they gave us beautiful vestments and 20 gold drachmas and honoured our pilgrimage exceedingly for Christ's sake. We remained with them two and a half months . . .[87]

We are aware that these splendid textiles that Gonçalo Gil (and earlier, Gama) had distributed so generously came in part from the spoils of the *Mîrî*; some of them were used, ironically, to dress up the hunchbacked pilot of that unhappy ship, who was given over to the charge of Vicente Sodré in the *Esmeralda*. This information comes to us from a series of receipts signed by Gama himself at Cannanur in late February 1503, the first surviving documents written by the Portuguese in India.[88] Gama himself set sail now for Portugal, leaving behind the two Sodrés and their sub-fleet (three *naus* and two caravels), as the first permanent attampt at a Portuguese naval presence in the Indian Ocean. Gama wished to sail by a new, and more direct, route to Mozambique, but eventually was able to reach that port only on 12 April, having passed through a part of the Seychelles, and a

[87] Schurhammer, 'Three letters of Mar Jacob', p. 336.

[88] AN/TT, CC, II–7–1, II–7–19, II–7–20, and II–7–21, all documents dating 21 and 22 February 1503, published in Brito Rebello, 'Navegadores e Exploradores Portuguezes', Docs. 63 to 66, pp. 311–13, and reproduced with photographs in Keil, *As Assinaturas de Vasco da Gama*, pp. 13–17. For a discussion, see Francisco Leite de Faria, 'Os mais antigos documentos que se conservam, escritos pelos Portugueses na Índia', in Luís de Albuquerque and Inácio Guerreiro, eds., *Actas do II Seminário Internacional de História Indo-Portuguesa* (Lisbon: IICT, 1985).

series of minor islands close to them (the so-called Amirante Islands, that thus bear Gama's name). From here, the fleet was able to leave for Portugal only 'two by two and three by three' (to quote the anonymous Portuguese account), because of a shortage of supplies for the long Atlantic voyage, and because several of the ships had to be careened. The first to leave were the *São Gabriel*, and the *Santo António*, already on 19 April. They reached Lisbon in the second half of August, bringing with them news of Gama's triumph in the Indian Ocean: 'in every place that he has been, either through love or through force, he has managed to do everything that he wanted', reports Affaitati on 20 August 1503.[89] But Gama himself set sail from Mozambique with the last ships only on 22 June, after several false starts. By mid-June, they had crossed the Cape, and then sailed north until on 27 September, they made landfall at the island of Terceira in the Atlantic. The return voyage was especially hard, and the anonymous Portuguese reports that in the Atlantic, those on board his ship were forced to eat two dogs and two cats on board their vessel, since the biscuits were full of insects. Eventually, by 24 October, at least thirteen ships had made the return, of which the largest number (nine, including Gama's *São Jerónimo*) entered the Tagus on 10 October. The return cargo, according to reports sent from Lisbon by the Cremonese merchant Affaitati in mid-September, may have been as high as 26,000 *quintais* of pepper and 6,000–7,000 *quintais* of assorted spices, especially cinnamon.

The Venetian letters also carry an interesting rumour that may shed light on the aftermath of the expedition. In one of them, Affaitati notes that Vasco da Gama himself and the other captains brought back pearls of considerable value on private account; the Admiral himself is reported to have returned rather rich, carrying between 35,000 and 40,000 ducats-worth of diverse high-value goods for himself.[90] Of particular significance is the fact that this repeats a pattern from the first voyage. In Sernigi's letter, written just after the return of Nicolau Coelho in 1499, in a discussion of the jewel-trade at Calicut, we already find an aside that states: 'And they say that the captain has brought back some great riches because he bought them with the silver he had for his own

[89] Letter from Affaitati in Lisbon to Pietro Pasqualigo in Spain, dated 20 August 1503, in *Diarii di Marino Sanuto*, Vol. V, pp. 130–1, written after the arrival of the *São Gabriel*. The same letter reports that the fleet carries '35,000 *quintali* of spices of every sort, but for the most part pepper, and a little cloves'.

[90] See the letter from Affaitati in Lisbon to Pietro Pasqualigo, dated 17 October 1503, in *Diarii di Marino Sanuto*, Vol. V, pp. 841–3.

service, and since he has not yet arrived, it is not known how much he has brought.'[91] A month later, the letter by Detti reiterates this, even adding a few details. The ships on their return carried few spices on Crown account, he reports, because the Portuguese did not carry much gold or silver. On the other hand, Gama had, on board his own vessel, silver cups and other objects 'which served for his own use in the ship as captain'; these he had sold and bartered against spices in Calicut, and thus had a personal cargo on his return of pepper, ginger, cinnamon, cloves, nutmeg, lac and rather fine stones.[92] Gama thus accumulated not only fame on these voyages, but quite literally a fortune.

In the first flush of triumph after his return on 10 October 1503, it is quite possible that this was at first overlooked. Once in Lisbon, so the chroniclers report, with great pomp and circumstance, Gama carried the gold tribute of the ruler of Kilwa to the palace, in a procession with drums and trumpets. This was the first major tribute (*páreas*) brought back from Asia, and signified the submission moreover of a Muslim ruler to the Portuguese Crown. From the gold, D. Manuel had an elaborate monstrance (*custódia*) made, designed by Gil Vicente; this object, finished in 1506, was destined for use in the Jeronimite monastery, where it remained until its transfer to the Museu Nacional de Arte Antiga.[93] Once more the glory of the Admiral, and his dramatic re-entry, led him to jostle for centre-stage with Dom Manuel himself.

But can the expedition really be counted a success from the Portuguese point of view? From a purely commercial viewpoint, the Crown had a major return cargo on its hands, rivalling any of the later years of the century, even if the captains (and Gama himself) had also done rather well out of private trade.[94] But what of larger strategic questions? As we have seen, Gama's most violent actions were directed at Calicut, and this must have had the broad concurrence of the Portuguese Crown, even if the tenor

[91] Biblioteca Riccardiana, Florence, Codex 1910, letter from Sernigi, 10 July 1499, fl. 63r (a), in Radulet, *Vasco da Gama*, p. 173.

[92] Biblioteca Riccardiana, Florence, Codex 1910, letter from Guido di messer Tommaso Detti, 10 August 1499, fl. 69v (a), in Radulet, *Vasco da Gama*, p. 193.

[93] Keil, *As Assinaturas de Vasco da Gama*, p. 37, reproduces a photograph of the monstrance, which had been modified in the late sixteenth century, but restored in 1929.

[94] Compare the cargo brought back by Lopo Soares in 1505, in Bouchon, 'L'Inventaire de la cargaison rapportée de l'Inde en 1505', pp. 102–3; this study is based on a cargo-list of the epoch, rather than on Italian reports, which are naturally less accurate. In this fleet, Gama, as Admiral of the Indies (and entitled to deploy 200 *cruzados* a year in Indian goods), had a small cargo of pepper on his account on the ship *Batecabelo*.

of a number of his actions may have been more extreme than the Crown wished. Bartolomeo Marchionni had already written, in September 1502, that Gama had gone with the intention of 'subjugating all of India'; Affaitati for his part had stated in a letter of the same month that since the Portuguese did not have enough ready cash (*danari contanti*) to pay for spices, they would have to terrorise the ruler of Calicut into either lading their ships, or else fleeing his country.[95] In Cochin, advised by Gonçalo Gil Barbosa, he behaved with relative tact and patience with the ruler, though it was later claimed by the Mappila merchants Cherian and Mammali Marikkar that he and his relatives Vicente Sodré and Estêvão da Gama had extorted cinnamon from them without payment.[96] In the case of Cannanur, his actions in breaking off negotiations in early October 1502 were distinctly ill-advised, but in the final analysis, he was able to recover lost ground and actually firm up the earlier tenuous basis of the factory. There is good reason to believe, then, that in February 1504, when D. Manuel made Gama two major additional grants, including an annuity of 400,000 *reis* (to be drawn from the Lisbon salt-tax), he was not wholly discontented.[97] The grant-letter began by citing the circumstances of the first voyage, while praising Gama for going far beyond the Ancients; then, explicit mention is made of the tribute extracted at Kilwa (whose ruler is described as 'a King of great power and riches', and who has in his power the gold mine at Sofala, the richest in the world); the letter goes on to cite with approval Gama's attack on 'the Moors from Mequa, enemies of our Holy Catholic Faith', and his successful negotiations with other rulers; and finally, mention is made of the fleet that he returned with, 'well-laden and with great riches'.[98]

[95] These statements may be found in letters dated 20 and 26 September 1502, in *I Diarii di Marino Sanuto*, Vol. IV, pp. 544–5, and 664–5.

[96] Letter from Gaspar Pereira in Cochin, dated January 1506, cited in Bouchon, *Mamale de Cananor*, p. 65.

[97] This grant is in addition to another of the same month, drawn from the Casa da Índia e Guiné, by means of a tax on caravels returning from São Jorge da Mina; cf. Biblioteca Nacional de Lisboa, Reservados, Mss. 244, No. 4, parchment document dated 20 February 1504, transcribed in Luciano Cordeiro, 'De como e quando foi feito Conde Vasco da Gama', in Cordeiro, *Questões Histórico-Coloniais*, Vol. II (Lisbon: Agência Geral das Colónias, 1936), pp. 206–7. For payments on this account between November 1504 and July 1505, see AN/TT, Livro da receita e despesa do tesoureiro da Casa de Guiné, 1504, fl. 227, in Brito Rebello, 'Navegadores e Exploradores Portuguezes', Doc. 89, p. 32.

[98] Biblioteca Nacional de Lisboa, Reservados, Mss. 73, No. 50, parchment copy dated 1682, of the grant-letter of 1504; another copy is in AN/TT, Chancelaria D. João III, Livro 3, fl. 167; transcription in Brito Rebello, 'Navegadores e Exploradores Portuguezes', Doc. 72, pp. 508–13.

All in all, then, the total impression that one gathers is that Gama's political position had evolved considerably from 1497. This is not entirely surprising, for the overall political situation in D. Manuel's court had changed considerably, necessitating new alliances all around. It is obvious that, by 1504, the monarch's own position was considerably stronger than at the start of his reign. His second marriage with a daughter of the Catholic Monarchs had meant that a direct succession was now assured, for with the birth of his son João (later to be D. João III), the risk that the Crown would pass to the unstable Duke of Bragança no longer existed. A consensus had at last been arrived at in court circles that the enterprise in the Indian Ocean would be pursued, even if the precise modalities were open to discussion. In the dispute between messianists like Duarte Galvão and commerce-oriented 'pragmatists' like the Baron of Alvito, it is clear that Gama would have rather thrown his lot in with the latter. But the Crown was too powerful to ignore, especially for a small-to-middling member of the service nobility, who depended on Crown grants for his social mobility; besides, the fate of D. Jorge, now destined to be forever marginal to court politics, had been more or less sealed by 1504. Even men like D. Francisco de Almeida were prepared to compromise with the Crown now, although their own conception of the relationship between trade, war, and messianic schemes differed radically from that of D. Manuel and his close advisers. As for Vasco da Gama, we know from his actions where he stood on the trade issue; he was all in favour of trade and profit for the nobles and captains who went to Asia, and far less enthusiastic about D. Manuel's monarchic capitalism. Still, in early 1504, it may have seemed that an accommodation was still possible between this vision and that of D. Manuel; the monarch needed to flaunt the glory of his agents in the discovery, even as his agents like Gama needed the monarch's patronage. But things were soon to go badly wrong between Gama and the monarch.

POSTSCRIPT TO THE EXPEDITION: THE SODRÉS

The key to the problem almost certainly lies in what transpired in the Indian Ocean after Gama's departure from Kerala in February 1503, though it is possible that D. Manuel came eventually to be displeased by reports concerning the affair of the *Mîrî*, or by Gama's private trading fortune, or even by his excessively violent comportment in Cannanur and Calicut. But the affair of the Sodrés is not to be neglected either. We have seen that, in keeping with the Crown's instructions, five (and not, as in some versions, six) ships were left behind, commanded by Gama's maternal uncle and close confidant, Vicente Sodré. Obviously, a part of Sodré's commission was to capture hostile shipping, and to make prizes of them (*andar às presas*, as it was termed in the epoch); in this fashion, he was supposed to be a perpetual thorn in the flesh of shipping between the Red Sea and Kerala (more particularly Calicut). But there was also another purpose to leaving such a fleet behind in the Indian Ocean, namely to protect the fledgling factories at Cochin and Cannanur from hostile action. This required Sodré to remain in the broad vicinity of the Indian south-west coast, even as he carried out his first commission and made a private fortune in the process; Gama had in fact promised the Cochin ruler that Sodré would protect his port.[99]

Now, it was amply clear, even after the engagement between Gama's fleet and that of the Mappilas of Calicut in February 1503, that the Samudri was not done with the Portuguese. The Italian Matteo da Bergamo, writing while at Mozambique at the end of March 1503, explicitly mentions current rumours that the Calicut ruler had the intention of destroying the Cochin factory, and the kingdom as well; it is hard to believe that the same rumours escaped the ears of Sodré.[100] Nevertheless, he and his brother Brás Sodré decided to leave the Kerala coast, seeking better targets for their corsair activity in the vicinity of the Red Sea itself. By April, they had already captured and raided a number of Muslim-owned vessels, largely in the neighbourhood of Cape Gardafui. It is almost certainly to these activities that the South Arabian chronicler Muhammad bin ᶜUmar al-Taiyib Ba Faqih refers in the following passage from his *Târîkh al-Shihrî*:

> In this year (Rajab 908/1503) the vessels of the Frank appeared at sea *en route* for India, Hurmûz, and those parts. They took about seven vessels, killing those on board and making some prisoner. This was their first action, may God curse them.[101]

[99] Letter from Diogo Fernandes Correia to Afonso de Albuquerque, dated 25 December 1503, AN/TT, CC, I-4-43, in Bulhão Pato (ed.), *Cartas de Afonso de Albuquerque*, Vol. III, pp. 211–12.

[100] Cf. Bouchon, *Mamale de Cananor*, p. 68. The author also quotes Gaspar Correia, *Lendas*, Vol. I, pp. 306–8, 345–9, to demonstrate that Vicente Sodré had already proved a bloodthirsty tyrant at Cannanur, pillaging the houses of Muslim merchants, and mutilating a rich merchant of Cairo who owed money to the Kolattiri.

[101] Cited in R.B. Serjeant, *The Portuguese off the South Arabian Coast: Hadrami*

A powerful eyewitness account is available to us, by a Portuguese who participated, namely Pêro de Ataíde.[102] In a letter to D. Manuel, written at Mozambique in February 1504, he notes that after Gama's departure, Sodré took his fleet out towards Cape Gardafui, capturing *en route* a *sambuq* with pepper and sugar, and then four other ships, with textiles, cloves, benzoin, camphor and rice on board. The goods on these ships were monopolised however by the Captain-Major's brother, Brás Sodré, who emerges clearly as the villain of the piece. Only after he had helped himself, would he let the other captains on board, and they for their part usually found only huge bales, that were much too large for them to transport on to their vessels. Tensions thus grew apace within the Portuguese fleet, further exacerbated with the capture of a large ship from Chaul carrying textiles, including silk. Ataíde explicitly claims that Brás Sodré stole some of these goods, 'without entering them in the books of Your Lordship, besides many others that he took when he wanted, for no one dared to go against him, since his brother permitted him to do whatever he wanted'. On 20 April, Sodré's fleet arrived at the Khurian-Murian islands off the south-east Arabian coast of Masqat and Oman, instead of making for the ports of Berbera or Zeila (near the Horn of Africa), as others preferred. These islands were inhabited for the most part by poor Bedouin pastoralists, and the Portuguese were able to exchange textiles and rice, against the local sheep. One of the caravels was brought ashore, to be careened, and matters now took a turn for the dramatic. Ataíde writes:

> Some days later, the westerly winds calmed, and the Moors of the island told him [Sodré] that he should leave there, for another very strong wind was about to commence, and he responded that he had cables of iron (*amaras de fero*), and the Moors replied to him that he should leave at once, and that even if they were made of steel (*aço*) they were not powerful enough to resist, and that the

Chronicles (Oxford: Clarendon Press, 1963), p. 43. For a discussion of Ba Faqih and his sources, also see R.B. Serjeant, 'Materials for South Arabian History: Notes on new MSS from Hadramawt', *Bulletin of the School of Oriental and African Studies* 13, 2 (1950), 281–307, especially pp. 292–5. For other mentions of this incident in Arabic sources (viz. Abu Makhrama's *Qilâdat al-nahr*, and Abu ᶜAbd Allah Wajih al-Din al-Daybaᶜ's *Qurrat al-ᶜuyûn fî akhbâr al-Yaman al-maymûn*), see Lein Oebele Schuman, *Political history of the Yemen at the Beginning of the 16th Century: Abu Makhrama's Account of the Years 906–927 H (1500–21 AD)* (Groningen: Druk V.R.B. 1960), pp. 5, 54.

[102] Letter from Pêro de Ataíde at Mozambique to D. Manuel, 20 February 1504, AN/TT, CC, I-4-57, in R.A. de Bulhão Pato (ed.), *Cartas de Afonso de Albuquerque seguidas de documentos que as elucidam*, Vol. II (Lisbon: Academia Real das Sciências, 1898), pp. 262–7.

wind brought strong waves with it, and that he would do well to leave at once. So that, my lord, the next day, the wind rose so high and the sea became so rough that the ship of Vicente was dashed against the shore, and after it that of Brás Sodré with its mast broken, each of them having six cables for the prow, and Our Lord saw to it that at that time I was saved with my mast broken, and only two cables for the prow, for the remedy and salvation of as many of those wrecked people as were able to escape the ships.

What followed next is shrouded in mystery. Though Vicente Sodré was killed, his brother Brás Sodré apparently survived the wreck, came ashore, and promptly executed the pilot of the ship from Chaul, and the unfortunate hunchbacked pilot of the *Mîrî* as well, who had been given to the Sodrés by Gama in order to serve them as adviser. Ataíde particularly regrets the death of the latter, 'the best [pilot] that there was in all of India, and who was most necessary for Your Lordship'. Thereafter, Brás Sodré himself died, perhaps of natural causes, perhaps not.[103] Ataíde is coy on the matter, noting:

My lord, after the loss of the ships and the death of Vicente Sodré, many things transpired that Your Lordship will come to know of through many persons, and through me at greater length, if God brings me there [to Portugal].

With the hundred and fifty men who remained (including Sodré's factor, Rodrigo Reinel), Ataíde claims to have taken charge of the three surviving caravels, and the meagre supplies that were left to them of bread and rice. It was only two months later that he (in the *São Paulo*), and the two other captains (one of whom is identified as Fernão Rodrigues) were able to leave for Cannanur. Ataíde's letter ends with self-praise, for having brought back the ships to India, and recovered some of the artillery of the wrecked vessels, 'with such risk to my person, in order to serve Your Lordship as I desire'; he asks for a reward, in the form of the post of *alcaide-mor* of Tomar, a post that Vicente Sodré had held. But he cannot resist attacking his enemies as well at the end, the deceased Brás Sodré obviously, a certain António do Campo, and even the Admiral himself in passing. Naturally, Ataíde does not dilate overly on the fact that the Cochin and Cannanur factories

[103] For an inadvertently comic account, see the romantic recreation by Manuel Pinheiro Chagas, 'O Naufrágio de Vicente Sodré', in *Brinde aos Senhores Assignantes do Diário de Notícias em 1892* (Lisbon: Typographia Universal, 1892), pp. 5–80.

remained vulnerable through the period of his absence from Kerala, until late August 1503.

In the case of Cannanur, this lapse turned out not to be terribly serious. The canny Gonçalo Gil Barbosa was relatively well informed of happenings around him; besides his nephew Duarte Barbosa, he could also count on the assistance in 1503 of the son of Gaspar da Gama, who was converted to Christianity, and under the name of Baltasar (another of the Magi) entered the service of the Cannanur factory.[104] But Cochin was another matter. From the departure of the Sodrés, the Samudri mounted pressure on the Cochin ruler to hand over Diogo Fernandes Correia, and the other Portuguese established there to him; Correia for his part is even reported to have asked the Cochin ruler for a vessel to allow him to transfer his factory to the safety of Cannanur. The latter refused, trusting in his own abilities to protect the Portuguese, a confidence that was in the event misplaced. The Portuguese chronicles then report the following letter, sent from the Samudri to the Cochin Raja:

> Samorim, to you Trimumpara [Tirumalpad], King of Cochin, I make it known to you that I have tried hard to avoid making war on you. Had you not wanted to be so contumacious and only did as I asked you, because it is just, and even of mutual profit, not to offer a place to the Christians, from whom we have received so much harm; and so that this may go no further, in your wishing to continue in your ways, I have come here to Repelim [Eddapalli], determined to enter your land and to destroy it, and to seize the Christians with all their things. However, I pray you to hand them over to me, and avoid seeing your own destruction and that of your land, which you may take as certain, and I shall divest myself of the hatred I have for you on account of the discourtesy of your letters, even though I have many complaints against you. And if you do not do this immediately, I promise the Gods that I will destroy you and depose you as king.[105]

When Unni Goda Varma refused to comply, military action brusquely followed, and the much weaker Cochin ruler hastily had to abandon his devastated capital, taking refuge from the Samudri in the island of Vaipin. The Portuguese too fled with him

[104] Letter from Gaspar da Gama (Gaspar da Índia) to D. Manuel, n.d., in Bulhão Pato (ed.), *Cartas de Afonso de Albuquerque*, Vol. III, p. 202. Gaspar appears to have acompanied Vasco da Gama on his voyage of 1502–3, but his activities on this occasion are somewhat obscure; cf. Lipiner, *Gaspar da Gama*, pp. 145–55.

[105] *Crónica do Descobrimento*, ed. Albuquerque, p. 93, and to be found with minor variants in Castanheda's chronicle.

to safety, so that their trading activities came abruptly to a halt.[106] It was only in late August 1503 that the Portuguese could recover some ground, with the arrival in Kerala of Francisco de Albuquerque and Nicolau Coelho, then of Pêro de Ataíde (returning from the Khurian-Murian islands), and finally in mid-September of ships commanded by Afonso de Albuquerque. A wooden fortress was rapidly constructed at Cochin, and a truce (temporary in the event) was signed between Francisco de Albuquerque and the Samudri in mid-December.[107] But the damage had been done. A letter of Gonçalo Gil Barbosa, probably directed to Francisco de Albuquerque, and dated early October 1503, recounts the new conjuncture. He reports the arrival of letters (*olais*) from the Samudri and from the Calicut-based Mappila merchant Koya Pakki, proposing a truce that would allow the *paradesi* merchants a certain latitude in trade. The letter continues:

> Yesterday, that is 3rd October, there arrived in this port [Cannanur] a ship of Coje Çamecedim [Khwaja Shamsuddin al-Misri], a Moor resident here, which came from Adem and which brought three horses and fifteen or twenty *bahares* of opium. The news that it brought saddened all these Moors, and what could be known from him was that in Mequa and all those parts there was great famine and wars, and that the King of Cayro had captured that of Mequa and had made himself master of that land, and that there were great divisions there. It is also said that five ships left Adem for these parts, of which two were lost, and one with its mast broken made for Barbaria or some other nearby port, and that one made for Dabull and that another was separated to Callecut.[108]

Thus, nature (in the form of unseasonal storms) had partly interfered with the shipping from Aden, which otherwise would

[106] For a near-contemporary account, see the letter from Diogo Fernandes Correia to Afonso de Albuquerque, dated 25 December 1503, AN/TT, CC, I–4–43, in Bulhão Pato (ed.), *Cartas de Afonso de Albuquerque*, Vol. III, pp. 211–12.

[107] For details of these activities, see Jean Aubin, 'L'apprentissage de l'Inde: Cochin, 1503–1504', *Moyen Orient et Océan Indien* 4 (1987), 1–96; a summary in Geneviève Bouchon, *Albuquerque, le lion des mers d'Asie* (Paris: Editions Desjonquères, 1992), pp. 47–51.

[108] AN/TT, CC, II–7–173, Gonçalo Gil Barbosa at Cannanur, 4 October 1503; letter addressed to 'Vossa Mercê'; transcription in Aubin, 'L'apprentissage de l'Inde'. For dealings between Barbosa and the fleets of Francisco de Albuquerque and Pêro de Ataíde, see AN/TT, CC, II–7–159, and II–7–163, orders dated late August 1503. The reference to conflicts between the Mamluks and the ruler of Mecca probably refers to the troubled succession of the Sharif Barakat II, which was resolved only in 1505; cf. Serjeant, *The Portuguese off the South Arabian Coast*, p. 6.

have reached the Indian west coast, completely defeating the purpose for which Sodré had been left behind in India. News of these events was sent back to Portugal through Pêro de Ataíde, who died *en route* while at Mozambique, in early 1504. When the information concerning the ill-judged activities of the Sodrés reached D. Manuel's court, there was a general condemnation of the brothers, which survives into the later chronicles. Thus, Castanheda notes that 'it seemed that the loss of the two brothers was on account of the sin that they committed in not helping the king of Cochin, and leaving the Portuguese in the peril in which they were'; Damião de Góis later in the century repeats the same censure, adding that the brothers were destroyed by the 'judgement of God'.[109] As late as 1508, in the instructions given to Diogo Lopes de Sequeira, explicit mention is made of the carelessness of Vicente Sodré and how it had cost him and the Crown so dear; Sequeira was thus advised to keep an eye out for storms, and a close lookout night and day on the weather and the local people.[110] All of this reflected badly not merely on the brothers, but on their nephew and ally, Vasco da Gama, with whom the brothers Sodré may well have discussed their raiding plans before his departure from Cannanur for Portugal in late February. If, from an earlier time, Gama was known to have been less than enthusiastic with respect to D. Manuel's plans in the Indian Ocean, the irresponsibility of the Sodrés may have appeared to be no less than an act of insidious sabotage, clearly privileging the trade and raiding of the nobility over the well-being of the Crown's factories and personnel. Thus, by late 1504, we find Gama under a cloud, struggling to retain that which he had been granted by the Crown in the five preceding years.

[109] Cf. Aubin, 'Como trabalha Damião de Góis', p. 110.

[110] 'Regimento de Diogo Lopes', 13 February 1508, AN/TT, CC, I–6–82, in Bulhão Pato (ed.), *Cartas de Afonso de Albuquerque*, Vol. II, pp. 403–19 (clause on p. 418).

CHAPTER FIVE

THE WILDERNESS YEARS

King Manuel has not infrequently been charged with a niggardly disposition, but whatever his conduct may have been in other instances there can be no doubt that he dealt most liberally with the navigator who was the first to sail a ship from a European port to India. This liberality had been called forth by the sensation produced by the discovery of an ocean highway to India, and the expectation that great wealth would pour into Portugal as a consequence; it was kept alive by the persistent importunities of the discoverer.

E.G. Ravenstein, *A Journal of the First Voyage of Vasco da Gama* (1898)[1]

ADMIRAL-IN-WAITING

WE NOW ENTER A PHASE OF ROUGHLY TWENTY YEARS, FROM 1504 to 1523, in which Vasco da Gama's career has traditionally been rather neglected by historians.[2] But these years, though poorly documented in terms of Gama's activities, are in fact crucial ones, in which Gama struggles to use his legend to defend his career, struggling against royal authority, and eventually constructing a triumph of sorts. In early 1504, as we have noted, the returning Admiral had received further cash-grants from the King, and besides he had made a considerable fortune in the course of the voyage itself. At home, he consolidated his lineage, and seven children (including his heir D. Francisco) were born in

[1] E.G. Ravenstein, ed., *A Journal of the First Voyage of Vasco da Gama, 1497–1499* (London: The Hakluyt Society, 1898), p. 225.

[2] For a flat account, recounting the 'principal events' of these years, see, for example, the official biography of J. Estêvão Pinto and Maria Alice Reis, *Vasco da Gama* (Lisbon, 1969), pp. 71–7. Teixeira de Aragão, *Vasco da Gama e a Vidigueira*, is remarkably discreet concerning this interim period; so is José Maria Latino Coelho, *Vasco da Gama* (Oporto: Lello e Irmão, 1985; reprint of the 1882 text).

quick succession, five of whom – Estêvão da Gama, Pedro da Silva da Gama, Paulo da Gama, Cristóvão da Gama, and Álvaro de Ataíde – had India-related careers. But it remained for him equally to consolidate his claim over his 'home-base', the *vila* of Sines, which he had been theoretically given in 1499, but the control of which eluded him. We have already noted the conflicts in 1500–1 between his men and those of the *alcaide-mor* and *comendador* of the place, D. Luís de Noronha; thereafter, matters were further complicated when Dom Jorge himself took direct control of the town as *alcaide-mor*.[3] Nevertheless, it would seem that between late 1503 and 1507, Gama was at Sines, where he had sumptuous houses built for himself, and also patronised the construction of the hermitage of São Giraldo and Nossa Senhora das Salas. Documents of the Order of Santiago from later in the century, on the occasion of inspections done by either the Master or his representative, list the objects presented by the Admiral and his relatives (his mother Isabel Sodré, his brother D. Aires da Gama, and his brother-in-law, Lopo Mendes de Vasconcelos) to various establishments, as well as minor properties gifted by them.[4] The impression one retains is of a family of considerable wealth and property, anxious to make an impact on the provincial town that was Sines. Gama had, in all probability, also purchased property elsewhere, notably in Lisbon. Though references are hard to come by, we may imagine that he sought to retain an attachment to the court and its politics, observing the succession of fleets that set out from the Tagus after the Albuquerques: Lopo Soares de Albergaria (or Alvarenga) in 1504, and then D. Francisco de Almeida in 1505.

We may briefly sum up what we can gather concerning Lopo Soares's voyage, which is of some indirect significance for our purposes. Let us note that the Captain-Major, later to distinguish himself as a sort of 'Albuquerque-in-reverse', was an important member of the middle nobility, whose father had been *chanceler-mor* of Portugal under D. Afonso V; his clan was linked by marriage to the Almeidas, amongst many others, and he had served as captain of São Jorge da Mina for four years, from 1495 to 1499. He set out from Lisbon with a fleet of eleven vessels, in

[3] Unpublished *Livro das menagens* in an unnamed private collection, cited in Leite Faria, 'Pensou-se em Vasco da Gama', p. 33, n. 112.

[4] 'Visitaçam da villa de Sines feita per Dom Jorge', 9 and 15 November 1517, in AN/TT, Cartório de Santiago, Livros 160 and 164, transcribed partially in Brito Rebello, 'Navegadores e Exploradores Portuguezes', Doc. 38, pp. 158–63; also see Docs. 39, 40 and 70 of the same collection.

late April 1504, and one of the captains to accompany him was Vasco da Gama's future brother-in-law, Lopo Mendes de Vasconcelos. The instructions (*regimento*) given to Lopo Soares have survived, and clearly refer him to the lessons learnt from Vasco da Gama's experiences in the course of his second voyage; the news of the disastrous affair of the Sodrés had not yet reached Portugal.[5] There was some private participation in this fleet, as in that led by Gama: one of the vessels here (as in the earlier fleet of the Albuquerques in 1503), belonged to the rather enigmatic figure of Catarina Dias de Aguiar, influential in the court, and a sometime trading-partner of Tristão da Cunha. Doubling the Cape towards late June after a few minor mishaps, the fleet touched at Mozambique in late July, where they found a letter from Pêro de Ataíde, who (as we have seen) had survived the unhappy end of the Sodrés in the Khurian-Murian Islands, returned to Kerala, but had eventually succumbed to an illness in late February or early March 1504 in East Africa, while on his way back to Portugal. Lopo Soares appears not to have touched at Malindi, despite instructions to the contrary; rather, being anxious to reach Kerala, he made directly for the Anjedive Islands, where his fleet encountered the ships of António de Saldanha and Rui Lourenço Rodrigues Ravasco, left over from the *armadas* sent in 1503.[6] These captains, after a series of exploits on the coast of East Africa, had spent a good part of early 1504 off the coast of south Arabia, capturing Muslim-owned vessels (including one from al-Shihr) that were for the most part on their way to or from the Red Sea, before meeting up with Lopo Soares's fleet in August 1504. The entire formation now made for Cannanur, where they spent a brief period in September, before setting out for Calicut, from where emissaries had in the meantime arrived to open negotiations once more with the Portuguese.

Now, it is worth noting that Lopo Soares's instructions explicitly ordered him *not* to strike a compromise with the Samudri Raja. The *regimento* thus stated: 'If by chance the king of Calicut sends a message to state that he seeks our friendship and wishes to make up for the damage that we have suffered, listen to him, but your response should be as follows: that you have no orders to

[5] 'Regimento de Lopo Soares' (n.d. 1504), AN/TT, Leis e regimentos sem data, Maço 1, No. 20, in Bulhão Pato (ed.), *Cartas de Afonso de Albuquerque*, Vol. III, pp. 185–93.

[6] Cf. Alexandre Lobato, *Da época e dos feitos de António de Saldanha* (Lisbon: CEHU, 1964), pp. 15–25, for an account based on the chronicles; also Avelino Teixeira da Mota, *A viagem de António de Saldanha e a rota de Vasco da Gama no Atlântico Sul* (Lisbon, 1971), for a contemporary Italian account.

make an agreement, but rather to do all the harm you can, to him and to all those who pertain to him.' Still, Lopo Soares did make for Calicut, partly in response to overtures from the Mappila chief, Koya Pakki, but there soon lost patience, and used the usual violent methods. His opening demands not having been met, the Portuguese fleet once more bombarded the port (which had now suffered a whole series of such attacks since 1500), and then departed for Cochin. Between the merchants of this port and those of the more southerly lying centres of Kollam, and Kayamkulam, a good sized cargo of spices was made up. Lopo Soares also took the time to mount a rather substantial attack on the town of Cranganor (where a Calicut force had gathered), lying at the frontier between Calicut and Cochin, before leaving for Portugal in late December 1504. Four vessels were left behind to guard the Portuguese establishments at Cochin and Cannanur against further attack; the departing Portuguese fleet also fought a naval engagement with a Calicut squadron (rather like Gama on his departure in 1503), but this time off Kappatt, which is to say between Calicut itself and Pantalayini-Kollam.[7] The fleet then returned to Lisbon, for the most part in late June and July, with a few stragglers coming in by mid-August 1505. The cargo was a particularly good one, though not as good perhaps as that of Gama's fleet in 1503: nearly 21,000 *quintais* of pepper, and minor shipments of ginger, cinnamon and cloves. In many respects, the fleet followed the pattern of Gama's second expedition, and was met with considerable triumph in Portugal. The rupture with Calicut, which had reached a new pitch with Gama, was still further deepened by Lopo Soares's activities. But there were two differences between the expedition of 1504–5 and that of 1502–3, one major, the other minor. First, the fleet of four vessels left behind in Kerala by Lopo Soares, commanded by Manuel Teles de Vasconcelos did the task it was meant to perform, unlike the *armada* of the brothers Sodré; it is worth noting that Vasconcelos was, after all, the nephew of Duarte Galvão, and hence perhaps closer in spirit to D. Manuel's conception than the Sodrés. Second, the fleet of 1504–5 brought back to Portugal the 'hero of Cochin', Duarte Pacheco Pereira, who had led the Portuguese forces in the defence against the Samudri in 1504, helping the Portuguese retain a foothold in Cochin.[8] Thus, already, on the

[7] For a detailed account, see Geneviève Bouchon, 'Le premier voyage de Lopo Soares en Inde (1504–1505)', *Mare Luso-Indicum* 3 (1976), 57–84.

[8] For a caustic look at the career of this Portuguese nationalist hero, see Jean Aubin, 'Les

arrival of the first ship in early July 1505, D. Manuel ordered that public celebrations be organised, to mark 'the great victories and destruction that our captains have visited on the King of Calecut and the Moors who are in his city and land after the departure of Afonso d'Alboquerque and Françisco d'Alboquerque, and that which was carried out by the said Lopo Soarez who burnt thirteen large ships of his, and some of them laden with spices in which many Moors also died'.[9]

As was usual, the India fleet of 1505 had already been sent off without awaiting Lopo Soares's return. This fleet is again a relatively well-documented one, that led by D. Francisco de Almeida. However, the despatch of Almeida, officially named Captain-Major, but given orders to assume the title of viceroy once in the Indian Ocean, is itself shrouded in some confusion. It would appear that the first person actively considered (and even perhaps named) to the post was the powerful trading *fidalgo*, Tristão da Cunha, who had earlier sent substantial goods to India on board the *Leitoa Nova*, a ship of Gama's fleet of 1502–3. It is reported by the chroniclers, however, that on the eve of his departure for India in early 1505, Cunha suffered a sudden and mysterious loss of vision, which rendered him incapable of voyaging.[10] He was hence substituted by D. Francisco de Almeida, but in the next year, 1506, was well enough to undertake the arduous voyage to India, where he was then involved in a bitter conflict with the viceroy, who even went so far as to make sarcastic remarks about his private trade.[11] Almeida, quite probably, had moved from the Order of Santiago to the Order of Christ in January 1505, and received the grant-letter naming him 'captain-major of all the said fleet and armada, and to remain in the said India for three years', on 27 February 1505.[12] The chroniclers' accounts of the replacement of Cunha by Almeida, like that of

frustrations de Duarte Pacheco Pereira', *Revista da Universidade de Coimbra* 36 (1991), 183–204.

[9] AN/TT, CC, I–5–31, letter from D. Manuel to the Vicar-Provincial of the Dominicans, Lisbon, 10 July 1505, published in Bouchon, 'L'inventaire de la cargaison rapportée de l'Inde en 1505', pp. 135–6.

[10] António Alberto Banha de Andrade, *História de um fidalgo quinhentista português: Tristão da Cunha* (Lisbon: Instituto Histórico Infante D. Henrique, 1974), pp. 50–4, *passim*.

[11] Cf. the letter from D. Francisco de Almeida to D. Manuel, December 1507, AN/TT, Fragmentos, Caixa 4, Maço 1, No. 67, published in António Dias Farinha, 'A dupla conquista de Ormuz por Afonso de Albuquerque', *Studia* 48 (1989), 464–5.

[12] AN/TT, Gavetas, VII/10–14, Papal Bull from Julian II; also AN/TT, Gavetas, XIV/3–14, grant-letter to D. Francisco de Almeida, dated 27 February 1505, Silva Rego (ed.), *Gavetas*, Vol. III, pp. 599–602.

earlier substitutions (Cabral by Gama, in 1502, for example) hence should probably not be taken too literally; what they conceal is in all probability another shift in the court-politics of D. Manuel, leading to the easing out of men like Cunha and his brother-in-law and ally Lopo Soares de Albergaria (whose primary interest was trade), and their substitution by Almeida, who – at least at the outset of his viceroyalty – promised to implement a vigorous policy of fortress construction on behalf of the Crown, and even carried out a part of these intentions at Kilwa in 1505. It is another matter that, once in India, he came increasingly to see matters otherwise (perhaps under the influence of advisers who knew the Asian scene better than he), so that his viceroyalty represents in the final analysis a long spell of relative inactivity. By late 1506, he was already beginning to drag his feet, and the next year he reacted in an ambiguous fashion to the rumours of the submission of Hurmuz to Afonso de Albuquerque: while in fact probably not greatly enthused (among other reasons because Albuquerque had stolen the thunder of his son, D. Lourenço), he did however assure D. Manuel in flattering terms, in a letter of late 1507 cited above, that he should no longer hesitate or seek counsel, but instead 'call yourself Emperor, for never has a prince had such just reason to so become'. The viceroy remained during most of the period of his government at Cochin (which now definitively became the *de facto* Portuguese headquarters), in a situation of some tension with the Cochin ruler, who was later to complain bitterly against him to D. Manuel.[13] He left only once for an extended spell, to avenge the death of his son, D. Lourenço de Almeida, killed in a celebrated engagement with a Mamluk fleet led by Amir Husain Mushrif al-Kurdi, at Chaul. He halfheartedly pursued the opening up of commerce with Melaka (an express priority in his *regimento*), failed to pursue the alliance offered by the rulers of Vijayanagara, and soft-pedalled the idea of setting up a fortress in Sri Lanka, where his instructions had suggested he might even make his headquarters (*principal assento*).[14] He did however send a fleet, under his son D. Lourenço, to the west coast of Sri Lanka

[13] AN/TT, Gavetas, XV/21–30, 'Carta que el-Rei de Cochim escreveu a el-Rei de Portugal a respeito da pimenta etc' (n.d.), in Silva Rego (ed.), *Gavetas*, Vol. V, pp. 530–3; the letter was probably written in very late 1509 or early 1510, before the disastrous Portuguese attack on Calicut.

[14] AN/TT, Gavetas, XX/4–15, 'Sumários das cartas que vieram da Índia' (1506–7), summary of a letter from Dom Francisco de Almeida, dated 27 December 1506, in Silva Rego (ed.), *Gavetas*, Vol. X, pp. 356–72.

in the latter half of 1506.[15] His detractors in India, notably one Pêro Fernandes Tinoco, pointed out to D. Manuel on more than one occasion that Almeida was more concerned with commerce than with political alliances, or the spread of Christianity for that matter.[16] We see his conception, as it had evolved by late 1508, in a celebrated letter to D. Manuel, in which after announcing the death of his son, and his intention to lead the naval force against Amir Husain, he adds:

> As for the fortresses that you order be built, the more fortresses you build, the weaker your power will be, for your entire force rests on the sea, and if we are not powerful there, we will easily lose all the fortresses, for the past war [in Kerala] was with people who did not know much about it [the sea], whereas now we are at war with Venetians, and Turks and Mamluks of the Soldão.[17]

These differences in conception also led to a bitter process of succession in the government in India: as is well known, D. Francisco refused in late 1508 to allow Afonso de Albuquerque to succeed him, delaying the latter's accession by nearly a whole year, to December 1509. Of some significance in all this is that D. Francisco was always held up for respect by Gama and his family, and he for his part made it a point to listen closely in India to Gama's chosen clients, as also to his wily and opportunistic godson, the Jewish convert, Gaspar da Gama.

It is clear that despite the fact that Almeida had apparently gained a measure of royal favour in 1505, Gama himself remained thoroughly marginalised. The conflict between the King and the Admiral was, moreover, no secret. The excellent and detailed account of the Venetian Lunardo da Cà Masser, resident at Lisbon in these years, is particularly illuminating, not so much for its description of the first nine Portuguese voyages to the Indian Ocean, but for its analysis of the political situation in Portugal in 1506. After noting the importance of the *corregedores* and, above

[15] Geneviève Bouchon, 'Les rois de Kôttê au début du XVIe siècle', *Mare Luso-Indicum* 1 (1971), especially pp. 74–6. For an earlier discussion, see Donald Ferguson, 'The discovery of Ceylon by the Portuguese in 1506', *Journal of the Ceylon Branch of the Royal Asiatic Society* 19 (1906–7), 284–400; and for an overview, Flores, 'Os Portugueses e o Mar de Ceilão', Part II, ch. 2.

[16] Letters from Pêro Fernandes Tinoco to D. Manuel, dated 21 November 1505 and 15 January 1506, AN/TT, CC, I-5-59, and AN/TT, Cartas dos Vice-Reis da Índia, No. 72, in Bulhão Pato (ed.), *Cartas de Afonso de Albuquerque*, Vol. II, pp. 341–4; Vol. III, pp. 170–3. Also see AN/TT, Cartas dos Vice-Reis da Índia, Doc. 53, an undated letter from 'El Rey de Narcinga' to D. Manuel (perhaps from 1505), offering a marriage alliance between the two crowns.

[17] *Crónica do descobrimento e primeiras conquistas*, ed. Luís de Albuquerque, p. 337.

all, the *vedores da fazenda* in politics (stressing the positions of the Baron of Alvito, Dom Martinho de Castelo-Branco and Dom Pedro de Castro), he goes on to write of the positions that were next in prestige, namely the *secretário-mor*, António Carneiro, and the *escrivão da puridade*. Sandwiched between the *vedores* and Carneiro, in terms of importance, is the following figure:

> An Admiral, which is to say a Captain-General of the Sea, who is Don Vasco da Gamba, the one who discovered India; this is a most highly honorable office, which office was given by this Most Serene King to the said Don Vasco, making him Admiral; even if he is not very grateful to His Highness, because he is an intemperate man, without any reason (*homo destemperado, senza alcuna ragione*); he has done many things in India during his voyage which have not endeared him to His Highness: but being the one who illuminated this route to India, and its discovery, this Most Serene King made him Admiral, and gave him a castle, from which he has an income (*intrada*) of 1500 ducats; he has at present an income of 4000 ducats, and also he has this privilege from His Highness that he can employ 200 ducats on the route to India, which he can invest in any sort of spices that he likes without paying any duties whatsoever; so that this is a great income, even if he were to have no other: he is of a low condition (*bassa condizione*), but at present has been made a Fidalgo, which is to say a gentleman (*gentiluomo*), and lives honorably, and is considered among the greatest grandees of that Kingdom.[18]

The 'castle' in question was evidently Sines, which still remained problematic, with D. Jorge (whom Cà Masser still refers to in 1506 as a pretender of sorts) refusing to relinquish it either to the Crown or to Gama. Did some further violent altercations take place in this respect in 1506 or early 1507? At present we have no documentation on the question, but we are aware that matters took a turn for the worse in 1507. Dom Jorge now petitioned the ruler, to the effect that he could no longer have the Admiral interfering in the affairs of this town which after all still pertained to his Order; he wished him expelled forthwith from his residence there. Dom Manuel now performed a volte-face, and decided that he would not support Gama in the affair at all. A

[18] Cf. the 'Relazione de Lunardo da cha Masser', published in Prospero Peragallo, 'Carta de El-Rei D. Manuel ao Rei Cathólico narrando as viagens portuguezas à Índia desde 1500 até 1505 reimpressa sobre o prototypo romano de 1505, vertida em linguagem e anotada. Seguem em appêndice a Relação análoga de Lunardo cha Masser e dois documentos de Cantino e Pasqualigo', *Memórias da Academia Real das Sciências* (2ª Classe, Lisbon) 6, 2 (1892), 67–98, especially p. 89.

Plate 20 Blue-tiled murals at the altar of an Alentejan church, in Alvito, reflecting overseas influence.

royal order (*alvara*), issued at Tomar on 21 March 1507, and made public by Dom Jorge at Santiago de Cacém on 26 June of that year, gave Gama a mere thirty days to leave Sines. It ran:

> We, the King, make it known to you Dom Vasco da Gama, Admiral of the Indies and of Our Council, that we consider it well and to our service that for certain reasons which move us, within thirty days of the making of this order, you take your wife, and all your household from the Villa of Sines where you at present maintain them. And neither you nor your wife and household may return to or enter the said Villa or its limits save with the permission of my well-beloved and prized nephew the Master [Dom Jorge]. And if either of you does enter without his permission, either with or without your household, we consider it well that you pay 500 *cruzados* in fine for the captives [ie. as ransom for hostages on the Barbary coast], and besides it will remain for us to give you the castigation due to those who do not obey the orders of their King and Lord . . .[19]

Explicit mention is made in a postscript, to the 'construction works' (*obra das casas*) that Gama has begun in the town (and which were still continuing in mid-1507), which should stop at once. Nineteenth-century historians like Teixeira de Aragão have seen the matter as motivated by the malevolent Dom Jorge, who is portrayed by him (in a devastating footnote to his work) as the worst sort of libertine and wastrel. The possible political significance of this falling out escaped his attention; what we see here is Dom Manuel using a convenient complaint from a rather unexpected quarter to put Gama in his place. Aggravating the matter from the point of view of D. Jorge was the fact that sometime in the first half of 1507, Gama had decided to follow the footsteps of D. Francisco de Almeida, and definitively abandon the Order of Santiago for the Order of Christ. Whether he hoped by these means to reinstate himself in royal favour is unclear; but it certainly did not help to endear him to D. Jorge. Gama was hence obliged to return the *comendas* of Mouguelas and Chouparria that he had from the Order of Santiago; the latter was already reassigned on 9 June 1507 to a certain Francisco de Lemos.[20]

[19] AN/TT, Livro dos Copos, fl. 257, in Teixeira de Aragão, *Vasco da Gama e a Vidigueira*, Doc. 18, pp. 250–2. Ironically, the order was publicly presented by Gama's own uncle, João da Gama, himself a member of the Order of Santiago.

[20] AN/TT, Ordem de Santiago, Livro de Registo (1505–1507), fl. 130v, *provisão* dated 9 June 1507, published in Teixeira de Aragão, *Vasco da Gama e a Vidigueira*, doc. 19, p. 252. This document too was drafted by João da Gama.

Expelled from his erstwhile residence, the Admiral seems to have moved to Évora, where he attempted without success in 1508 to buy himself another minor landed position, this one in the Ribatejo – that of *alcaide-mor* of Vila Franca de Xira. But though permission was granted in 1508 by D. Manuel to Luís de Arca, the *alcaide-mor* of the town, to proceed with the sale, the transaction does not appear to have been prosecuted.[21] Local tradition in Évora tends to associate Gama with a series of houses with frescoes in that town, which the Admiral reputedly had painted during this phase of 'exile', particularly one in the Rua de Valdevinos.[22] He continued through this phase to send an agent to collect the fiscal dues he had assigned to him in Sines, Santiago de Cacém, and Vilanova de Milfontes, though – as we see from an admonitory letter of D. Manuel of November 1511 – not without a certain difficulty. His old order, the Order of Santiago, not content with the Admiral's humiliation, appears to have placed obstacles in even this process.[23]

THE LIMITS OF ROYAL CENTRALISATION

Now, we should bear in mind that these years of eclipse for Gama, and particularly those from about 1505 to 1514, were also those when Dom Manuel's domestic political position was at its strongest. We are still woefully lacking in detailed studies at a monographic level that would enable us to gain a true measure of the relationship between royal power, state and civil society in Portugal in the epoch of D. Manuel.[24] However, there are sufficient indications that the Crown in this period attempted to continue some essential aspects of the centralising line of develop-

[21] *Alvará*, dated 18 November 1508, in Biblioteca Nacional de Lisboa, Reservados, Mss. 237, No. 32, published in Teixeira de Aragão, *Vasco da Gama e a Vidigueira*, Doc. 20, p. 253.

[22] Cf. the study by António Francisco Barata, *Vasco da Gama em Évora, com várias notícias inéditas*(Lisbon: Typographia B. Dias, 1898). But also see more recently, João Paulo de Abreu Lima, 'Vasco da Gama e os Frescos das "Casas Pintadas" da Cidade de Évora', *Panorama* (Revista Portuguesa de Arte e Turismo, 4th Series) 31 (1969), 51–63. This special number of *Panorama*, commemorating the fourth centenary of Gama's birth, is a fascinating piece of nationalist propaganda, from the last years of the *Estado Nôvo*. Not the least amusing aspect is the comparison between 'argonauts and astronauts', for we are in 1969, the year that Neil Armstrong walked on the moon!

[23] Biblioteca Nacional de Lisboa, Reservados, Mss. 244, No. 3, dated Lisbon, 19 November 1511, published in Luciano Cordeiro, 'De como e quando foi feito Conde Vasco da Gama', in Cordeiro, *Questões Histórico-Coloniais*, Vol. II, p. 208.

[24] However, Jean Aubin is preparing such a study (tentatively entitled *D. Manuel, 1495–1521: Le Portugal des Découvertes*), which will undoubtedly change our conception of the Manueline 'landscape'.

ment set out by D. João II, while considerably enlarging its scope. D. João's major target had been the titled and landed aristocracy, allied either overtly or covertly with the Catholic Monarchs; he also resisted the expansion of the numbers of titled nobility, and created only one title for a Portuguese during his reign. At the level of international relations, too, the rivalry with Spain occupied him in large measure, whether with respect to west Africa, or the Atlantic. However, in the process of implementing his vision, D. João II came to symbolise a form of personalised, highly autocratic rule in Portugal, remembered with awe mixed with a degree of terror as the 'Perfect Prince'.

Psychologically, but also in terms of the instruments he deployed, D. Manuel was rather different, as his posthumous official sobriquet of the 'Fortunate King' (*O Rei Venturoso*) already suggests. A fairly large number of new titles (eight new Counts) were created during his reign, a measure of his variable situation in respect of the aristocracy. Far less sure of his legitimacy than D. João II, he appears to have attempted to rule using a modified council-system, which however tied his hands at various moments. One possible response to these situations of impasse was to claim that the king had direct divine inspiration, a tactic that had to be used sparingly in order for it to be effective, and which is unlikely on the face of it to have impressed the king's sceptical opponents. An institution that gradually came to assume importance in this period was the *desembargo do paço*, which accompanied the king, and met in council once a week to review appointments of magistrates, and discuss the political and economic situation; the *desembargo* had had a merely advisory position under D. João II. Another much-discussed institution is the Royal Council itself; and while about sixty members of the Council under D. Manuel (including Vasco da Gama) have been identified so far, no attempt appears to have been made to date to sort them out into varying political tendencies. We are also aware that influence in the court could be exercised in other ways than through the Royal Council. For the record, we may note that under a half of the council had the right to use the prefix *Dom*; the few titled aristocrats amongst these included the powerful *vedor da fazenda* D. Diogo Lobo, the Baron of Alvito, and D. João Coutinho, Conde do Redondo. For the period of D. Manuel as much as D. João III, we have the following general formulation from the authoritative pen of Jean Aubin:

> We should note that the Council was far from a homogenous body. Great seigneurs, hostile to the royal bureaucracy, sat side by side with their critics, the *letrados*, and with *fidalgos* of very different lineages and life-styles, called upon to attend on account of their merits. The position of member of the Council, in any event, was essentially honorary. The King governed with a very small number of counsellors, four or five more often than not, and at times even fewer. Political affairs were dealt with in small committees. When a large issue was under discussion, the King consulted a broader group. Thus it was during the discussions concerning the maintenance or the abandonment of the garrisons in Morocco [in D. João III's reign].[25]

We may imagine that the same would have been the case when the decision to abandon or continue with the presence in the Indian Ocean was discussed under D. Manuel. But there were also other forces in play. It would appear that the first decade of the reign was occupied with a hesitant (at times downright confused) sounding out of the various elements in balance. Thus, D. Manuel sought from his accession to ally himself by marriage to the Catholic Monarchs, by marrying one of their daughters, preferably the oldest one, Dona Isabel, the 'virgin widow', in some contemporary reports, of D. Afonso (d. 1491, the son and heir-apparent of D. João II), but failing that their third daughter, D. Maria. The alliance was opposed by one group in the Portuguese court (including the Marquês de Vila Real and his son D. Fernando, and D. João Manuel), which quite rightly feared the influence of Castile on the wavering D. Manuel, but it was supported by others, notably D. Fernando Mascarenhas, D. Álvaro de Portugal, and also D. Francisco de Almeida (who after his sojourn in Spain, was well regarded by the Catholic Monarchs). In the first half of 1496, it appeared that D. Manuel would have to settle for the younger and less prestigious of the two sisters, but eventually the reluctant and mystically minded D. Isabel was brought around. In turn, she (perhaps acting on her parents' behest) set down a number of political preconditions to D. Manuel, who appears to have been willing to compromise as a way of gaining proximity to the Spanish succession.[26]

The contract of marriage was signed by proxy (through D. Álvaro de Portugal) at the end of November 1496, but it was only

[25] Jean Aubin, 'La noblesse titrée sous D. João III: Inflation ou fermeture?', *Arquivos do Centro Cultural Português* 26 (1989), 418.

[26] The whole question is discussed in detail in Jean Aubin, *D. Manuel*; I am grateful to the author for permitting me access to the relevant sections of his manuscript.

in late September 1497, three months after the departure of Gama's fleet for India, that D. Isabel crossed the frontier. The conjuncture was a curious one, for a few days later, the heir-apparent, the Infante D. Juan of Castile died, leaving the succession open to D. Manuel and his new bride. Thus, for a brief period of eight months, D. Manuel emerged in an unusual role, not merely as King of Portugal, but as heir to the kingdoms of Castile and Aragon. However, with the death of D. Isabel in 1498 in childbirth, this possibility was lost, to the relief of a good many in the Portuguese court.[27] In 1500, D. Manuel then married the other unattached daughter of the rulers of Spain, D. Maria, through whom he had his sons the Infantes D. João (later to be D. João III), D. Luís, D. Fernando and D. Henrique. In this sense, his policies were quite different from those of D. João II, for D. Manuel sought time and again to use his in-laws to shore up his own prestige. This meant in turn that, with each marriage-negotiation, he was susceptible to influence from Spain. It has been much debated whether the decree of expulsion of the Jews and Muslims from Portugal promulgated late in 1496 reflects the influence on D. Manuel of the Catholic Monarchs, or whether he decided upon this *suo motu*. Whatever may be the case, there is indisputable evidence that D. Manuel was under considerable pressure in 1496–7 from his bride D. Isabel, who apparently saw Portugal as a hotbed of heretics and apostates. Besides, it is clear that this decision of D. Manuel could not have displeased his future in-laws; later, too, they used their influence to deflect Portuguese policies in this or that direction, in keeping with their own convenience. Indeed, as a Spanish chronicler was to put it, D. Fernando offered his daughter to D. Manuel in 1500, 'in order to ensure that the Portuguese did not begin to turn their thoughts towards novelties'.[28]

The aid of the Spanish Crown was essential in order for the Portuguese to maintain the oldest of their overseas enterprises, that in North Africa. In D. Manuel's scheme of things, despite the growing importance that the Indian Ocean came to have, Morocco was nevertheless given a role of key significance, as D. Manuel sought to profit from the fragmentation of the power of the Wattasid dynasty to consolidate Portuguese control over

[27] His son by his first wife, D. Miguel, died in infancy, in mid-1500.

[28] Zurita, *Historia del Rey don Hernando el Católico*, quoted in Jean Aubin, 'Le Portugal dans l'Europe des années 1500', in *L'Humanisme Portugais et l'Europe* (Paris: Centre Culturel Portugais, 1984), p. 220, n. 4.

the western Maghreb. A series of projects of fortification were taken up in the early sixteenth century, on sites that looked out on the Atlantic, such as Santa Cruz, Mogador, and Safi, as well as later Azemmour and Mazagon, even though some of these were completed only relatively late in D. Manuel's reign. A key player in these activities was the captain of Safi, Nuno Fernandes de Ataíde, who orchestrated a campaign with indigenous allies, the purpose of which was to attack the inland centre of Marrakesh. In order to be able to manage these activities better, D. Manuel even signed an agreement with a representative of the Catholic Monarchs at Sintra in 1509, to readjust the division of zones of influence and future expansion. The underlying idea was very clearly to attack the southern underbelly of Morocco, with the perspective in the long run of opening up the possibility of a 'western front' on Islamic power centred in Egypt and Palestine.

Here then are the key elements of D. Manuel's crusade against Islam: the salient into North Africa, the fleets into the Indian Ocean, and the hope of a Christian ally (perhaps Prester John) who would attack the Mamluks from the rear. Besides, in 1500, he offered the Pope 8,000 men for an eventual joint Christian force in the Mediterranean, but soon realised that this meant that he cut a ridiculous figure, poorer than even the ruler of Poland. Desirous of being a more major player in European politics, in 1505 he offered as many as 15,000 men. But where was this force to be financed, with so many other commitments in North Africa and the Indian Ocean? For at this time, of the overseas enterprises managed by Portugal, only Mina afforded regular and substantial revenues. Two voyages to the Indian Ocean (those of Gama and Lopo Soares de Albergaria) had afforded substantial returns, but the presence in the Indian Ocean could by no means be counted on to finance itself at this stage.

Partly on this account, and partly on account of the inertial political trajectory that we have noted, the Portuguese Crown began to make moves from 1504 onwards (but which bore concrete fruit in 1505) towards a major internal reform, involving both civil and military government. The keystones of the civil reform that was contemplated were two in number. First, a substantial reorganisation of internal fiscal resources, by redefining the duties and jurisdictions of the *contadores*, by regularising weights and measures, and by strengthening the hands of centrally directed magistrates and inspectors (*juizes de fora, desembargadores* and *corregedores*) over local fiscal and administrative authorities, who derived their position either from autochthonous collectives (*concelhos* and *câmaras*) or from landed aristocrats (*senhorios*). The authority of the *corregedor*, which extended over a district (*comarca*, of which there were twenty-one in Portugal) was extensive; he was to visit each town and village under his jurisdiction every year, investigating misdemeanors and supervising local officials.[29] As Stuart Schwartz puts the matter:

> The presence of the *juíz de fora* and the *corregedor* in the towns and villages of Portugal signaled an attempt by the monarchy to limit the control of local elements of power. One contemporary observer of Portugal noted that it was also the duty of the *corregedor* to 'quiet factions and dissension and to restrain the powerful of the province'. Both the *corregedor* and the *juíz de fora* were mainstays of royal government at the local level.[30]

Second, the middle years of D. Manuel's reign witnessed the imposition of a more or less uniform juridical regime on Portugal, by the regularisation and redefinition of charters (*forais*) of all sorts of local and regional institutions, and the creation of an orderly system of records, revised in order to be rid of inconsistencies of interpretation. The documents thus produced were to be kept in multiple copies, one with the local seigneur, one with a collective representative body (like the *câmara*), and the third in the central record-room (Torre do Tombo). From this dual project outlined above emerge the great works of the Manueline regime, the *Ordenações Manuelinas* (1514), as well as the extensive texts of the so-called *Leitura Nova*, including both new documents concerning the juridical position of notable individuals, families and institutions, and a careful selection of older documentation.[31] For the first time, some form of uniform fiscal order was sought to be imposed on Portugal, which was to that

[29] This is not to say that conflicts of this sort had not existed earlier at all; cf. Humberto Baquero Moreno, 'A presença dos corregedores nos municípios e os conflitos de competências (1332–1459)', *Revista de História* (Oporto) 9 (1989), 77–88.

[30] Stuart B. Schwartz, *Sovereignty and Society in Colonial Brazil: The High Court of Bahia and Its Judges, 1609–1751* (Berkeley: University of California Press, 1973), pp. 5–18, for a useful general discussion (quotation on p. 7). Nevertheless, resistance to royal centralisation continued well into the seventeenth century; cf. António Manuel Hespanha, *As vésperas do Leviathan: Instituições e poder político, Portugal – século XVII* (Coimbra: Almedina, 1994).

[31] Cf. for example, *Ordenações do Senhor Rey D. Manuel*, 5 vols. in 4 (Coimbra, 1797); also the discussion in Henrique da Gama Barros, *História da administração pública em Portugal*, 11 vols. (Lisbon, 1945–54), especially Vol. I.

point a palimpsest of variant institutions, partly as a result of the evolving methods of the Reconquest.

This project, as we may well imagine, had its supporters among the *letrados*, which is to say the caste of university graduates who served as scriveners, record-keepers, and legal experts (and who had risen into prominence after the Cortes of Coimbra in 1385); it also had a large number of detractors among other groups, who feared this ominous and advancing shadow of royal centralisation. Those who were put in charge of this enterprise were men such as the *chanceler-mor* Rui Boto, the *desembargadores* João Façanha and Rui da Grã, and a cavalier of the royal household, and later *fidalgo*, Fernão de Pina (a brother of the chronicler Rui de Pina), who served D. Manuel in his internal reforms for almost the entire duration of his reign.[32] Their task included the sending out of royal representatives to various cities and towns, to brotherhoods, chapels, hospitals and charitable institutions, to gather authenticated copies of the assets and revenues possessed by these. Consider, for example, the task given in July 1516 (somewhat late in this process) to a certain Brás de Ferreira, sent out by Fernão de Pina as D. Manuel's representative to Colares, Chileiros and Arruda; he was required to gather together, in a sitting, the local representatives of the Municipal Chamber (*juizes, vereadores, procurador e o esprivam*), in the presence of an accredited official (such as an *almoxarife* or *mordomo*), and also summon the revenue-farmers (*rendeiros*) both past and present. Besides, some respectable citizens who were aware of revenue affairs were to be invited too, and in their presence enquiries were to be made concerning revenues, but also investigating certain specific complaints that had been received in respect of revenue-collection. These investigations were to be done in all detail: tithes, cattle-taxes, taxes on artisans, all of these were to be considered in depth.[33]

Further research is obviously required before one can precisely delineate how this task of centralisation and regularisation was put into practice, beginning in the Estremadura region and spreading eventually layer by layer, from the largest centres to secondary and tertiary levels, and from the south and centre, to the regions of the distant north-east. But in a sense, we are

[32] For a first approach to these materials and the men behind them, see Maria José Mexia Bigotte Chorão, *Os Forais de D. Manuel, 1496–1520* (Lisbon: Arquivo Nacional da Torre do Tombo, 1990).

[33] AN/TT, Gavetas, XX/11–27, text in ibid., pp. 10–12.

witnessing a particular manifestation of a familiar phenomenon of the early modern world: the attempt by the central state to redefine in a radical fashion its relationship with other, alternative centres of fiscal and political power in the realm. Accompanying this is the creation of imposing visual manifestations of royal power, particularly the huge churches and monasteries in the so-called Manueline style, with a particular iconography (often reflecting the ruler's messianic pretensions), and symbolism.[34] There was a dialectic here of fiscal power and the visual imagination, for the Crown not only intruded into the landscape by its construction activity, it also needed to raise money through the fisc for this very purpose. Despite its small size, the case of Portugal is rendered complex by the layers of political institutions that were involved. The Crown thus came into conflict not only with the large territorial nobility (indeed, though present, this was probably not the primary conflict in D. Manuel's time), but with the military orders (such as the Order of Santiago), and above all with some members of the middling and lower nobility, the town councils, and even with sections of the well-off gentry.

But there was a second dimension to this conflict, a military one. For D. Manuel's reforms extended into that sphere as well, where he wished to implement a vigorous policy of 'modernisation'. In this he was not alone, for even D. Jaime, the Duke of Bragança, with whom he did not see eye to eye on some other matters, saw the need for such a reform. Once more, there was an eastern influence, that of Spain and northern Italy, where a new style of infantry warfare, and salaried professional soldiery armed with pikes, had been perfected in the late fifteenth and early sixteenth centuries. A key figure was *el Gran Capitan*, D. Gonzalo Fernández de Córdoba, who had enabled the armies of the Catholic Monarchs at Naples to become a formidable instrument, more efficient even than the legendary Swiss infantry. In contrast, the primary battlefield experience of the Portuguese in the early sixteenth century was derived from North Africa, where *fidalgos* disembarked with their somewhat disorganised clientele, engaged in cavalry skirmishes and raids, and still conceived warfare in a rather 'medieval' mode. It was a tradition that

[34] Cf. among numerous other references, Ana Maria Alves, *Iconologia do Poder Real no Período Manuelino* (Lisbon, 1985), and Sylvie Deswarte, *Les enluminures de la "Leitura Nova", 1504–1522: Etude sur la culture artistique au Portugal au temps de l'humanisme* (Paris, 1977).

they were to carry over into the Indian Ocean in the early sixteenth century as well.

The attention devoted in the literature on military history to the spread of the use of firearms in this period, and its social and political consequences, has rather obscured the significance of other organisational changes. In Portugal itself, the decision to introduce companies of halberdiers (*halabardeiros*, or more generically *gente da ordenança*) appears to have been taken in 1505, and the formation of these corps finds mention in Cà Masser's valuable relation of 1506, cited above.[35] Instructors for these companies were actively sought among Portuguese who had done service in Naples or elsewhere in Italy; in late 1506 or early 1507, for example, Cristóvão Leitão, who had served with D. Gonzalo de Córdoba as late as 1505 (in the campaign at Pisa, against the Florentines), returned to Portugal with a salary of 200 ducats (according to Cà Masser). By early February 1508, D. Nuno Manuel had been named Captain-General of all the *gente da ordenança* in the service of D. Manuel; later in the same year, in August, the ruler embarked a group of 2,500 infantry to participate in an unsuccessful attack on Azemmour, led by D. João de Meneses, Conde da Tarouca, and in the later engagement at Arzila. Amongst these men were such experienced hands as Leitão himself and Gaspar Vaz; they were later to be joined by other veterans from Italy, such as João Fernandes, Pêro de Morais, João Fidalgo and Rui Gonçalves, the last two eventually going on to serve in Asia.

Despite the lack of success of the 1508 Morocco campaign, the recruitment of experienced former mercenaries and the training of infantry continued. In 1513, a large expeditionary force of 400 vessels and some 18,000 men was sent out under D. Jaime, Duke of Bragança, to push through the capture of Azemmour, and the Duke took every care to have companies of soldiers trained in the new tactics by the veterans of Italy. For once, it seemed that he and D. Manuel saw eye to eye, but it is not advisable to see this congruence of views on military tactics as a larger meeting of minds. D. Jaime (1479–1532) is after all a rather enigmatic character from the history of the Manueline reign. Having fled with his family to Spain at the age of four, when his father was executed by D. João II, he returned only in 1496, and was

[35] My discussion here draws heavily on Jean Aubin, 'Le capitaine Leitão: Un sujet insatisfait de D. João III', *Revista da Universidade de Coimbra* 29 (1983), 87–152.

thereafter named heir-presumptive to D. Manuel, a position he retained till the birth of the latter's sons. In 1502, a political marriage was contracted for him with D. Leonor de Mendoza, daughter of the Spanish Duke of Medina Sidonia; shortly thereafter, he fled rather mysteriously to Rome, and asked the Pope to allow him to become a member of the Order of Piedade. Permission was refused, and he was persuaded to return. Ten years later, he accused his wife of adultery with a young *fidalgo* in his service, and stabbed her to death in Vila Viçosa in November 1512, locking himself into a cistern in his palace thereafter, as a daily form of penitence. The expedition to Azemmour was a more public form of divesting himself of guilt, by which the most important nobleman in Portugal risked his life on a military expedition.[36] However, once this task had been performed successfully, and Azemmour captured in early September 1513, D. Jaime retreated to his home-base in the Alentejo, from where he continued to express scepticism concerning D. Manuel's global programme and pursued a maverick career. In January 1517, for example, he obtained a Papal Bull from Leo X, permitting him to transform fifteen churches under his jurisdiction into *comendas* of the Order of Christ and give them to cavaliers as he pleased, without reference to the Portuguese Crown. And as late as 1529, he still expressed unhappiness to D. João III concerning some of the policies that the latter's father had followed overseas.[37]

If the Duke reformed the soldiery in his own employ, training them 'in the Swiss style' in 1513, his example was scarcely followed by others. Whether on account of sabotage or incompetence, other commanders in Morocco refused to use the pikemen appropriately. Recruits too resisted the discipline that was required, and often deserted at the first opportunity. Since the programme required drilling, and further required men used to being on horseback to mix with their social inferiors on foot, this was objected to by members of the lower nobility (*escudeiros* and *infanções*), as we see from the widespread unrest that attended the attempt to revive the system of *ordenança* in 1526–7. Rather like the Mamluks of the military caste described by David Ayalon in his celebrated monograph, the Portuguese nobility resisted the social demands of the new technology, and resented the Crown

[36] Cf. Anselmo Braamcamp Freire, *Brasões da Sala de Sintra*, 3 vols., intro. Luís Bivar Guerra (Lisbon: Imprensa Nacional, 1973), Vol. III, pp. 63–6, 343–4.

[37] Letter dated Vila Viçosa, 12 February 1529, AN/TT, Gavetas, XVIII/10–10, in Silva Rego (ed.), *Gavetas*, Vol. IX, pp. 536–40.

that wished to impose it.[38] After 1515, as D. Manuel's position weakened, he was hence obliged to give up this military reform, and even dissolved his company of halberdiers.

But while the great wave of Manueline reform lasted, in the years from roughly 1505 to 1514, Portugal was projected with force on the European stage, as the returning fleets from India continued to bring in substantial (if not astonishing) return cargoes. Venice was now distinctly shaken, with its commercial position in the East threatened by Portugal, and its relations with Egypt under a shadow on account of the erratic comportment of Sultan Qansawh al-Ghawri (r. 1501–16). Besides, a major change was emerging in the politics of West Asia and North Africa, a shift in the political conjuncture which the Portuguese Crown was quick to seize. Besides the usual players, the Ottoman Sultan Bayezid II and the Mamluks, in a situation of uneasy equilibrium at the turn of the sixteenth century, a disturbing new force was to be discerned, namely the Shi[c]i Safavid polity that Shah Isma[c]il had established, with dramatic military successes after 1500, in first Azerbaijan, then in eastern Iraq and the province of Fars. As Europe grappled with the mystery of the Great Sofi, as Shah Isma[c]il was usually termed, D. Manuel made use of the unease of the Mamluks to mount pressure on them through Rhodes, and the Order of St John of Jerusalem. By 1507, in the face of these multiple threats, the Mamluks had entered a temporary alliance with the Ottomans, who supplied them with war-materials for the readying of vessels in Alexandria and Suez, to be used both in the Mediterranean and against the Portuguese in the Indian Ocean.

That the Mamluks did not react immediately to the Portuguese arrival in the Indian Ocean may seem surprising; but it can probably be explained by the succession struggle that emerged in Egypt at the end of the long reign of Sultan Qa'it Bay (r. 1468–96). A series of four short-lived Sultans followed, before the emergence in 1501 of Qansawh al-Ghawri. Indeed, one of the early Egyptian reactions to the Portuguese in the Indian Ocean is that of that celebrated but rather obtuse chronicler, Ibn Iyas, who states (probably literally, but perhaps metaphorically) that the Portuguese had broken down the wall (*sudd*) that Alexander the Great had built in ancient times, in order to circumnavigate

[38] David Ayalon, *Gunpowder and Firearms in the Mamluk Kingdom: A Challenge to a Medieval Society* (2nd edn London, 1978); also Geoffrey Parker, *The Military Revolution: Military Innovation and the Rise of the West, 1500–1800* (Cambridge University Press, 1987).

Africa.[39] This statement can be read as flattering to Portuguese pretensions (Duarte Galvão himself, we have seen, was not averse to the Alexandrine parallel), but it must also be recalled that, in Islamic lore at least, Alexander built the wall to protect civilisation from the depredations of Gog and Magog. At any rate, though slow off the mark, the Mamluk state did stir in time. Besides, it would appear that already in 1505 rumours had begun circulating of an imminent Portuguese attack on Jiddah. By November of that year, Amir Husain Mushrif al-Kurdi was hence sent out from Cairo to the Red Sea, with the task in the first instance of fortifying the port that gave access to Mecca. Thereafter, he set out with a fleet, probably in August or September 1507, his first destination being the Gujarat port of Diu; Hadrami sources speak of Amir Husain as a Holy Warrior (*mujâhid fî sabîl Allâh*), who had set out 'to engage the Franks who had appeared in the Ocean and cut the Muslims' trade-routes'.[40] The Egyptian fleet, reputedly comprising some twelve ships and 1500 men-at-arms, had an initial and rather dramatic success against a Portuguese fleet in Chaul in March 1508, led by D. Lourenço de Almeida, the viceroy's son, who was killed in the engagement. The news was brought back to Cairo late that year, and the Sultan reputedly had drums of victory beaten in the city for three days in a row, in particular since it was reported that Amir Husain had taken a 'considerable booty'.[41] But Amir Husain also called for reinforcements, apparently mistrusting his Indian allies, in particular the governor of Diu, Malik Ayaz who – fearing the larger intentions of the Egyptians – had meanwhile opened negotiations with the Portuguese viceroy. But before Cairo could act, a Portuguese fleet led by the viceroy himself left Cannanur for Diu, via the Konkan port of Dabhol. Abandoned by his erstwhile Gujarati allies, Amir Husain saw his fleet destroyed on 3 February 1509, and eventually returned to Cairo after many tribulations, only as late as December 1512.[42] In late 1509, Sultan Qansawh al-Ghawri,

[39] Cf. Serjeant, *The Portuguese off the South Arabian Coast*, p. 25n. This statement is typical of the common millenarian vocabulary of the epoch, used by Muslims and Christians alike. It recurs, interestingly, in a modern poem by Jorge Luis Borges ('Los Borges'), where his Portuguese forbears are described as the people who 'broke down the wall of the Orient'.

[40] Serjeant, *The Portuguese off the South Arabian Coast*, p. 44, quotations from the *Târîkh Shanbal*, and the *Târîkh al-Shihrî*.

[41] Cf. Gaston Wiet, trans., *Journal d'un bourgeois du Caire, Chronique d'Ibn Iyâs*, 2 vols. (Paris: Armand Colin, 1955–60), Vol. I, p. 138.

[42] The best discussion of this episode is in Jean Aubin, 'Albuquerque et les négociations de Cambaye', *Mare Luso-Indicum* 1 (1971), 12–19.

incensed at this reversal, wrote letters to Diu, threatening Malik Ayaz with reprisals for his treachery; ironically, in mid-1510, ambassadors arrived once more at Cairo from Sultan Mahmud of Gujarat and some other unspecified Indian rulers, asking again for Egyptian aid against 'the European pirates . . . who had inflicted a defeat on Husain, the commander of the Egyptian expedition'.[43]

The Sultan may have been tempted, for he now had at his disposal a secret trump card, in the form of a Portuguese turncoat, Álvaro Vaz da Fonseca. Fonseca had served several years in Cochin, as scrivener of the factory, and had even participated in the defence of Cochin in 1504 against Calicut forces.[44] Thereafter, he had been rewarded with the important post of factor of the Portuguese factory at Antwerp in 1506, but two years later, absconded with the factory's liquid resources and – after living it up for a time with a courtesan in Rome – , eventually appeared in Alexandria at the same time that the ambassadors from Gujarat had their audience in Cairo. The Sultan appears even to have toyed with the idea of sending the experienced and well-informed Fonseca out at the head of a fleet into the Indian Ocean; after all, another of his high officials, the *tarjumân* Taghribirdi, was a renegade from Valencia.[45] But nothing eventually came of the matter. An Egyptian fleet, which the Sultan was having constructed in the Gulf of Ayas, was partly destroyed and partly captured in late August 1510 by the Knights of St John from Rhodes, led by D. Manuel's agent Frei André do Amaral. With the expansion of the Spaniards in the Maghreb to add to his difficulties, Qansawh al-Ghawri (d. 1516) was to remain largely on the defensive for the rest of his reign. Instead of sending fleets into the Indian Ocean, he would see the Portuguese capture Hurmuz and Socotora, enter the Red Sea on an expedition, pose a threat to Aden, and even to the very Hijaz.

GAMA VERSUS ALBUQUERQUE?

If D. Francisco de Almeida's arrival in India in 1505 marks the formal beginnings of a continuous Portuguese government in

[43] Wiet, *Journal d'un bourgeois*, Vol. I, pp. 176–7.

[44] Letter from Álvaro Vaz to D. Manuel, dated 24 December 1504, AN/TT, Gavetas, XV/2–36, in Silva Rego (ed.), *Gavetas*, Vol. IV, pp. 132–40.

[45] For a discussion, see Jean Aubin, 'La crise égyptienne de 1510–1512: Venise, Louis XII et le Sultan', *Moyen Orient et Océan Indien* 6 (1989), 123–5. Also see the anonymous letter from a Florentine merchant to António Carneiro, dated 10 August (1510), AN/TT, Gavetas, XV/19–4, in Silva Rego (ed.), *Gavetas*, Vol. V, pp. 240–1.

India, the years from 1509 saw the true creation of a Portuguese maritime empire in the Indian Ocean, once the governorship came into the hands of Afonso de Albuquerque. As we have had occasion to mention, Gama and Albuquerque represent two poles in the Portuguese nationalist historiography.[46] There is, of course, a whole panoply of 'heroes' for nationalists to play with in these early decades of the sixteenth century, from the mysterious Cabral, to Duarte Pacheco Pereira, to D. Francisco de Almeida himself, but Portuguese romantic nationalism has nevertheless devoted the greatest attention to Gama and Albuquerque, posing the two as visionaries each in a different mould, struggling to rise above the petty machinations of their contemporaries, and burdened by the erratic and even downright mean policies of D. Manuel. Now, it is abundantly clear from recent historiography that Albuquerque represented a particular tendency in Portuguese politics, and a rather special imperial vision. Close in spirit to ideologues like Duarte Galvão, and supported by able courtiers like the *vedor da fazenda* D. Martinho de Castelo-Branco (named Count of Vila Nova de Portimão in 1514), Albuquerque represented a form of royal centralism in the Indian Ocean, allied to the universalistic and messianic ideals that motivated D. Manuel and Galvão. In this view of the world, the blockade of Egypt was a pressing priority, and the military alliance with Prester John of Ethiopia its logical prerequisite. Thus, the Sultanate of Babylonia would be destroyed, and Mecca along with it, if possible. At the same time, the royal spice trade had to be developed, be it in western India, or in Melaka; Crown ships were to be extensively deployed everywhere, rather than ships owned in large measure by wealthy trading nobles. Fortresses were to be built at a good number of key points, protecting the Crown's goods, and ideally to be manned by a professional soldiery (in the 'Swiss style', as it was termed in the epoch), rather than by the clients of individual *fidalgos*. Above all, in the Indian Ocean, authority was to be centralised in the governor and a few chosen subordinates around him (such as, in Albuquerque's case, the Florentine Francesco Corbinelli), rather than divided out amongst a number of Crown factors, each of whom corresponded independently with the Crown in Portugal, and concomitantly claimed a largely autonomous domain of action. Disdaining these men, Albuquerque

[46] For a recent biography of Albuquerque, which struggles without great success to escape the burden of Portuguese nationalism, see Geneviève Bouchon, *Albuquerque, le lion des mers d'Asie* (Paris: Editions Desjonquères, 1992).

Plate 21 Inscription with the arms of D. Diogo Lobo da Silveira, Baron of Alvito, *vedor da fazenda*, and the great opponent of Manueline imperial policy.

trusted the advice of his Florentine allies (though by no means all Florentines!), converted Jews such as Francisco de Albuquerque and Alexandre de Ataíde, and even local corsairs like Timoji.

It was inevitable that this vast, and rather ambitious, conception would find opponents within the ranks of the Portuguese in Asia. To pragmatists, the messianic flavour of the enterprise to destroy Egypt and recover Jerusalem (as well as the alliance with Prester John) made little sense. The self-interest of the independent Crown factors was in turn directly threatened by the new centralised conception of decision-making. As for the question of Crown trade versus private trade, here Albuquerque found himself at odds not only with the trade-oriented (and corsair) *fidalgos*, but with Florentines like Girolamo Sernigi, whose ship Albuquerque confiscated in 1510, while it was *en route* to Melaka.

Opponents to Albuquerque were to be found not only in Asia

but in Portugal. Leading them was D. Diogo Lobo, the Baron of Alvito, and a powerful – if relatively low-ranking – titled noble from the Alentejo. It is interesting to remark the congruence of views that emerges between the Baron and his Alentejan neighbour, D. Jaime, Duke of Bragança, and which eventually extends to another resident of the region, Vasco da Gama, expelled from Sines and installed now in first Évora and then Niza. The Admiral for his part continued, even in this hour of distress, to keep an active interest in India affairs. In April 1511, his brother Dom Aires da Gama went out to Asia as captain of the *nau Santa Maria da Piedade*, notionally a part of the fleet commanded by Dom Garcia de Noronha.[47] In point of fact, however, Dom Aires's ship and another one (the *Belém*, commanded by Cristóvão de Brito) left some days after the main body of the fleet, but managed to make a far more rapid voyage, arriving on the Indian west coast by late 1511 (while the rest of the fleet missed the monsoon at Mozambique and failed to cross the Arabian Sea). Gama and Brito appear to have been closely allied, and once in India, proceeded to take advantage of the absence of the governor Afonso de Albuquerque (at that time engaged in the conquest and subsequent pacification of Melaka) to create serious difficulties in Cannanur for its captain Diogo Correia. Now, Correia had been named captain of Cannanur by Albuquerque on his own initiative, in March 1511, despite the fact that he had no royal grant-letter to this effect. He was put there in place of a certain Manuel da Cunha, very briefly captain of the fort since December 1510, and a close ally of the 'Cochin coterie' (comprising men like Lourenço Moreno, Diogo Pereira, Gaspar Pereira and António Real), the sworn enemies of Albuquerque and his plans. Real, Moreno, Pereira and their allies naturally sought by all possible means to discredit Correia, and in this it appears that they were aided by D. Aires da Gama in his brief stay in India. According to Albuquerque in one of his letters to Dom Manuel, Gama had used his own position and his family's prestige to discredit the governor, and in his absence rounded up a good number of Portuguese who were reluctant to aid Albuquerque, and shipped them back with him to Portugal in 1512, in view of the governor's continued absence in Melaka. He and Cristóvão de Brito also

[47] AN/TT, CC II–25–5, published in H. Lopes de Mendonça (ed.), *Cartas de Afonso de Albuquerque*, Vol. VI, pp. 415–16, *mandado* dated February 1511, concerning supplies for Dom Aires's ship, on its way to India. For basic details of the voyage, see the account in Barros, *Da Ásia*, Década II, Livro 6, Ch. 10, pp. 148–51.

apparently carried back letters of complaint from Albuquerque's opponents to the court, obliging Albuquerque later to defend himself against attacks from Portugal. Also of passing interest is the fact that Gama and Brito made it a point on their return voyage to stop *en route* in south-east Africa, to pay their respects – significantly enough – at the spot where D. Francisco de Almeida had been killed in early March 1510, in a disastrous and, from a Portuguese viewpoint, altogether humiliating sortie against local black tribesmen, while on his voyage home.[48] More curiously, it is claimed by Albuquerque that rumours were spread by D. Aires to the effect that his own position was precarious (we may note that in 1512, in point of fact, his first three-year term as governor was to end), and that a successor was contemplated in the court. Diogo Correia, it is stated by Albuquerque, was contemptuously accosted one day by the minister of the Kolattiri, the *wazîr* (*alguozil*) of Cannanur, and told 'that he was no captain, and nor could I [Albuquerque] name a captain, and that Dom Aires had said that the Admiral [Vasco da Gama] was due to come that year, and that I would have to leave'.[49] This persistent rumour is referred to by Albuquerque elsewhere as well, eventually being displaced in 1514 by the rumour that it was Tristão da Cunha who would come to India as Albuquerque's successor.

It is interesting to note that the two best-known Portuguese texts of the period describing the trading networks in Asia, the *Suma Oriental* of Tomé Pires, and the *Livro* of Duarte Barbosa, are authored by men on the opposite sides of the Albuquerque–Gama divide.[50] Pires, probably born about 1465 in Lisbon or Leiria, reached India only in 1511 (coincidentally, with D. Aires da Gama and Cristóvão de Brito), when Albuquerque was already governor. He was an apothecary, and had already served the Infante D. Afonso before the latter's death in 1491; in India, where he was initially named *feitor das drogarias* in Cannanur, he was very quickly to make a minor fortune, perhaps selling medicines. He seems to have been very well regarded by Albuquerque, despite his tendency to enrich himself, and by late 1512 had been sent to Melaka to assist the captain and factor there, Rui

[48] AN/TT, CC, I–14–12, dated 3 December 1513, in Bulhão Pato (ed.), *Cartas de Afonso de Albuquerque*, Vol. I, pp. 181–2.

[49] AN/TT, CC, I–14–11, letter from Cannanur dated 2 December 1513, in ibid., Vol. I, pp. 176–7.

[50] For an earlier analysis along these lines, see Luís Filipe Barreto, *Descobrimentos e Renascimento: Formas de ser e pensar nos séculos XV e XVI*, 2nd edn. (Lisbon: Imprensa Nacional, 1983), pp. 143–68.

de Brito Patalim and Rui de Araújo. He remained in and around Melaka to 1515, the year in which he appears more or less to have completed the *Suma Oriental*; he did however make a trading voyage as factor on a Crown fleet to Java in 1513. In 1516, Pires was designated leader of a disastrous embassy to Ming China, and ended his life there, probably in 1523 or soon after, when he and his companions were sentenced to death by the Chinese authorities, on account of the serious misdemeanors of some of their compatriots.[51]

The prologue of the *Suma Oriental*, dedicated to D. Manuel, makes it clear enough where Tomé Pires's political sympathies lie, for it celebrates the fact that the bellicose Portuguese flag flies high, so far and wide in the lands and waters of Asia. As for Asian rulers, 'those who are vassals live peacefully, and those who are rebels are fearful and tormented'. He continues:

> And all of this is caused by the great power that Your Highness has, exercised and managed in war by the most magnificent and great cavalier Afonsso Dalboquerque, your captain-general, spirited, astute, capable in war, and in human terms most prudent, who continuously and with such effort, now in upper India now in Arabia, and in between, ceases not to make war on the name of Mafamede.

All the expenses ('which no other Christian king has ever had') incurred in war by the Portuguese state in India are thus justified and exalted by Pires, in the name of the crusade that the Holy Catholic Faith must carry out against the false and diabolical Moorish religion of 'the most ignominous and false Mafamede'.

In contrast, we have Barbosa, who had arrived in Kerala as early as 1500 along with his uncle Gonçalo Gil Barbosa, and remained there for nearly five decades, with only brief absences; he died at Cannanur, probably between September 1546 and May 1547.[52] If Pires, with the mentality of a botanist and classifier, in his *Suma Oriental* is primarily concerned with the idea of the mercantile network, and the manner of controlling intra-Asian trade through the mastery of key points, Barbosa is far more interested in a type of global anthropological observation, re-

[51] For a comprehensive biographical sketch, see Armando Cortesão, 'Introdução', in *A Suma Oriental de Tomé Pires e o Livro de Francisco Rodrigues* (Coimbra: Universidade de Coimbra 1978), pp. 10–65. However, Cortesão probably exaggerates the length of Pires's life after his imprisonment in China.

[52] Cf. Georg Schurhammer, 'Doppelgänger in Portugiesisch-Asien', in Schurhammer, *Orientalia*, pp. 121–3.

sulting from his detailed knowledge in particular of Kerala society. Along with his uncle, he was at first based at Cochin (where he returned as late as 1503, at the time of Francisco de Albuquerque's sojourn there), but then moved on as interpreter (*língua*) to Cannanur. In 1506, both Barbosas returned to Portugal, in keeping with royal orders, and Gonçalo Gil was actually rewarded with the status of *fidalgo*, and made a *comendador* in the Order of Christ. In 1511, Duarte Barbosa returned to India, to serve still in the relatively modest post of scrivener of the factory at Cannanur; he was on the ship of D. Aires da Gama, with whom he appears to have maintained rather close relations. Barbosa's relations with Albuquerque very rapidly deteriorated, as his association with the other opponents of the governor became clear. In late 1511, he was associated with D. Aires (and the Cannanur factor, Gonçalo Mendes) in trying to undermine the position of Albuquerque's appointee as captain of that port, Diogo Correia.

Barbosa's position, and conception of what the Portuguese enterprise in Asia ought to be is made clear in a letter addressed by him to D. Manuel in early 1513.[53] Barbosa opens the letter by noting that he knows Kerala very well indeed on account of his long stay earlier, and his familiarity with the language; besides, he has an intimate acquaintance with the history of dealings with local rulers, 'the trade and customs, and conditions decided on by the Admiral, and confirmed by other Captains-Major who have been there till now, and which we have to guard [from respect] for the king and people of the land'. Constituting an explicit attack on Albuquerque and his policies, the letter then goes on to argue that the factories set up between 1502 and 1506 should be preserved, but that everything else (the fortresses created by Albuquerque in Goa and Melaka, the project to set up a new factory at Calicut) should be abandoned. Barbosa identifies himself here quite explicitly as a spokesman for local interests, suggesting that he was closely associated with certain Mappila merchants of Cannanur. But Barbosa's opposition to Albuquerque went beyond mere words, for he was also substantially implicated in the curious episode by which the Ethiopian ambassador to D. Manuel, Abraham (or Mateus, as he was also known), was apprehended on charges of being a Muslim, a pederast, and a

[53] Letter from Duarte Barbosa, dated Cannanur, 12 January 1513, AN/TT, CC, I–12–56, in Bulhão Pato (ed.), *Cartas de Afonso de Albuquerque*, Vol. III, pp. 48–51; also see a later letter by the same, dated Cannanur, 15 January 1527, AN/TT, CC, I–37–76.

Mamluk agent, in January 1513. In 1507, after several failed attempts, the Portuguese Crown had finally sent three emissaries to the Ethiopian court, including a Tunisian Muslim, a Morisco converted to Christianity, and a Portuguese. Disembarking in April 1508, near Cape Gardafui, these men made their way (the Morisco separately, the other two together) to the Ethiopian court in circumstances that remain obscure; we are aware however, that in late July 1508, the ruler Na'od had just died, and the political situation was fluid. The regent, the aging convert from Islam, Queen Eleni, showed some interest in pursuing an alliance with the Portuguese, in order to push back Mamluk control over the Red Sea.[54] However, the vicissitudes of the embassy (with Abraham-Mateus being under a cloud of suspicion from 1513 right until his death, in 1520), meant that this proposed alliance could not bear fruit during D. Manuel's reign; it is of course another matter whether the rulers of Ethiopia would really have been able to make a significant difference to the balance of power in the Middle East. By his words and actions, Barbosa (like Gaspar Pereira, António Real, Lourenço Moreno, and other opponents of Albuquerque) made it amply clear that he was opposed to the whole crusading project of D. Manuel, to building an alliance with Ethiopia, and to destroying the Mamluks (rather than skimming off a part of their trade, through raids on ships bound for the Red Sea). Eventually exasperated with his attitude, Albuquerque took a series of measures against Barbosa; in late November 1513, he wrote to D. Manuel that he was removing him from Cannanur, 'because he is the interpreter (*lymgua*), and the cause of all these revolts'.[55] Sent instead to Calicut, where Albuquerque succeeded in installing a factory and fortress (after instigating the assassination of the Samudri, by his brother in early 1513), Barbosa appears quickly to have ingratiated himself with the ruler there.[56] When, in May 1514, Albuquerque ordered that Barbosa be sent in irons to Cochin (no doubt for some

[54] For a detailed discussion on this embassy, see Jean Aubin, 'L'ambassade du Prêtre Jean à D. Manuel', *Mare Luso-Indicum* 3 (1976), 1–56.

[55] AN/TT, CC, I-13-112, letter dated Cannanur, 30 November 1513, in Bulhão Pato (ed.), *Cartas de Afonso de Albuquerque*, Vol. I, p. 134.

[56] The early history of the Calicut factory, which was eventually wound up in 1525, would repay more detailed study; cf. AN/TT, Núcleo Antigo, No. 755 (56 ff.), comprising the 'Livro da receita e despesa' of supplies in the factory, held intermittently between March 1514 and October 1515 (and mistakenly catalogued to 1504). For similar materials on another Portuguese fortress, compare Jean Aubin, 'Ormuz au jour le jour à travers un registre de Luís Figueira, 1516–1518', *Arquivos do Centro Cultural Português* 32 (1993), 15–42.

further misdemeanours), he was actually protected by the Calicut ruler. Thus, Barbosa weathered out the period of Albuquerque's government, and continued in later years to build a career (and perhaps even a fortune) in Kerala.

If the adherence of Barbosa to Vasco da Gama, as a sort of model and an alternative to Albuquerque, was based in part on personal ties (through his uncle or otherwise), which are unknown to us, it was equally based, quite obviously, on a widely held view in the epoch of Gama's stand on the essentials of Portuguese expansion in Asia. It has been noted often enough in recent times that Albuquerque's opponents have received a consistently bad press, in view of the hero-centred world-view of Portuguese historiography. Men like Gaspar Pereira (secretary of the Portuguese government in Asia), or Diogo Pereira, have thus been presented often enough as mean-spirited intriguers, incapable of comprehending the grandeur of the conception of the Terrible Afonso. In questioning this view, however, we should not fall into another trap, that of seeing men like Barbosa, the Pereiras, or Moreno – and *a fortiori* Vasco da Gama – as representing the spirit of 'free trade' against a heavy-handed mercantilism, or as defending the rights of native rulers and merchant-princes (the Raja of Cochin, Mammali Marikkar of Cannanur) against the Portuguese state for high-minded reasons. The tensions were rather more complex, and the motives more sordid. What the 'Cochin coterie' wanted was a privileged position for their own class, the middling Portuguese nobility, both in trade in the Indian Ocean, and in prize-taking. To this extent, their conception was far more limited, and in a manner of speaking rather more 'pragmatic' than that of Albuquerque and D. Manuel, who wished to develop Crown trade on as many routes as possible, and to build a very extensive Asian empire based on an universalised Christianity. But violence was as much a part of the first strategy as the second, the principal distinction being between the form, and the level of systematisation, of the violence involved. The position of Albuquerque's opponents was eventually seemingly vindicated, with his fall from grace in 1515, and his substitution as governor by Lopo Soares de Albergaria, cousin and nominee of the Baron of Alvito. With him, the Cochin coterie too rode the crest to positions of renewed power and influence. Diogo Pereira, whom Albuquerque had sent back in disgrace to Portugal, returned as secretary to Lopo Soares (and went on rapidly to build up a private trading fleet), while

Lourenço Moreno was even briefly named to the post of captain of Cochin by Lopo Soares, in place of Aires da Silva. Silva, in turn, wrote a long and violent letter from Melaka in August 1518, denouncing the governor, and stating that he 'wishes me much harm, and if he could destroy me he would'; he also suggested that Lopo Soares had a good deal of private trade at Pulicat on the Coromandel coast, perhaps through the Florentine Piero di Andrea Strozzi.[57] From this and other letters, it is clear that in Lopo Soares's time, Crown shipping was pulled back to an extent; already, in the fleet that accompanied Lopo Soares to India, several were privately equipped, and even had royal licences 'to go trading in India wherever they might wish, carrying goods from one part to the other'.[58] Also, attempts at territorial aggrandisement ceased for a time.

There are two exceptions to this latter rule for the period of Albergaria's government. The first is the decision, taken in the last year of his stay (and implemented in September 1518), to build the long-delayed fortress at Colombo; but this can be explained easily enough, for in view of the growing hostilities between the Portuguese and the Mappilas, the need for a fortress even to protect the private trading interests of the Cochin-based Portuguese (who were interested in cinnamon from Sri Lanka) had become pressing. The second exception proves the rule: this is the case of Lopo Soares's bizarre expedition to the Red Sea in 1517. Organised at a singularly auspicious time for the Portuguese, when the Ottoman Sultan Selim I had just finished his conquest of Mamluk Egypt (and the Red Sea region was especially vulnerable), this expedition bore no fruit whatsoever. The huge fleet, carrying some 1,900 Portuguese, 600 'malabares', and between 600 and 1,000 mariners, left Goa in early February 1517, and arrived before Aden on 14 March 1517. Here, to their surprise, the Portuguese were welcomed, and given water and supplies, as well as pilots to help them navigate as far as Jiddah; for unbeknownst to them, the rulers of Aden had only a few months earlier beaten off a Mamluk attack, led by Amir Husain (who was

[57] Unsigned letter from Melaka, 15 August 1518, AN/TT, Gavetas, XV/21–16, in Silva Rego (ed.), *Gavetas*, Vol. V, pp. 472–86; the author's identity is clear, though, from a reading of the text. The letters has gaps, and the passage on Coromandel states that Lopo Soares has 'cargos e tratos que tem em Paleacate per mão de hum . . . estrangeiro'.

[58] AN/TT, CC, I–17–66, letter from Tristão da Cunha in Lisbon to D. Manuel, 31 January 1515, published in Banha de Andrade, *História de um fidalgo quinhentista*, pp. 219–21; for a discussion, see ibid., pp. 46–8. Cunha himself sent a ship (one of four such), probably the *Santo António*.

executed shortly thereafter) and Salman Ra'is, who though in Mamluk service was in fact a *Rûmî*, by the definitions of the time. Lopo Soares however turned down the chance to set up an establishment at Aden, if we are to believe a series of contemporary accounts. Thus, in the version of the German merchant, Lazarus Nürnberger, in Asia at the time:

> It is believed, the first time Lopis Suaris came before Aden, when those of Aden asked if he wanted to attack the town, or what his intention was, that if he had said yes, that he wished to have the town, or that they should hand over a tribute to the king of Portugal, they would readily have agreed to it, for they were in great fear of the Rumis (*Romern*), and they would have preferred to submit to the Portuguese rather than to the Rumis, because the Rumis treat the Moors [Arabs] very badly.[59]

Instead, Lopo Soares sailed off to Jiddah, and after weathering a storm, anchored off the port in mid-April. The Portuguese captains consulted, and finding the port strongly defended, decided to abandon the effort, instead choosing to attach softer targets at Kamaran and Zeila. By the time they left the Red Sea, the governor at Aden, ᶜAmir Marjan, was no longer interested in treating with them, and the fleet thus made for Hurmuz. After a storm near Cape Gardafui which wreaked havoc on the fleet, Lopo Soares arrived in Hurmuz by the beginning of November 1517, and made his way back to Cochin at last by 8 December. Between losses and desertions, the formidable force that he had taken with him had been very substantially depleted.[60] Thus, while making a show of following instructions from Portugal that he was to take stern action in respect of the Red Sea, Lopo Soares in fact insidiously helped sabotage the whole process envisaged by D. Manuel. A sense of this is conveyed even by Lazarus Nürnberger, who prefigures Gaspar Correia's criticism of Lopo Soares's style of government by some decades. The German merchant states baldly that 'at this time, there is a very poor government in India under the Portuguese (*ser bosz regiment in India untter den portugalesern ist*)', with badly paid soldiers, who are hence

[59] Anne Kroell, 'Le voyage de Lazarus Nürnberger en Inde (1517–1518)', *Bulletin des Etudes Portugaises et Brésiliennes* 41 (1980), 68.

[60] For an account, see Jean-Louis Bacqué-Grammont and Anne Kroell, *Mamlouks, Ottomans et Portugais en Mer Rouge: L'Affaire de Djedda en 1517* (Cairo: Annales Islamologiques, Supplément, 1988), pp. 21–42. For the encounters at Aden, see the letter from Dinis Fernandes de Melo at Cochin to D. Manuel, dated 2 January 1518, in ibid., pp. 61–3. Compare Serjeant, *The Portuguese off the South Arabian Coast*, pp. 50–1, for the account of Ba Faqih.

constantly deserting to the enemy. He adds that 'if the Rumis arrive in India before the arrival of the ships from Portugal, the greater part of the Portuguese will give themselves up to the Rumis'.[61]

D. MANUEL IN RETREAT

The government of Lopo Soares has been described as a phase of 'decompression' after the strict regime that Albuquerque attempted to maintain. Aided by his clan, notably his nephews D. João da Silveira, and D. Aleixo de Meneses, the latter highly active throughout the governor's triennium as *capitão-mor do mar*, Lopo Soares sought to settle scores with those suspected of being sympathetic to Albuquerque. Matters in Asia reflected in good measure the situation in the metropolis. Between late 1514 and 1518, the political position of D. Manuel touched a low within the Portuguese court. We can discern this fact from a number of indices, but to explain it in fact proves rather more difficult. The rapidity of Albuquerque's conquests, combined with the complaints that came in the form of letters every year from India, raised questions concerning the nature of the Asian empire that was being created; the degree of autonomy that the governor (who was accused by some of wishing to become 'Duke of Goa') enjoyed was particularly viewed with hostility. The failure of the huge Portuguese maritime expedition to Mamora in North Africa, in 1515, in which over 4,000 Portuguese were killed, could only have done further damage to the ruler's prestige, while opposition to his policies of internal centralisation has already been noted. The most obvious sign of D. Manuel's fragility was of course the decision to displace Albuquerque, taken in late 1514, against the wishes of the king himself, of the queen D. Maria, and of the aging ideologue of messianic Christianity, Duarte Galvão. Instead, Galvão himself was sent out in early April 1515, in the fleet that carried Lopo Soares to India, in order to accompany the Ethiopian ambassador Abraham-Mateus, back to his country in a return embassy. The aging Galvão by now proved irascible and unstable in his judgements, quarrelling bitterly with the Ethiopian envoy, and allowing himself to be influenced by the critics of Albuquerque. He even wrote to D. Manuel in early January 1516,

[61] Kroell, 'Le voyage de Lazarus Nürnberger en Inde', p. 69.

with a thinly veiled attack on the recently dead Albuquerque, accusing him in effect of having harboured excessive ambitions.[62]

Things had obviously come a long way from early 1514, when D. Manuel had sent a hugely expensive embassy to the Medici Pope Leo X in Rome, to show off his oriental riches, and to make a point of the power that Albuquerque's campaigns had allowed him to accumulate. This embassy, headed ironically enough by Tristão da Cunha (and which counted as its secretary Garcia de Resende), made its way via Alicante and Majorca, arriving at the outskirts of Rome by mid-February 1514. Somewhat slowed down by the stately progress of the *pièce-de-resistance* of the embassy, an elephant, Cunha and his companions were able to enter Rome only on 12 March 1514, many of them in fancy dress, in an 'Indian' style.[63] Besides the elephant, which obviously made a great impression, the embassy comprised forty-two other beasts (including a cheetah from Hurmuz), and all of one hundred and forty persons, which added up to a quite considerable expense. By April, Cunha had begun to run out of money, forcing him to borrow locally, and reminding him of his friend, 'the Baron [of Alvito], who told me in Lisbon that the embassy to Rome was not good, because a lot of money was being spent on it, as he well knew'.[64] Nevertheless, the exercise was gone through with to the finish; on 29 April, the Pope signed the Bull that the Portuguese sought. The Pope also sent back lavish gifts, to which D. Manuel responded in 1515 by sending him a ship laden with spices, and later a rhinoceros, offered to the Portuguese monarch in 1514 as a gift by Sultan Muzaffar of Gujarat. Appropriately enough, the ship carrying the rhinoceros, which left Lisbon in October 1515, went down off Genoa in early February 1516, rather accurately reflecting D. Manuel's own situation at that time.

Problems of a political, financial and diplomatic character, all were beginning to weigh heavily on the Portuguese Crown. A troubling issue was the recrudescence of the rivalry with Spain, which had been held in check since the 1490s by D. Manuel's successive marriages (his second wife, D. Maria, dying in 1517). The death of Ferdinand of Castile in January 1516 meant that the Habsburg prince, Charles V, became co-ruler of Castile and

[62] Letter cited in Jean Aubin, 'Duarte Galvão', *Arquivos do Centro Cultural Português* 9 (1975), 84–5.

[63] Banha de Andrade, *História de um fidalgo quinhentista*, pp. 119–31; also see Salvatore de Ciutiis, *Une ambassade portugaise à Rome au XVIe siècle* (Naples, 1899).

[64] AN/TT, CC, II-266-60, Tristão da Cunha at Rome to António Carneiro, dated 11 April 1514, in Banha de Andrade, *História de um fidalgo quinhentista*, pp. 212–14.

Aragon with his mother, the mentally unsound D. Juana. This succession had notionally been on the cards since 1504–5 (since the death of the Catholic Monarch, Isabel), but the situation a decade later was significantly different. Arriving in Spain in September 1517, Charles V went about setting his house in order with a certain determination. He also effectively sought to reopen the whole question of the 'partition of the world', consummated at the Treaty of Tordesillas, aided by a proposal brought before him by the Portuguese, Fernão de Magalhães, who is better known to posterity as Ferdinand Magellan.[65]

Magalhães, born about 1480 in the north of Portugal, had probably gone to Asia in 1505, in the company of D. Francisco de Almeida. He remained there for eight years, participating in campaigns on the west coast of India, and east Africa, besides accompanying Diogo Lopes de Sequeira on the first, unsuccessful, Portuguese mission to Melaka in 1509. Here, he became closely associated with Francisco Serrão, later to be Portuguese factor in the Moluccas. In 1510, Magalhães was in Goa, participating in its capture, and the next year again in Melaka, together with Albuquerque. In 1513, he returned to Portugal and took part in the Duke of Bragança's expeditionary force against Azemmour. Despite this rather hectic activity, Magalhães found himself unable to get the rewards he desired at the court of D. Manuel, perhaps because of rumours that he had rather freely helped himself to loot in the capture of Azemmour. Whatever may be the case, between 1514 and 1516, he made at least two petitions for career advancement to the court, and both were turned down.[66] Thus, gathering together the information he could from men who had experience of the Moluccas, such as Francisco Serrão, and supplementing it with the aid of the cosmographers, Rui and Francisco Faleiro, and the cartographer Jorge Reinel, Magalhães determined to present a project in Spain, to show that the Moluccas were in fact within the Spanish half of the division made at Tordesillas.

[65] For earlier attempts to reach the Moluccas from the west under Ferdinand of Castile, see Max Justo Guedes, 'Estreito de Magalhães', in Albuquerque and Domingues (eds.), *Dicionário de história dos descobrimentos portugueses*, Vol. II, pp. 640–4.

[66] A somewhat different version, but similar in its essentials, may be found in the text of Fernando Oliveira, put together in the mid-sixteenth century from an earlier account; cf. Pierre Valière, *Le voyage de Magellan raconté par un homme qui fut en sa compagnie* (Paris: Centre Culturel Portugais, 1976), pp. 27, 31. For the authoritative accounts of Magalhães, see Visconde de Lagoa, *Fernão de Magalhães: A sua vida e a sua viagem*, 2 vols. (Lisbon, 1938), and Jean Denucé, *Magalhães, la question des Moluques et la première circumnavigation du globe* (Brussels, 1911).

Geographically, of course, Magalhães had got matters wrong. But the political potential of his project was at once realised in Spain, not least of all by a certain Diogo Barbosa, an important Portuguese merchant based in Seville, where Magalhães arrived in late October 1517, having publicly renounced his status as a subject of D. Manuel. With the aid of Barbosa (whose daughter he married), of Juan de Aranda (factor of the *Casa de Contratación*), and of the Bishop of Burgos, Magalhães managed to get his project accepted by Charles V, and a contract between them was signed at Valladolid on 22 March 1518, in the teeth of strong opposition from other groups in the Spanish court; the expedition was financed by the Fuggers and by the influential Cristobál de Haro. The rest of the story is well known, Magalhães's departure with five ships from Seville on 10 August 1519, his arrival in Rio de Janeiro late that year, the revolts against Magalhães by other members of the expedition, his difficult penetration of the Pacific in late November 1520, and his eventual death in the island of Cebu on 27 April 1521. Of the five ships, only the *Victoria*, captained by Sebastian de Elcano, actually managed to circumnavigate the globe, returning to Spain via the Cape of Good Hope on 6 September 1521, with a mere eighteen crewmen.

Vasco da Gama in the meantime still waited, as it were, in the wings. However, in the years immediately following on the return of D. Aires da Gama from India (1512), we do see a revived interest on the Admiral's part in the India trade. Documents from the Casa da Índia suggest that he was able to ingratiate himself to some extent with D. Manuel again, perhaps through the powerful António Carneiro, secretary to the King. Thus, in June 1513, Gama was granted the privilege of receiving money and goods from India, either by employing his own funds, or on account of persons who sent goods to him, all this without paying freight or duties in Lisbon. Naturally, those spices on which a royal monopoly was exercised were excepted from this. Two years later, in August 1515, Gama was allowed to send one of his men each year to India, 'to collect and bring back goods for him from there'; this agent was to be paid the same as a man-at-arms.[67]

In 1514–15, Gama had shifted residence to the region of Portalegre, near the Spanish border, as we see from a letter written by him in one of these years, at Niza, to António

[67] Royal grant-letters, published by Luciano Cordeiro, 'O prémio da descoberta: Uma certidão da Casa da Índia', in Cordeiro, *Questões Histórico-Coloniais*, Vol. III (Lisbon, 1936), pp. 129–30.

Carneiro. The letter refers to three of Gama's clients, António Lopes, and Francisco and Fernando Anes, who had been accused of poaching, and killing a boar (perhaps from royal reserves, or *coutos*) in the process. Gama thus writes:

> My lord: Antonjo Lopes and Fernand Anes and Francysco Anes, squires (*escudeyrros*) resident in Benevente are men for whom I desire to do a great deal: His Highness accuses them of having killed a boar (*porco*), and they are in great terror of the ire of His Highness: I would be very grateful to you, my lord, if you could get an order signed by His Highness in which he lets them go free, and they give over 100 *cruzados* each in bond, for this [accusation] is not true, and they wish to show they are not guilty, and as they are honourable men and do not want to go to prison, you my lord would do me a great favour (*muyta merce*) in preparing this very just order. Trusting in your lordship, in Njsa, on the 30th of December. The Admiral.[68]

Again, in August 1515, Gama had a change made in the 1504 grant of 400,000 *reais* to him, once more suggesting his location at Niza: for henceforth, half of this sum was to be paid to him from the salt-tax at Niza, or failing that from the *almoxarifado* of nearby Portalegre.[69]

We are aware that shortly thereafter, in 1517–18, Vasco da Gama had moved to Lisbon, for a document of minor interest (which does carry his signature though) confirms this fact. This is a royal order from Dom Manuel to a certain Nuno Vaz, the customs-collector of the region between the Tejo and Odiana, asking him to pay a sum to the sister-in-law of the Admiral, Dona Ana de Ataíde, wife of D. Aires da Gama. This sum was the last third due from her dowry of 352,000 *reais*, which the Crown had given her; but a note from Vasco da Gama on the document, dated 31 July 1518, confirms that he had received the sum on his brother's behalf, and the letter of procuration (dated 22 March 1518) notes that it was written 'in the houses of the residence of the lord Admiral, D. Vasco da Gama' in the city of Lisbon. A minor sidelight of the document is that it identifies, among the witnesses, the Admiral's chaplain Pêro Vaz, and another of his

[68] AN/TT, Cartas dos Vice-Reis, No. 98, published in Keil, *As Assinaturas de Vasco da Gama*, pp. 17–18.

[69] Biblioteca da Sociedade de Geografia, Lisbon, Colecção Vidigueira, Maço I, Doc. 20, parchment; for a discussion of this collection, acquired in 1892 from D. Tomás da Gama, Count of Vidigueira, see Rosalina Silva Cunha, 'Sociedade de Geografia: A Colecção Vidigueira', *Boletim Internacional de Bibliografia Luso-Brasileira* 1, 1 (1960), 65–99.

servants (*criado*), João Cordeiro. This document is also proof of the close and continuing relations between Dom Vasco and his brother Aires da Gama, which we have already remarked above.[70]

By late 1518, D. Aires himself was back in India, this time as captain of Cannanur. He returned there with Diogo Lopes de Sequeira, who had been named governor of Portuguese India in February that year, to replace Lopo Soares de Albergaria. Diogo Lopes arrived in Goa in mid-September 1518 after a rapid voyage, and then went on to Kerala the next month, in order among other things, to visit Cannanur.[71] D. Aires was one of the major *fidalgos* to be sent out that year, and it would appear that Albuquerque's old complaints against him had either been forgotten or set aside. A letter written by him to D. Manuel within a few months of his takeover of the captaincy of Cannanur survives, and carries his first impressions of India after his return. The letter is also interesting as a collective articulation, so to speak, of the Gama family's perspective on how India ought to be governed. D. Aires begins by noting that there were too many large Portuguese ships in the Indian Ocean; these should be replaced by small vessels, and oared ships (*navios de remos*). He then goes on to note, in what is implicitly a criticism of Diogo Lopes, that the governor of India ought not to have too many arbitrary powers. Rather, he should govern in a consensual fashion, together with three or four other persons of a certain weight and seniority, even if the governor (*capitam mor*) remains superior to them, and is not obliged to take their advice. The letter now goes on to the spatial organisation of Portuguese India, beginning with Goa, which he treats with considerable scepticism. Goa is no doubt a 'very noble thing', and 'a very honoured thing', but, according to D. Aires, it represents an enormous expense to the exchequer because of the very large number of officials, scribes and others, all of whom live in style ('eating betel at your expense'), so much so that 'there are more officials there than in two Lisbons'! One solution to this, suggests D. Aires, would be to expand the revenue base, by negotiating with the ᶜAdil Shahs of Bijapur ('o Çabajo') for rights over Bardes, and one or two small islands; however, on no

[70] AN/TT, CC, II-76-157, published in António Baião, 'Vasco da Gama e as suas expedições à Índia (com documentos novos)', in *Duas conferências no Paço Ducal de Vila Viçosa* (Lisbon: Fundação da Casa de Bragança, 1956), pp. 40–2.

[71] Cf. the interesting letter from Pedro de Bastroni Corço to D. Manuel, dated Cochin, 10 November 1518, AN/TT, Gavetas, XV/12–13, in Silva Rego (ed.), *Gavetas*, Vol. IV, pp. 386–90. The author, who accompanied the new governor, condemns Lopo Soares's government, and praises Diogo Lopes in fulsome terms.

account should the Portuguese attempt to gain a foothold in the mainland (*terra firme*).

D. Aires da Gama's purpose in this exposition soon becomes clear as he moves on to the next item on his agenda: Cochin. In continuity with the point of view of his old allies, such as António Real and Lourenço Moreno, D. Aires wished to denigrate Goa, and stress the central importance of Cochin in the Portuguese scheme of things. 'Cochin is the thing in India of which Your Highness has the greatest need', he declares unequivocally, stressing the pepper trade there, and its role in supplying the return cargoes to Europe. The letter now goes on to put forward a rather interesting project for trade within the Indian Ocean, but only after dealing with the questions of Cambay and Diu. D. Aires, not surprisingly, declares himself opposed to the setting up of factories or fortresses in Gujarat (Cambay), arguing that any trade with the region can be done through the existing establishment at Chaul. On the other hand, he notes, Diu must be dealt with firmly; the governor there, Malik Ayaz, in particular, who has taken to constructing armed ships in the Portuguese style (*navios e artelharia à vossa usança*), must be obliged to desist. The practice of blockading the port by a patrolling fleet, and capturing its ships that go to the Red Sea, appears to him to be a sound one, not least of all because it yields revenues. However, he warns against an attack on the port itself, preferring a maritime blockade. Once Diu has been reduced, and the 'Rumes' have been discouraged from their Indian projects, the real plan can be put into operation: which is to displace the Muslims entirely from the trade of the Indian Ocean, substituting them by Christians and Gentiles. The Muslims, he declares, 'are your mortal enemies, and will always try to do you all the harm they can'. The Gentiles, on the other hand, 'naturally do not wish to do you harm'. Indeed, he argues, in earlier times, Gentiles had been converting in India to Islam, 'in order to benefit from this navigation'. D. Aires offers, as a first step, to put this plan into action at Cannanur, by offering the ruler there an increased share in the horse-trade from Hurmuz, if only he favours the Christians and Gentiles. But a further master-stroke is in store. Since, in D. Aires's opinion, as many as two hundred Moorish ships go every year to Hurmuz alone, the competition from local Christians and Gentiles will not suffice to displace them. The real solution lies in allowing 'as many Portuguese ships as so wish', to come to India each year, to participate in the intra-Asian trade. A division is thus proposed.

The Crown will trade on the Cape route, and will sell European goods in Asia; the private traders, in ships 'well armed with artillery and weapons', naturally enough, will take over the intra-Asian trade. The Portuguese Crown will thus be saved expenses and trouble, and the risks of trade will be passed on to private traders.

We may pass rapidly over the rest of the letter, which is concerned with the conservation and strengthening of Cannanur as a centre of Portuguese trade and settlement. Cannanur, D. Aires reminds the king, is in fact the first real centre at which the Portuguese established themselves in India, and should hence not be neglected; it was, according to him, in a deplorable state of repair by early 1519.[72]

THE ADMIRAL STRIKES BACK

The naming of Diogo Lopes de Sequeira as governor of Portuguese Asia in February 1518 has been seen as a reversal of the trends set by Lopo Soares de Albergaria, and a return in part to the policies of Albuquerque. Of course, matters were rather more complex than they appear at first sight. As Luís Filipe Thomaz has shown, even during the government of Lopo Soares, D. Manuel managed to push through some of his own measures, creating for example the post of *vedor da fazenda* (financial intendant) in India in 1517, and naming a certain Fernão de Alcáçova to the post.[73] This post was obviously designed to check the governor's power, since the *vedor* was theoretically answerable to the Crown alone, and also reserved the right to remove erring financial officials in Asia; however, since Alcáçova arrived in India in September 1517, and was promptly shunted back to Europe in January 1518 by Lopo Soares (with whom he entered into conflict), it is in fact clear that the monarch did not really succeed in imposing his will. Similarly, the political position of Diogo Lopes *vis-à-vis* projects of royal centralisation in Asia is not wholly clear. To be sure, he pursued D. Manuel's line in respect of Hurmuz, Gujarat and the Red Sea (where he led an expedition in 1520, successfully establishing contact once more with Ethiopia). But from certain other points of view, the situation is more ambiguous. Either on account of his limited

[72] AN/TT, Gavetas, XV/9–11, letter from Aires da Gama to D. Manuel, Cannanur, 2 January 1519, in A. da Silva Rego (ed.), *Gavetas*, Vol. IV, pp. 213–21.
[73] Thomaz, *De Ceuta a Timor*, p. 454.

powers, or because of a lack of will, Diogo Lopes permitted corsair activities in the Maldives and in the Bay of Bengal (for example, the case of João Moreno), and maintained rather good relations with men such as Diogo Pereira, noted proponents of the intra-Asian private trade of the nobility. This was particularly curious, because a new set of regulations promulgated by the Crown in February 1518 reaffirmed the importance of Crown monopolies in Asia, and also explicitly forbade the payment of salaries to soldiers and officials in goods, in order to discourage them from trading.[74]

If we are to suppose that Diogo Lopes represented an instrument of Manueline policy (as opposed to Lopo Soares, closely associated with the Baron of Alvito), we also need to resolve some knotty problems of chronology. As we have noted, the years from late 1514 to mid-1518 were marked by a considerable retreat on the part of D. Manuel, to which a number of factors contributed. In this period, the opposition began to gather around the heir-apparent, the young Infante D. João; in mid-1517, the monarch while making his will seemed gloomily resigned to the shelving of his grandiose messianic plans.[75] Then, in 1518, D. Manuel struck back, to restore the balance as it were, by a powerful symbolic gesture. Negotiations had commenced with the Habsburgs for a double marriage alliance: for D. Manuel, Guillaume Chièvres de Croy, Charles V's powerful mentor in this early period, proposed the hand of D. Margarida (governor of the Low Countries, and Chièvres de Croy's political rival), while Charles's own sister D. Leonor was in view for the Infante D. João. D. Manuel in mid-1518 quickly changed that, taking the much younger D. Leonor as his bride, an act that Damião de Góis disingenuously notes was very much resented by those around the Infante for reasons of pride. More recent historians too have at times seen this essentially political act in neo-Oedipal terms, which is scarcely helpful. However, this domestic drama took place some weeks *after* Diogo Lopes's departure for India, and

[74] These documents, from 1518, are published in J.H. da Cunha Rivara, *Archivo Portuguez-Oriental*, 6 Fascicules in 9 Parts (Goa, 1857–76), Fascículo V , Docs. 6 to 8, pp. 9–11, from late copies in the Historical Archives, Panaji (Goa). As Thomaz (ibid.) notes, these regulations were later systematised by a royal decree of September 1520, as the *Ordenações da Índia*, for the text of which see Luís Fernando de Carvalho Dias, 'As Ordenações da Índia', *Garcia da Orta* (Special No.) (1956), 229–45. For a discussion of other restrictions, as late as 1521, also see Flores, 'Os Portugueses e o Mar de Ceilão', pp. 169–70.

[75] AN/TT, Gavetas, XVI/2–2, Testamento de el-Rei D. Manuel, 7 April 1517, in Silva Rego (ed.), *Gavetas*, Vol. VI, pp. 111–63.

thus at the time when the governor was named (in February 1518), D. Manuel was still on the defensive. Taking one thing with another, then, it seems best to see Diogo Lopes de Sequeira not as an agent of D. Manuel against internal opposition, but rather as a compromise candidate. If, on the one hand, he acted at times in accordance with D. Manuel's expectations, on the other hand, he was a cousin of Lopo Soares himself, and continued to trust considerably in men like D. Aleixo de Meneses (Lopo Soares's relative and close ally), whom he left in charge on the Indian west coast when he went to the Red Sea in early 1520.

But the governor's factional politics run deeper than one might have imagined. Thus, for example, Diogo Lopes's dislike for D. Aires da Gama comes through clearly in several of his letters. In one of these, addressed to D. Manuel, and written in late December 1519, he accuses the Admiral's brother and a certain Rui de Melo of untrammelled greed (*cobyça*), on account of which he wishes to keep them from positions of real financial responsibility. Two earlier captains, Dom Guterre de Monroy and Simão da Silveira were rumoured respectively to have taken back fortunes to Portugal of 40,000 and 30,000 *cruzados*; but D. Aires and Rui de Melo wish to outdo them, he states, 'without regard for anything of the service of Your Highness nor even of their honour, and I believe that they are not very content with me because I have not given them a hand in this'.[76] Elsewhere, noting the physical improvements he has brought about in the fortress of Cannanur (which by D. Aires's own estimation was in bad shape in 1519), Sequeira states that despite his actions, D. Aires is annoyed (*agastado*) with him, hinting broadly moreover that it is because he has not let the latter be 'fattened' in the process.[77] In sum, if Diogo Lopes was not wholly averse to the entourage of Lopo Soares (and thus to a certain tendency in the Portuguese court), he was on the other hand no friend of the Gamas. It will thus not do to reduce the Manueline court and its nobility to a simple binary conflict between those ranged for and against the centralising line.

In mid-1518, despite his internal *coup d'état*, D. Manuel found himself still caught in a cleft stick. By means of his new matrimo-

[76] AN/TT, CC, I–25–83, letter from Diogo Lopes de Sequeira sent through Simão de Alcáçova, published in Ronald Bishop Smith, *Diogo Lopes de Sequeira* (Lisbon: The author, 1975), p. 35.

[77] Also see AN/TT, Fragmentos, 4–4–3, 'Hobras que mamdey ffazer de pedra e caal' by Diogo Lopes (n.d. 1520), in ibid., p. 53.

nial project, he had no doubt strengthened himself at home, but laid himself vulnerable abroad to the Habsburgs; thus, in 1519, he was prompt to support Charles V's candidature to be Holy Roman Emperor, and later, in 1520–1, supported him in the revolt of the *comuneros* in Castile.[78] Besides, it was no secret in Portugal by 1518 that Charles V had signed an agreement with Fernão de Magalhães, and that an attempt by the Spaniards to establish rights on the Moluccas would soon follow. Thus, even if, in the second half of 1518, D. Manuel's position was stronger than it had been in the previous three years, the monarch was still vulnerable. It was this vulnerability that Vasco da Gama sought to exploit.

Our analysis rests on an extraordinary document from August 1518, first published by Luciano Cordeiro in the late nineteenth century, namely a letter addressed to Gama by Dom Manuel. The letter runs:

> Admiral, friend. It seems to us that this petition (*requerimento*) that you present to us for the title of Count, which you say we had promised you, you have presented it as you saw fit, and we, on account of the services that you have done for us do not wish to give you the permission that you ask from us for you to leave our Kingdoms, but by this [letter], we order you that you remain in our Kingdoms until the end of the month of December, this next one that comes in the present year, and we hope that in this time, you will see the error that you are committing and will wish to serve us as is due, and not go to such an extreme, and as soon as the said period is finished, if you wish to remain unmoved in your intention to leave our kingdoms, even though we would feel this heavily, we would not prevent you from going or from taking your wife and children and your moveable goods. Written in Lisbon on the 27th of August by the Secretary 1518. The King.[79]

Let us examine the text in detail. It is the reply to an earlier petition from Vasco da Gama (which has unfortunately not come down to us), stating that D. Manuel had to grant him the title of Count, as allegedly promised on an earlier occasion. If not, Gama asked for permission to leave Portugal, and it is obvious (if unstated) that his intention was to go over to Charles V. We may imagine the effect that this would have had, if, besides Magalhães

[78] Aubin, 'Le Portugal dans l'Europe des années 1500', p. 221.

[79] Luciano Cordeiro, 'De como e quando foi feito Conde Vasco da Gama', in Cordeiro, *Questões Histórico-Coloniais*, Vol. II, Doc. 7, p. 209 (the text was first published in 1892). The document is reproduced in Teixeira de Aragão, *Vasco da Gama e a Vidigueira*, Doc. 23, pp. 257–8.

(who was yet to depart on his voyage), the Habsburg ruler of Spain had the Portuguese Admiral of the Indies in his service. Dom Manuel was obviously caught in a quandary. To refuse to concede anything at all to the Admiral was dangerous, when so many other powerful sources of internal and external opposition already existed; besides, the myth of Vasco da Gama already existed in embryonic form, both in Portugal and abroad (as we see from the writings of Garcia de Resende, for example) and the Admiral simply could not be treated as just another noble, or commander of an expedition. Gama was simply not Pedro Álvares Cabral.

A compromise was thus hammered out, through a process that we can only barely glimpse from the documents. By mid-April 1519, a general order (*alvará*) had been issued, to the effect that any one of D. Manuel's subjects could sell properties to the Admiral, thus permitting him (after the failed affair of Vila Franca de Xira) to try and acquire a landed base once more.[80] But the crucial intervention seems to have been that of D. Jaime, Duke of Bragança, who was himself (as we have noted) no great admirer of the Asian policies of D. Manuel; the flavour is once more implicitly that of a political alliance between the Duke and the Admiral. By November 1519, the terms of the arrangement were agreed upon. From the lands of his Dukedom, D. Jaime would hive off the towns and territories of Vidigueira and Vila de Frades, which he would give over to Vasco da Gama; these would then form the territorial basis for his title of Count, which D. Manuel would grant him. Since the territories formed a part of the Bragança properties, Gama's rights would be particularly extensive, and he would also be shielded in some measure from the intrusive wedge of the *Ordenações Manuelinas*, which did not apply to the Duke's lands. The entire transaction took place in the Alentejo, between Évora (where Vasco da Gama and D. Manuel were at the time), and Vila Viçosa, the usual residence of D. Jaime. The first legal step was taken in Évora on 24 October, when D. Manuel issued an *alvará*, permitting the Duke to sell the two *vilas* to D. Vasco, and the latter to hand over the annual pension revenues of 400,000 *reais* he had received from the Crown in 1504, and drawn from the revenues of the Casa da Mina, to the Duke in exchange. Next, the Duke signed a letter at Vila Viçosa on 4 November, giving powers of attorney to a certain João

[80] AN/TT, CC, III–7–20, copy of an *alvará de licença*, dated Almeirim, 16 April 1519.

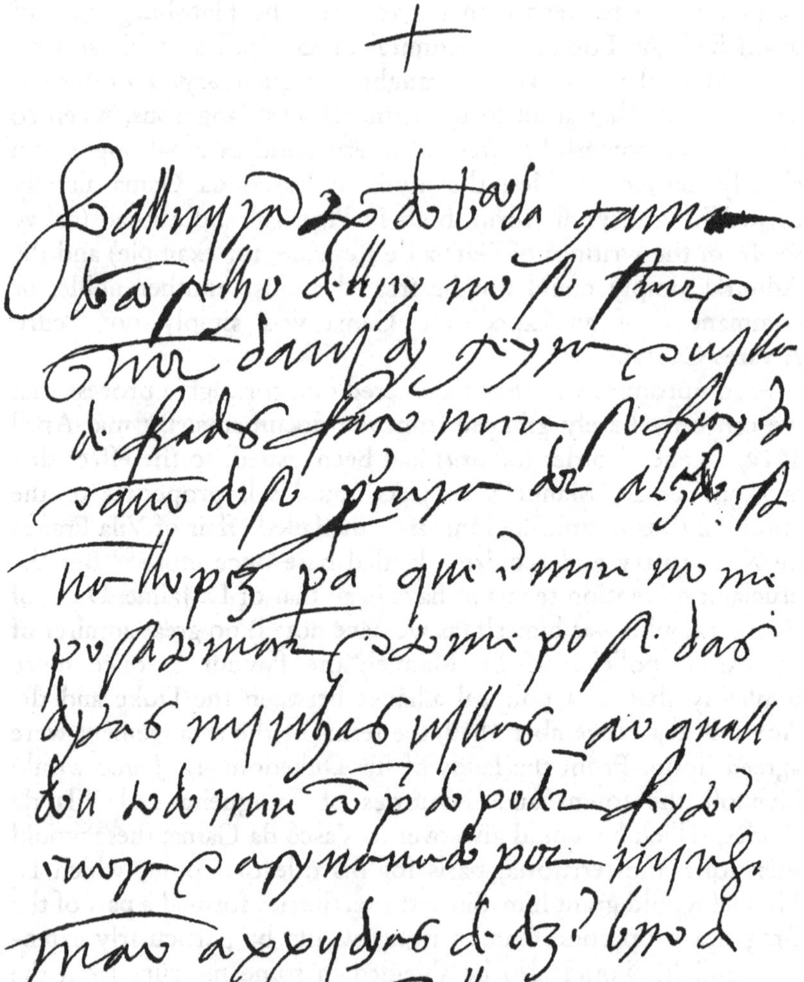

Plate 22 Facsimile of a letter of power of attorney from D. Vasco da Gama to Estêvão Lopes, 22 December 1519, concerning possession of Vidigueira and Vila de Frades.

Alves, actually to carry out the transaction. On 7 November, Alves was in the (probably temporary) residence (*pousadas omde ora pousa*) of Gama at Évora, with the latter's wife D. Catarina de Ataíde, and his son and heir, D. Francisco da Gama. Here a detailed act of sale (and 'renunciation') was set out, with the proviso that besides the amount of the pension (due from January 1520 onwards), Gama would pay the Duke 4,000 gold *cruzados* in cash. The sale was held to be binding on the heirs of both Gama and the Duke, even though they were minors at the time.

Subsequently, a royal letter, dated 17 December 1519 at Évora, confirmed the transaction, adding formulae of praise for the Admiral, who had rendered 'so many and such insign services . . . especially in the discovery of the Indies, and the settling of them, from which there resulted, and results great profit not only to us and the Crown of our kingdoms and lordships, but generally universal profit to their residents and to all of Christianity, on account of the exaltation of Our Holy Catholic Faith . . .' And finally, on 29 December 1519, the second part of the transaction also was completed, this one between the King and the Admiral. D. Vasco's representative, a certain Estêvão Lopes, was sent to Vidigueira and Vila de Frades, to take possession of the towns (receiving and handing back their keys) and the Admiral was publicly acclaimed by the local representatives. On the very same day, in Évora, D. Manuel at last named him Count 'of the vila of Vidigueira . . . with all the honours, preeminences, prerogatives, authority, grace, privileges, liberties and freedoms that the Counts of Our Kingdoms have and possess'.[81]

The compromise had something in it for all parties. The new Count of Vidigueira was considerably beholden to the Duke of Bragança. D. Manuel, for his part, had created a Count at a small financial cost, since Vasco da Gama had effectively bought the revenues of his title, rather than received them from the Crown. On the other hand, for Gama, the title did represent an enormous leap in terms of social advancement, for in Portugal at this time, there were relatively few titled noblemen (nineteen at the death of D. Manuel), amongst whom were two Dukes, two Marquises, a Count-Bishop, and twelve other Counts. Thus, in the two

[81] The entire documentation, with several interesting facsimiles of texts, may be found in Ribeiro, 'De como e quando foi feito Conde Vasco da Gama', pp. 209–23. One of these texts, the 'auto de posse e entrega', is in the Biblioteca da Sociedade de Geografia de Lisboa, Colecção Vidigueira, Maço I, Doc. 21. Also see Teixeira de Aragão, *Vasco da Gama e a Vidigueira*, Docs. 24, 26, and 27, pp. 258–9, 261–70.

decades since his return from India, Vasco da Gama had come a long way from his initial niche in the lower nobility. He became a major figure in the court in the last years of D. Manuel. Thus, in Garcia de Resende's account of the departure of the Infanta D. Beatriz for Savoy in August 1521, we find the Conde Almirante and his sons, D. Francisco and D. Estêvão, conspicuously mentioned: it is also noted that Gama on this occasion distinguished himself with the public display of his riches, through his extravagant expenditures.[82] In Vidigueira too, at this time a small centre, with barely 327 households in the Census of 1527, Gama made an effort to construct in keeping with his status, and in continuation of what he had begun in Sines; of this, one of the few survivals is an inscription in the clocktower (*Torre do relógio*), which mentions that he had a bell placed in it in 1520. Among the distinguished inhabitants of the town (possibly from this time on, but certainly somewhat later), was a certain Lourenço Moreno, who had the chapel of Nossa Senhora da Piedade built there; it would bear investigation if this was the same person as the former opponent of Afonso de Albuquerque.[83]

THE END OF THE MANUELINE DREAM

Until his death in December 1521, it is clear that D. Manuel continued to cherish certain mystical notions in relation to his Asian empire, and in general harboured a conception of Portuguese expansion that was rather too grandiose for Portugal's meagre resources. Vasco da Gama was not one of those who shared this vision, which was instead one that was fostered by men like Duarte Galvão and D. Martinho de Castelo-Branco, and supported thereafter by men like Galvão's son-in-law, Duarte Pacheco Pereira, in his *Esmeraldo de situ orbis*. The years after 1518 see the recrudescence of some of these ideas, which had fallen into disuse after about 1514, in particular the notion of an alliance with Prester John of Ethiopia. For their part, orthodox theologians in Portugal were quite doubtful about the Christianity of the Ethiopians, which seemed to them excessively tainted by Judaism. Still, on his expedition to the Red Sea, the governor Diogo Lopes de Sequeira had, in April 1520, made his way to the coast of Erythrea, and once more made contact with

[82] Garcia de Resende, 'Hida da Infanta Dona Beatriz pera Saboya', in *Crónica de Dom João II e Miscelânea*, ed. Veríssimo Serrão, pp. 323, 328.
[83] Teixeira de Aragão, *Vasco da Gama e a Vidigueira*, pp. 160–1.

the subjects of the Ethiopian monarch, Lebna Dengel. An embassy, under the leadership of D. Rodrigo de Lima, had been deposited to go to the court itself.

News of these successful contacts appears to have reached Lisbon by the spring of 1521. D. Manuel once more saw, within the matrix of his millennialist vision, the possibility of destroying Mecca, and with it 'the evil sect of Mafamede'. A brief document, entitled 'Letter with the News that came to the King Our Lord, of the discovery of Prester John', was thus almost immediately published, in May or June 1521, with summary details of Diogo Lopes's actions, and the imminent alliance with Ethiopia.[84] At much the same time, D. Manuel wrote a letter to Pope Leo X (also published in 1521, in Latin), with the familiar projects of the union of Eastern and Western Christianities, the imminent destruction of Mecca (and the 'tomb' of Muhammad there!), all under the patronage of the Papacy. In keeping with this optimistic spirit, D. Manuel also named Martim Afonso de Melo Coutinho in March 1521, to command a fortress that he had decided should be contructed in China. After a brief expedition to the Ming domains in August 1522 (in which three ships, and a great part of the crew of the fleet were lost), Coutinho returned to Melaka and thence to India, sadder and wiser. He was to complain bitterly to D. João III in 1523, that he was 'so blinded by the information that had been given there [in Portugal] to the King, your father – may he rest in Holy Glory – that it seemed to me that at least half must be true, until I saw that it was all the contrary'.[85]

One major legacy remained, carried over from 1521 to the next year and reign. This was the Spanish threat to the Moluccas, against which Diogo Lopes de Sequeira was already taking precautions in 1519. He had sent word to Melaka, he reports to D. Manuel in his letter of December that year, and the captain there, Afonso Lopes da Costa, had been told to despatch D. Tristão de Meneses to the Moluccas in a caravel; a letter from Garcia de Sá in Melaka, dated August 1520, contains a mention of

[84] Armando Cortesão and Henry Thomas, *Carta das novas que vieram a el Rei nosso senhor do descobrimento do Preste João (Lisboa 1521): Texto original e estudo crítico, com vários documentos originais* (Lisbon, 1938). For a discussion, also see Jean Aubin, 'Le Prêtre Jean devant la censure portugaise', *Bulletin des Etudes Portugaises et Brésiliennes* 41 (1980), 33–57.

[85] AN/TT, CC, I–30–49, letter from Goa, dated 25 October 1523 in João Paulo Oliveira e Costa, 'Do sonho manuelino ao realismo joanino: Novos documentos sobre as relações luso-chinesas na terceira década do século XVI', *Studia* 50 (1991), 121–56 (quotation on p. 154). My discussion here closely follows that of Costa.

the sighting of the Spanish fleet.[86] The next year, 1520, D. Manuel is reported to have pressed for the construction of fortresses in the Moluccas and in Sumatra as a defensive measure, thus using the Spanish threat to justify military expansion in Asia. This rivalry with Spain would continue to figure as a major theme into the later 1520s, when most other elements of the Manueline dream had been abandoned, censored, or significantly modified. In one major respect at least, in the 1520s and 1530s, the pendulum would swing the other way; after the Portuguese expeditions to the Red Sea, it was now to be the turn of Portuguese India to live in fear of the impending invasion of the 'Rumes' (Ottomans). The dream of universal empire too would pass for a time from D. Manuel to the Ottoman Sultan, Süleyman the Magnificent.

[86] AN/TT, Gavetas, XV/10–2, Garcia de Sá to D. Manuel, letter dated 23 August 1520, in Silva Rego (ed.), *Gavetas*, Vol. IV, pp. 245–6.

CHAPTER SIX

A CAREER CULMINATES

Dom João III called the great Vasco. He gives him discretionary powers; he makes him almost his crowned emissary, himself over again, in such distant parts where the voice could hardly be heard, where the arm barely reached. And the great Vasco goes. It is no longer the bold firmness that nothing can break, nor the most bitter vengeance that reduces everything; it is Themis, implacable but serene, inflexible but tranquil.

José de Souza Monteiro, 'Vasco da Gama – A Psychologia d'um Heroe' (1898)[1]

SETTLING SCORES

WE HAVE SEEN THAT, IN THE LAST YEARS OF HIS REIGN, IT seemed that Dom Manuel might re-consolidate his political position, to the perdition of his foes, with the revival of his Ethiopian project being the most visible symbol of this change. But this revival stopped considerably short of fructification, and in December 1521, Dom Manuel died, little mourned by his successor – with whom his relations had steadily deteriorated. In turn, his son, Dom João III, found discontent rife against himself, as he attempted to press through his own policies. Rather than being the creator of a loose dispensation, favourable to the nobility and its ambitions both in Europe and in Asia, the new monarch turned out initially at least to have centralising ambitions of his own, albeit different ones from those of his father. But before entering into the thorny issues of the early Joanine reign, it may be useful to recapitulate the major steps of the last years of the reign of D. Manuel.

[1] José de Souza Monteiro, 'Vasco da Gama – A Psychologia d'um Heroe', *Revista Portugueza Colonial e Marítima*, (número comemorativo, 20 May 1898), 8.

It has already been noted that in the outgoing fleet of the spring of 1521, D. Manuel had sent out orders to construct a fortress in China, considerably underestimating the strength and vigour of the Ming state. Indeed, as mentioned in the previous chapter, the turning point appears to be not with the sending of the fleet carrying Diogo Lopes de Sequeira to India in 1518, but with that of the subsequent year. Let us note that this fleet of 1519 was a very substantial one, much larger than that of the years 1516 to 1518, irrespective of which of several divergent sources we choose to rely on. It was commanded by Jorge de Albuquerque, nephew of Afonso de Albuquerque, and carried on board a new *vedor da fazenda* (to replace Fernão de Alcáçova), namely Doutor Pedro Nunes (stoutly resisted as usual by the governor, Diogo Lopes de Sequeira, who masked his real reasons – conflict of powers – by denigrating the *vedor*'s professional competence), as well as Diogo Fernandes de Beja, who was apparently being sent out to assume the captaincy of a new fortress D. Manuel projected in or near Diu. The trend continued; the next year's fleet, led by Jorge de Brito, carried orders not only to oppose Fernão de Magalhães in the Moluccas with all possible vigour, but to construct fortresses at Pasai (in fact already begun by Jorge de Albuquerque), in the Maldives, as well as at Chaul on the Konkan coast, north of Goa. A new, post-Albuquerque, phase of fortress construction seemed to be beginning for the Portuguese, this time on royal initiative. Diogo Lopes went along with this shift in policy to some measure, and after failing to negotiate a position in either Diu itself or neighbouring Jafarabad from the wily Malik Ayaz, did manage to persuade Burhan Nizam Shah I of Ahmadnagar to permit him in 1521 to build a fortress at Chaul (termed São Dinis), in circumstances that are somewhat obscure. Later in the century, the Nizam Shahs were bitterly to regret this decision, and the fortress was to become a bone of contention between them and the *Estado da Índia*. Thus, with the fortress at Colombo having been completed a short while earlier, and another new fortress at Kollam (in southern Kerala) being organised by Heitor Rodrigues in 1519–20 despite stiff opposition from a part of the ruling house of Venadu, the rivals of the Portuguese in the Indian Ocean had good reason to feel a certain nervousness. This may well explain in part the fact that João Gomes (nicknamed 'Cheiradinheiro', or 'Smells Money'), sent to the Maldives to investigate the possibility of building a fortress there too, was killed in 1521, reportedly at the behest of 'Moors from Cambay'.

The fleet that left for India in 1521 carried a new governor, sent out to replace Diogo Lopes de Sequeira, whose three-year term was drawing to a close: this was the very powerful noble, D. Duarte de Meneses, son of the Count of Tarouca, D. João de Meneses, who, amongst his many accumulated distinctions, was also Prior of Crato, and *mordomo-mor* of both D. João II and D. Manuel. D. Duarte had long been associated with the city of Tangiers, where he was already captain from 1507 onwards, on behalf of his father, and he enjoyed a formidable reputation as a war-leader. It is difficult to discern why precisely he was chosen as governor of Portuguese India in 1520–1; but we can only guess that it was related to his authoritative personal position, and the need to prepare matters in Asia on a war-footing against the imminent Castilian threat. On his departure from Lisbon in April 1521, he reputedly was accompanied by a fleet of twelve vessels; his brother, D. Luís de Meneses, went along with him with the title of Captain-Major of the Indian seas.

On arriving in Asia in August 1521, after a relatively rapid voyage, D. Duarte found a situation where the immediate priorities were rather different from those that had been imagined. Added to this fact was D. Duarte's own vision, which was far more centred on the Middle East than on matters beyond Cape Comorin (or, for that matter, even on the Malabar coast). Goa was left largely in the charge of its captain, Francisco Pereira Pestana, a close associate of D. Duarte, who entered into conflict with a good proportion of its population of *casados,* or burgher-settlers. Besides, in 1522, even the return-cargoes sent to Portugal were very poor, partly because a very small fleet (of only three ships) was sent out that year from Portugal, but also reputedly because adequate attention was not paid to preparing even the sole ship that went back. Throughout the governor's triennium, Southeast Asia (and the Moluccas), while pressing, was given a distinctly lower priority than the Persian Gulf, where the control of Hurmuz had very rapidly turned into a major issue. We may recall that, in 1515, Albuquerque on taking Hurmuz had not displaced its ruler, Turan Shah (who had just succeeded to the throne in 1514), but rather reduced him to vassal status. Within the island-state, he continued with his courtiers to exercise a social and administrative influence, and the customs-proceeds were notionally shared between him and the Portuguese Crown, while the Portuguese were to protect him against external aggression and internal rebellion. Albuquerque had already sent an embassy to Shah Ismaᶜil in 1514, to reassure him, and after the construction of the Portuguese fort there, another emissary, Fernão Gomes de Lemos, visited the Safavid court, with munitions and war-materials, promising to help the Shah in his wars against the Ottoman ruler, Selim I. But relations remained uncertain: on at least two occasions, the Safavid ruler showed an interest in seizing Bahrain and Qatif, leaving the Portuguese distinctly nervous. Nevertheless, in 1517, two Portuguese, João Meira and António Gil, are to be found at Basra, at the head of the Persian Gulf, one of the main focal points of commerce with Hurmuz, in order to firm up commercial relations; thus, in the five years after the second Portuguese capture of Hurmuz, the Persian Gulf port appeared a relatively flourishing commercial centre.

But troubles were around the corner. Disputes arose with Shah Ismaᶜil over the payment of the *muqarrariya*, the toll that was to facilitate the passage of caravans with goods to and from Hurmuz, over the Iranian plateau; commerce to and from that direction was thus partially interrupted. Then, following difficulties between the rulers of Hurmuz and their notional vassals in Bahrain (the centre of a lucrative pearl-fishery), a Portuguese fleet under António Correia attacked the latter centre in late July 1521, reducing it to obedience, and summarily decapitating its ruler Muqrin bin Zamel.[2] However, such 'services' rendered by the Portuguese did not prevent a major rebellion at Hurmuz itself beginning in November 1521, instigated it would appear by the ruler Turan Shah himself and more particularly by his adviser and *wazîr* (in Portuguese, *alguazil*), Ra'is Sharafuddin Fali. This appears once more to have been inspired by resentment at the new face of Manueline government in Asia after 1518, for D. Manuel had decided in 1520 to send out orders, changing the statutes of the customs-house in Hurmuz, and the privileges of its ruler. Though the captain sent from Portugal for this purpose, a certain Diogo de Melo, failed to reach Asia that year, D. Manuel's letters and instructions did reach Hurmuz, where it fell on the captain, D. Garcia Coutinho, to implement the new orders. In effect, this required the substitution of several of the existing local officials by Portuguese, new weights and measures in the

2 For the situation in the Persian Gulf in this period, see the brief but pertinent discussion in Salih Özbaran, 'The Ottoman Turks and the Portuguese in the Persian Gulf, 1534–1581', *Journal of Asian History* 6, 1 (1972), 46–8, reprinted in Özbaran, *The Ottoman Response to Portuguese Expansion* (Istanbul: The Isis Press, 1994).

customs-house, and overall Portuguese supervision. Accepted with poor grace, these changes eventually helped provoke an attack a year after they were implemented, led by Ra'is Sharafuddin himself, with the ruler's approbation.

Heeding letters from Hurmuz, one of the first acts of D. Duarte was hence to send word to his brother (who was at the time overseeing the final touches to the Chaul fortress), to go out to the Persian Gulf at the head of a naval force, and remain there for a considerable period, raiding and trading at the same time. Before D. Luís could arrive with his naval force, the new governor also sent D. Gonçalo Coutinho (brother of the captain of Hurmuz) there in a galleon; this *fidalgo* appears to have been an able if unscrupulous diplomat, and managed to arrive at understandings separately with Turan Shah and Ra'is Sharafuddin, not least of all because dissensions had broken out meanwhile between them. Then, in early 1522, in uncertain circumstances, Turan Shah was strangled by a certain Ra'is Shamsher, acting on the orders of Ra'is Sharafuddin, and in his place, his nephew was placed on the throne, with the title of Muhammad Shah (r. 1522–34).

D. Luís de Meneses arrived in Hurmuz shortly after these events, and found that Ra'is Sharafuddin had decided to maintain a safe distance, residing in the island of Qishm along with the new ruler. Hurmuz was in a state of some confusion, and the priority was to provide merchants (who had apparently dispersed in some measure to other nearby centres) with a sense of security, and persuade them to return to the island of Jarun (the centre of the Hurmuz state) itself. D. Luís decided that the best solution was to negotiate with Ra'is Sharafuddin, to reassure him that his earlier misdemeanours would be forgiven, if he were to return to Hurmuz and take matters in hand. These overtures being met with very limited success, he hence resolved on an arrangement with the redoubtable Ra'is Shamsher, who now agreed to assassinate Ra'is Sharafuddin as soon as the occasion presented itself. In view of this assurance, D. Luís rested content, and after a rather elaborate series of raids on shipping and settlements in southern Arabia returned to India in the latter half of 1522. An elaborate list of slaves who were taken in these raids is available: 216 of them, from small children, to aged and infirm men and women, of various origins.[3]

[3] AN/TT, CC, II–102–19, document dated Hurmuz, 5 July 1522, concerning the

The solution that had been arrived at did not seem satisfactory to the governor D. Duarte de Meneses, who thus resolved to make a personal visit to Hurmuz in 1523, following it up with another in 1524. Ra'is Sharafuddin was contacted once more, placated, and restored to his former position with honours (a decision that created something of a scandal); on the other hand, his main rivals were expelled or put in their place.[4] On the first of these sojourns, a new treaty of submission was signed between the governor and Muhammad Shah on 15 July 1523, raising the annual tribute to 60,000 *xerafins* (from 15,000 in Albuquerque's time, and 25,000 in that of Lopo Soares); measures were also taken to regulate the use of arms by Muslims resident there.[5] On 1 September that year, the governor also sent out an embassy to resolve tensions with Shah Isma^cil. The moment chosen was an inappropriate one, for Shah Isma^cil died at Tabriz in May 1524, leaving the Safavid court in considerable uncertainty; the new ruler, the young Tahmasp, was unwilling or unable to make any firm commitments to the Portuguese envoy, Baltasar Pessoa. It is clear that these visits by the governor to Hurmuz had a commercial motive as well, for D. Duarte de Meneses, as an avid private trader, was particularly interested in the commerce between south-western India (in pepper, and sugar from the Kanara port of Bhatkal), and Hurmuz.

In the same year, 1523, before his own departure for Hurmuz from India, D. Duarte had sent his brother to the Red Sea and Massawa, with stoppages en route to mount a series of raids in southern Arabia. These raids, especially an attack on al-Shihr, find mention in the Hadrami chronicles, in particular since the governor of that town (*amîr al-balâd*), Amir Matran bin Mansur, was killed in the process by a stray bullet.[6] Castanheda too, in his chronicle, mentions this set of episodes in a fashion that cannot be described as complimentary to D. Luís. He notes the capture off Cape Gardafui of five ships, then a raid on Aden, where four

valuation of slaves from Sohar and Cabo Roçalgate; also AN/TT, Núcleo Antigo No. 592, 'Livro das presas da armada de D. Luís de Meneses'; both documents published in António Dias Farinha, 'Os Portugueses no Golfo Pérsico (1507–1538): Contribução Documental e Crítica para a sua História', *Mare Liberum*, 3 (1991), 55–65, 67–78.

4 In 1529, after further difficulties between him and the Portuguese, Ra'is Sharafuddin Fali was imprisoned and taken to Portugal; cf. the letter from him to D. João III, written at Viana in 1532, AN/TT, CC, III–11–89, in Dias Farinha, 'Os Portugueses no Golfo Pérsico', p. 101.

5 AN/TT, CC, II–109–13, 'Trelado do comtrato e comcerto que of governador Dom Duarte de Menesses fez com el-rei d'Armuz', in Dias Farinha, 'Os Portugueses no Golfo Pérsico', pp. 80–2.

6 Serjeant, *The Portuguese off the South Arabian Coast*, pp. 52–3.

others were burnt in the road. He then goes on to mention the attack on Shihr ('Xael'), described as a large town, well supplied, and plentifully provided with 'all the fruits there are in Spain'! Besides, there was a substantial horse trade there, that drew Muslim merchants from Malabar and Cambay, as well as those ships that missed the monsoon and were unable to enter the Red Sea in time. In Castanheda's view, the Portuguese raid was entirely futile, motivated solely by D. Luís de Meneses's chagrin, 'for he was annoyed that he had not yet done anything in India, and thought he could do something here'. However, the town was found partly deserted, and with very limited pickings for the Portuguese raiding party; nevertheless, it was sacked, 'by which some of them still became rich'.[7] These activities, and the corsair proclivities in the same region of a certain António Faleiro (reputedly a business partner of the captain of Goa, Francisco Pereira Pestana), did nothing to improve the Portuguese reputation, the more so since none of the raids on both places and ships seems to have distinguished too carefully between friends and foes.

Whatever else may have been the case, the Persian Gulf at least occupied D. Duarte's attention. The rest of Portuguese Asia, even the west coast of India, appears to have been of far less interest to him. Where Chaul was concerned, he removed its young captain Henrique de Meneses (who had been named by Diogo Lopes de Sequeira), and substituted him with that well-known (even notorious) *fidalgo*, Simão de Andrade, whose chief qualification to the post appears to have been that he had recently married the governor's illegitimate daughter, D. Brites.[8] In the hinterland of Goa itself, diplomatic efforts had been made by Diogo Lopes to preserve the status quo with respect to the ruler of Bijapur, Isma^cil ^cAdil Shah (r. 1510–34), even though the envoy, João Gonçalves de Castelo-Branco, who had been sent to his court in February 1520, returned with little achieved a year later. Two years later, in 1523, taking advantage of the governor's absence in Hurmuz, the Bijapur ruler sent in a force of roughly five thousand men and, overcoming some Portuguese resistance, took over the *tanadarias* of Ponda and Salsette, until that time in Portuguese control. The captain of Goa, Francisco Pereira Pestana, much too preoccupied with his own internal difficulties with a faction of

[7] Castanheda, *História*, Vol. II, Livro VI, Ch. 23, p. 188.
[8] Cf. João Paulo Oliveira e Costa, 'Simão de Andrade, fidalgo da Índia e capitão de Chaul', *Mare Liberum* 9 (1995), 99–116.

Plate 23 Pepper and other products arriving from upriver in Cochin by boat in the twentieth century.

Portuguese settlers, could offer no worthwhile resistance, and agreed to a truce.[9] Further south, in Kerala, the situation of the Portuguese in the fortresses of Calicut and Cochin also deteriorated. The prince called 'Nambeadarim', whom Albuquerque had encouraged in 1513 to assume the title of Samudri by assassinating his predecessor, appears to have died in 1522; a rather more bellicose successor was on the throne. This new Samudri, in a letter to D. João III written in January 1523, stated his complaints clearly enough. Contrary to what had been claimed at the time peace was made with Albuquerque in 1513–14 and the Portuguese fort in Calicut constructed, trade in his port had not developed compared to that in Cochin and Cannanur. Further, in the interior, one of his 'vassals', a certain 'Taquamayum' had rebelled against him; and when the Samudri had mounted a punitive expedition, the Cochin Raja had come to the former's aid. The Samudri's annoyance is scarcely concealed:

> In the time of our other ancestors, the kings of Cochym never had precedence over us or took our lands, and we in those other times always destroyed their ports and lands, and after Your Highness has taken possession of the land, with your Captains-Major and fortress-captains, we have never again made war on them, and further the King of Cochym, though not a king by rights but on account of the favour of Your Highness who has made him king, still came to do us ill.

In spite of this provocation, the Samudri writes that he has decided not to retaliate, merely to set his own house and that of his 'vassals' in order. But D. João III for his part should prevail on the Cochin ruler not to interfere in his affairs. The letter then recounts the situation of the envoys sent from Calicut to Portugal: the first ambassador who had been sent in 1513, converted and given the name D. João da Cruz in Portugal, had returned home in 1515, but had turned out to be a troublemaker, while the next envoy, the Mappila Koya Pakki, had turned out to be more interested in his own affairs than in those of the two rulers.[10] The

[9] AN/TT, CC, I–30–54, letter from the *Câmara da Cidade* of Goa to D. João III, dated 27 October 1523, reporting quarrels between the captain and a group of *casados* living on the outskirts of the city, men like Manuel de Sampaio, Jerónimo Coelho, Mateus Fernandes, Francisco Pinheiro, Pedro Homem and Jerónimo Rodrigues. In the quarrel, D. Duarte came down naturally enough on the captain's side. This letter is signed by Fernão Rodrigues, Fernão Álvares, Baltasar Moreira, and (Illegible) Sá.

[10] At least eight letters of D. João da Cruz exist, dating variously from about 1515 to 1537; for the references, see Georg Schurhammer, 'Letters of D. João da Cruz in the National Archives of Lisbon', in *Varia I: Anhänge*, ed. László Szilas (Lisbon: CEHU, 1965), pp. 57–9; for the text of one of these, AN/TT, CC, I–60–44, dated 15 December 1537,

letter ends with a plea for D. João III to seek advice from the captain of the fort at Calicut, D. João de Lima, and the factor Pêro Mousinho.[11]

By the next year, however, relations between the Samudri and D. João de Lima had become none too cordial. It is reported by Castanheda that 'when the Moors of Calicut saw the great carelessness of the governor, and that he did not punish them for anything that they did, they gained far more strength than they had had to make war on the Portuguese'. A decision was thus taken to send out eight large ships, laden with spices, to the Red Sea, and protected by a fleet of smaller vessels (*paraos*), all under the command of a Mappila notable and *ᶜâlim* from Tanur (a port between Calicut and Ponnani), a certain Kutti ᶜAli.[12] The Portuguese captain of Calicut catching wind of this, he asked D. Luís de Meneses to guard the Kerala coast, but the latter reportedly refused, preferring to remain in Cochin, and then leaving in October 1524 for Goa, to await the return of his brother, the governor, from Hurmuz. By this time, an overt conflict had begun between D. João de Lima on the one hand, and the Mappilas and the Nayar forces of the Samudri on the other.

This deterioration in Portuguese–Mappila relations thus represents a second phase in the conflict between the Portuguese and Muslim communities in the Indian Ocean. The first phase, which Vasco da Gama had been partly instrumental in bringing to a head, was directed at the Muslim merchants who traded to the Red Sea, whom the Portuguese assumed were also largely from the Middle East (the Maghreb, Egypt, Yemen, the Hadramaut and Iraq), and whom they termed generically as 'Mouros de Meca'. In the first fifteen years of the Portuguese presence in Kerala, they did considerable damage to the commercial interests of these merchants, terrorised them, and in part forced them out of the Kerala ports. Duarte Barbosa, who was not wholly unsympathetic to them, describes the community of *paradesis* ('a que chamam pardetis') in his extensive chapter on Calicut, noting that they included Arabs, Persians, Gujaratis, Khorasanis and Deccanis, and that they had a separate 'governor' to take care of their affairs. He

see Schurhammer, 'Dois textos inéditos sôbre a conversão dos Paravas, pescadores de pérolas na Índia (1535–1537)', in *Orientalia*, ed. László Szilas (Lisbon: CEHU, 1963), pp. 260–2.

[11] AN/TT, CC, I–4–2, dated Calicut, 15 January 1523 (a slightly different copy may be found in CC, I–4–3); both are wrongly catalogued to 1503.

[12] Castanheda, *História*, Vol. II, Livro VI, Ch. 67, p. 241.

continues (and we should recall that his text was written in around 1518):

> Before the Portuguese discovered India, they [the *paradesis*] were so many, and so powerful and free in the city that the Gentiles did not dare walk through it, on account of their arrogance; later, seeing the determination of the Portuguese, they tried to expel them from India, and being unable to do so, they themselves left little by little for their own lands, leaving India and its trade, so that there are thus very few who remained, and without any strength.[13]

However, with the new-found Portuguese interest in the Maldives and in the ports of the west coast of Sri Lanka, the commercial conflict soon extended to the Mappilas, many of whom (notably Koya Pakki of Calicut, but even Mammali of Cannanur to a limited extent) had not been hostile to the early Portuguese presence; indeed, the first Portuguese cargoes in Cochin were supplied to them by Mappila merchants. The wave of real and attempted fortress-building of the years 1518 to 1521, essentially at the initiative of the Portuguese Crown, and designed to control trade within the western Indian Ocean, brought forth new clashes of interest, and thus new enemies. This was further exacerbated by the quarrels between private Portuguese (the *casados* of Cochin, in particular) and the Mappilas over the control of particular resources such as the coastal trade between Kerala, the Pearl Fishery coast of Tirunelveli, southern Coromandel, and Sri Lanka.[14] It is these tensions, and the mounting anti-Portuguese sentiment of the Mappilas, that underlay a seventeenth-century chronicler's condemnation of the quality of government under D. Duarte de Meneses:

> And though it is true that he was greatly feared by the Moors of Berberia, he came to be so little respected by those of India, that on one occasion that he was in Cochin, they passed in sight of the city and him in their *paraus*, and launched flying fireworks towards the shore, which rose to the sky to show their scorn and despisal of us.[15]

[13] Neves Águas (ed.), *O Livro de Duarte Barbosa* (Lisbon: Publicações Europa-América, 1992), pp. 131–2; the same passage may be found in the standard edition, *Livro em que dá relação do que viu e ouviu no Oriente Duarte Barbosa*, ed. Augusto Reis Machado (Lisbon: Agência Geral das Colónias, 1946).

[14] Cf. the discussion in Jorge Manuel Flores, 'The Straits of Ceylon and the Maritime Trade in Early Sixteenth-Century India: Commodities, Merchants and Trading Networks', *Moyen Orient et Océan Indien* 7 (1990), 27–58.

[15] Frei Luís de Sousa, *Anais de D. João III*, ed. M. Rodrigues Lapa, 2nd edn (Lisbon: Sá da

Portuguese fears in the 1520s were that these conflicts with the local Muslim merchants ('Mouros da terra') would find an echo in a larger conflict, that with the Ottoman Empire; for the imminent threat of the 'Rumis' (be they the subordinates of the Mamluks or of the Ottomans), which had in fact abated after about 1510, nevertheless remained present in their minds. As has been convincingly demonstrated, at least during the last years of the brief and vigorous reign of Selim I (1512–20), no real thought was in fact given by the Ottomans to challenging the Portuguese in the Indian Ocean.[16] However, with the accession of his son Süleyman in 1520, the amplitude of Ottoman ambitions, and as a consequence their southern policy, changed somewhat. The expansion into Habesh by the Ottomans beginning in the 1520s is one measure of this, to be sure. But Ottoman plans can be seen far more clearly, for example, in the report presented by the celebrated Salman Ra'is to the Ottoman governor of Egypt, Ibrahim Pasha, in 1525. Salman Ra'is, born in Lesbos and formerly a corsair in Sicily, had in June 1515 been named admiral of the Mamluk fleet in the Red Sea, and continued to serve the Ottomans after the defeat of the Mamluks. His report begins with a list of ships and guns at Jiddah, to be used against the 'infidel Portuguese' (*Portakal-i bedîn*); it then goes on to state that with these weapons, 'it is possible to capture and hold all the fortresses and ports in India which are under Infidel domination'. These territories are then described, beginning with Hurmuz, then Diu (where it is incorrectly stated that the Portuguese have a fort), then Goa (their 'headquarters'), then Calicut and Cochin; finally, mention is made of the Portuguese forts at Colombo and Pasai (the report was thus somewhat out of date, as these forts no longer existed in June 1525). A long section then follows on the province of Yemen, where once again the Portuguese threat is highlighted.[17] Nevertheless, the point must be made that the tenor

Costa, 1951), Part I, Livro 2, Ch. 11. For a discussion of deteriorating relations with the Mappilas, also see Flores, 'Os Portugueses e o Mar de Ceilão', pp. 175–6.

[16] Jean Aubin, 'La politique orientale de Selim Ier', in *Res Orientales VI: Itinéraires d'Orient, Hommages à Claude Cahen* (1994), 197–216; this argument is directed in part at M.M. Mazzaoui, 'Global policies of Sultan Selim', in Donald P. Little (ed.), *Essays on Islamic Civilization presented to Niyazi Berkes* (Leiden, 1976), pp. 224–43, and more particularly at Jean-Louis Bacqué-Grammont, *Les Ottomans, les Safavides et leurs voisins: Contribution à l'histoire des relations internationales dans l'Orient islamique de 1514 à 1524* (Istanbul, 1987). Another recent attempt at synthesis, of somewhat limited interest for our purposes, is Palmira Brummett, *Ottoman Seapower and Levantine Diplomacy in the Age of Discovery* (Albany: SUNY Press, 1994).

[17] Salih Özbaran, 'A Turkish Report on the Red Sea and the Portuguese in the Indian Ocean (1525)', *Arabian Studies* 4 (1978), 81–8, reprinted in Özbaran, *The Ottoman*

of the document is not merely defensive; it also envisages an active, offensive campaign against the Portuguese, and their eventual expulsion from the Indian Ocean: for once Yemen is conquered, 'it would be possible to master the lands of India (*vilâyet-i Hindustân*) and send every year a great amount of gold and jewels to Istanbul (*Devlet-i Asitâne*)'. The Venetians caught echoes of this plan, and a letter from their consul resident in Istanbul, dated December 1525, states that he has met Salman Ra'is and discussed his plans with him. Mention is made of ships that are being constructed to this end, besides others that are already in Jiddah ('Alziden'); Ibrahim Pasha had also apparently indicated to the Venetians his support for the project, in view of the large potential tributes (perhaps from Gujarat?).[18] This thus forms the prelude to the Ottoman expedition to Gujarat actually mounted under the command of Salman Ra'is in 1527–8, and carried through with some difficulty, after his death in a rebellion, by his subordinates Mustafa Bairam and Safar al-Salmani.[19] The expedition was to have profound consequences in the medium term for the élite politics of the Gujarat Sultanate, since many of the Ottoman notables who took part in it became powerful magnates in western India.

THE NEW STRONGMAN

While these raids and expeditions were being planned or carried out in the western Indian Ocean (in a situation in which it appears that D. Duarte de Meneses had little contact with the metropolis), the situation in Portugal had evolved quite considerably. Diogo Lopes de Sequeira, who left India in December 1521 (the very month that D. Manuel died), found this out to his chagrin, immediately on his return to Lisbon. Here, he found himself rudely received by Doutor Fernão de Álvares de Almeida, who took charge of his personal possessions, declaring that he was

Response to Portuguese Expansion, pp. 99–109; also see Michel Lesure, 'Un document ottoman de 1525 sur l'Inde portugaise et les pays de la Mer Rouge', *Mare Luso-Indicum* 3 (1976), 137–60. Özbaran also cites another report, this one on Jiddah and Mecca, also by Salman Ra'is.

[18] Letter from Sier Piero Bragadin, *bailo*, dated 29 December 1525, at Pera, in *I Diarii di Marino Sanuto*, Vol. XL, ed. Federico Stefani, Guglielmo Berchet and Nicolò Barozzi (Venice: The Editors, 1894), pp. 824–5.

[19] The expedition and death of Salman Ra'is are mentioned by the Portuguese chroniclers; cf. Castanheda, *História*, Vol. II, Livro VII, pp. 398, 435; Livro VIII, pp. 585, 607, 616; also see AN/TT, Gavetas, XXIV/1–24, letter from Cristóvão de Mendonça, captain of Hurmuz, to the Duke of Bragança, 30 September 1530, in Dias Farinha, 'Os Portugueses no Golfo Pérsico', pp. 99–101.

under investigation, presumably for financial misdemeanours.[20] Diogo Lopes, at first inclined to believe that this was merely the act of some over-zealous official, complained to the Crown through his brother, then sent word to the *vedores da fazenda*, the Baron of Alvito and D. Martinho de Castelo-Branco, but was told that Almeida was in fact acting on royal orders. Enquiries revealed to him that the inquisitor was the relative and client of the Count of Penela, D. João Vasconcelos (brother-in-law of D. Francisco de Almeida), and was related as well to D. Lopo de Almeida, all of whom Sequeira did not count among his well-wishers. Old scandals regarding the returning governor's wife were dredged up, causing him embarrassment; he hence sought and received an interview with D. João III, but nothing was resolved there. When he retired to his house in Alandroal in the Alentejo, he was pursued there by a Crown agent, Paris Dias, who seized his goods, in connection with the accusation that he had brought back a secret private fortune from India of 150,000 *cruzados*. By mid-October 1522, Sequeira had more or less resolved to leave Portugal, but in order to conceal his intentions, wrote to the Crown that he was going on a pilgrimage to Guadalupe.

Sequeira claimed that he owed the Crown nothing, but that on the contrary it was the Crown that owed him the equivalent of 600 *quintais* of pepper (which he was permitted to trade at Hurmuz, but which he had refused, claiming that such trade was prejudicial to the Crown interest), and six thousand *cruzados*. He also set out at some length a counter-accusation, directed at a certain Jorge Dias, and other men 'who had been punished by me in India'. Nor was he the only victim of a new dispensation that was being put into place very rapidly in 1522–3. Duarte Pacheco Pereira, author of the *Esmeraldo de situ orbis* and flatterer of D. Manuel's messianic and imperial pretensions, had been rewarded with the plum post of captain of São Jorge da Mina in 1519, once D. Manuel had taken matters in hand after his retreat of the years 1514–18.[21] In 1522, Duarte Pacheco was brought back from Africa in chains, and replaced in his captaincy by the son of the Count of Penamcor, a certain D. Afonso de Albuquerque.

[20] AN/TT, CC, I-28-113, letter dated Alconchel, 14 October 1522, in Ronald Bishop Smith, *Diogo Lopes de Sequeira*, pp. 55–8.

[21] Cf. the posthumously published paper by A. Teixeira da Mota, 'Duarte Pacheco Pereira: Capitão e Governador de S. Jorge da Mina', *Mare Liberum* 1 (1990), 1–27; also Jean Aubin, 'Les frustrations de Duarte Pacheco Pereira', *Revista da Universidade de Coimbra* 36 (1991), 183–204.

Towards the end of 1523, after a period of imprisonment, he was however released. Other victims of the 'purge' of 1522 included D. Bernardo Manuel, who was imprisoned in Santarém, after being recalled from North Africa; the infantry captain, Cristóvão Leitão, too, claimed that he was a victim of the change of government, but more in the sense of unkept promises than actual persecution.[22]

In order to comprehend the changes of the period, our best source is the correspondence of the Habsburg ambassador Juan de Zúñiga, in Portugal from March 1523 and in close touch with a number of leading political figures. Zúñiga's mission was multiple: to gather information for his master, to ensure that Portuguese support was not extended to the revolting *comuneros* in Castile, and to prepare the ground for the negotiations on the Moluccas, held in April and May 1524 at Badajoz-Elvas.[23] In one of these letters to Charles V, he declared that he had rarely seen so much discontent in a kingdom, and much of it directed at none other than *el conde almirante*, Vasco da Gama, who had emerged apparently as the *éminence grise*, the strongman of the new regime.[24] In late July 1523, the first news had filtered through that Gama was being sent to India once more, and Zúñiga now wrote:

> Your Majesty would already have come to know from Dotor Cabrero how, for the coming year, the Conde Almirante has been named Captain-General, and that he will depart in March as is usual. I have worked to have some information from his household, and what is known from a trustworthy person is this: that he has been heard to say that he takes up this post to remedy past errors, as much those who have rebelled, as to defend that which this kingdom has won; and that in his time [there], no spices will be brought back from there for Your Majesty; and that he believes that the caravels that are awaited from there will not return here because orders have been given to take them, and to work in all possible ways to send them to the bottom, and in such a way that one cannot know what has been done to them; and that if he could do the same to Your Highness's people who have gone there, that

[22] Jean Aubin, 'Le capitaine Leitão: Un sujet insatisfait de D. João III', *Revista da Universidade de Coimbra* 29 (1983), 96–8.

[23] Some letters on the Moluccan issue from Zúñiga, written between May and July 1523, from the Archivo General de Simancas, are reproduced in an appendix to Luís de Matos, 'António Maldonado de Hontiveras e a "questão das Molucas"', in A. Teixeira da Mota (ed.), *A viagem de Fernão de Magalhães e a questão das Molucas* (Lisbon: IICT, 1975), pp. 548–77.

[24] Archivo General de Simancas, Estado 367, Doc. 87, Zúñiga to Charles V from Évora, 9 May 1524; for a discussion, see Aubin, 'Le capitaine Leitão', pp. 96–7.

> he will not leave it to his successor; and that even if there were an agreement between Your Majesty and the King of Portugal, so that Your Highness's caravels could continue [in] the possession [of the Moluccas], that he would not honour it, and instead work in all possible ways so that their possession is maintained in all that [region], and that of Your Majesty does not take effect, for he will have the apparatus to do so.[25]

Zúñiga, convinced of the soundness of Magalhães's calculations, also reported rumours that the navigational charts in Portugal were being altered, to ensure that the Moluccas fell in the Portuguese half of the world, and that a Castilian astronomer had been called in to aid them. The letter ends on an interesting note: Zúñiga reports that he has been contacted by Diogo Lopes de Sequeira, who wished to see him 'in great secret', perhaps about the charts; an appointment has been fixed.

In his next letter, barely two weeks later, the Spanish ambassador reports a somewhat changed situation. For one thing, he notes that the plague, very much present in Portugal that year, had begun to take a substantial toll. A discreet mention (in cipher) is made of Diogo Lopes, and how he desires to serve Charles V; no further news is available on Vasco da Gama, who 'is now at his residence', presumably in Vidigueira. A great part of the letter is concerned with the affair of D. Bernardo Manuel, who had been imprisoned in Santarém, but who had managed to escape and flee to Castile, taking advantage of laxity caused by the plague. Zúñiga had meanwhile been approached by one of his brothers, pleading his case before the Habsburgs.[26] Zúñiga's later letters, of August and September 1523, then continue with further details. It would appear that on the night of 18 August, the Castilian ambassador was paid a visit in his residence (*posada*) at Tomar by Diogo Lopes de Sequeira, who appeared alone and on foot. Diogo Lopes mentioned the grievances he had had against D. João III ever since his return from India, and stressed how useful he could be to the Habsburgs, as someone who had unparalleled knowledge of the spice trade, and of the navigational routes to India. By way of a first demonstration of his intentions, he handed over to Zúñiga recent letters that he had received from some of his clients and friends (*criados y amygos*) who were still in India; the

[25] Archivo General de Simancas, Estado 367, Doc. 115, Zúñiga to Charles V from Tomar, 21 July 1523.

[26] Archivo General de Simancas, Estado 367, Document 117, letter from Tomar, dated 6 August 1523.

Castilian ambassador promptly had copies made of them, and sent them on to Charles V.[27]

A good part of this letter is equally devoted to the preparations being made for the fleet that is to be sent out to India the next year. Some new ships were being constructed, bombardiers and other men-at-arms are being recruited, supplies (such as iron from Biscay, and munitions from Flanders) are being brought in. All in all, one gathers the impression of a great mobilisation, and indeed we are aware that Gama took as many as fourteen ships with him the next year, possibly the highest number that had been attained after Lopo Soares de Albergaria in 1515. At the same time, the frenetic activities of cartographers (men like Lopo Homem) and astronomers around the court find mention, in the struggle to define the real position of the Moluccas. Zúñiga naturally kept an eye on shifting fortunes in the court as well, noting that the stars of the Baron of Alvito and Luís de Silveira were on the decline, while those of the Count of Vimioso and D. António de Ataíde were on the ascendant; he also remarks that financial matters were rather badly organised (*en la hazienda ha mala orden*), causing concern all around. All in all, we gain the impression of an epoch of rumours and counter-rumours, of political agents negotiating positions based on real or imagined geographical and practical expertise, all of this the confused aftermath of the voyage of Magalhães and Elcano.[28]

A later letter, of 19 September, for its part largely concerns the affairs of D. Bernardo Manuel, who it turns out had a large number of enemies in the court, ranging from the Count of Vimioso to the Count of Vila Nova (D. Martinho de Castelo-Branco). Nevertheless, it had begun to appear that with the intercession in his favour by Charles V, the Portuguese king might actually overlook some of his earlier actions, including his unauthorised departure from Santarém. We may note too that, in closing this letter, Zúñiga mentions that a certain *vedor da fazenda* has been sent to aid the Count-Admiral in arranging all that he requires (presumably for the departure of the India fleet); we may suppose that this was Nuno da Cunha, whose name

[27] Archivo General de Simancas, Estado 368, Doc. 24, Juan de Zúñiga at Tomar to Charles V, 29 August 1523.

[28] After his return in 1522, Elcano was sent later in the 1520s as chief pilot of another expedition to the Moluccas, this one under the command of Fray García Jofre de Loyasa; the expedition failed, and some of its survivors wound up in Portuguese Asia; cf. J. Oyarzun, *Expediciones españolas al estrecho de Magallanes y Tierra de Fuego* (Madrid, 1976).

appears in several documents concerning the fleet, to which we shall refer below.[29]

We have already briefly noted the contacts between Zúñiga and another one of those who felt aggrieved with the new regime: Duarte Pacheco Pereira. He for his part went even further than Diogo Lopes, who at least held back some of his cards, rather than declare himself wholly pro-Castilian and anti-Portuguese. If the latter offered by January 1524 to come over to Castile with his troops (and we may note that Diogo Lopes had been named *almotacé-mor* in Portugal in late 1523, in spite of his distance from the centres of power), this was still a means of settling scores, rather than being rooted in a larger statement.[30] The rather more pretentious Duarte Pacheco, on the other hand, stated to the Castilian ambassador his willingness to head an expedition to conquer Arabia Felix, the area around Aden and Yemen, on behalf of Charles V, and also made no secret of his conviction that the Moluccas lay well within the part of the globe that pertained to Castile according to the Tordesillas treaty, perhaps by as much as five degrees longitude.[31] Indeed, so many and so varied were the verbal attacks on Gama and his power that the Castilian ambassador found himself in the piquant situation of having too many potential allies in élite Portuguese circles – an embarrassment of unexpected riches as it were. Nevertheless, though spurning some of the overtures made to him, he did make it a point to retain his good relations with Diogo Lopes, who for his part continued to supply Zúñiga with details of key matters (such as a *relacion de las naos*, perhaps those on their way to India the next year).[32] Since Diogo Lopes was one of the Portuguese delegates at Badajoz-Elvas, in the negotiations with Charles V, this places his whole participation there in a rather ambiguous light.[33] Of some importance is Zúñiga's stress that the former

[29] Archivo General de Simancas, Estado 367, Doc. 106, Juan de Zúñiga at Tomar to Charles V, 19 September 1523.

[30] A few useful elements for the last years of Diogo Lopes de Sequeira's life may be found in the privately published booklet by Ronald Bishop Smith, *Diogo Lopes de Sequeira: Elements on his office of Almotacé-mor* (Lisbon: Silvas, 1993).

[31] Cf. the discussions in Aubin, 'Le capitaine Leitão', pp. 96–7, and Aubin, 'Les frustrations de Duarte Pacheco Pereira', pp. 202–3. In the latter essay (p. 203, n. 65), Aubin notes the reluctance of a number of modern Portuguese authors, 'par réflexe de nationalisme traditionnel', to believe that Duarte Pacheco could have engaged in these negotiations.

[32] Archivo General de Simancas, Estado 367, Doc. 98, Zúñiga at Montemor-o-Nôvo to Charles V, 4 December 1523.

[33] For a general view of the discussion of the failed Luso–Castilian negotiations of April–May 1524, see Luís de Albuquerque and Rui Graça Feijó, 'Os pontos de vista de D.

governor of Portuguese India was motivated above all by the fact that 'he wishes great harm to the Count-Admiral', who was qualified as thoroughly unreasonable in temperament, besides being the 'greatest enemy of the Castilians' (*el mayor enemigo de Castellanos*) in Portugal.[34] Thus, Castile, not so long ago a potential place of exile for Gama, became his central target, at least in the view of Zúñiga and his informants.

But Zúñiga's portrayal, while evocative enough, is nevertheless a partial one, precisely on account of its excessive focus on the Moluccan question. For that was not Vasco da Gama's sole brief: rather, he took with him to India a far wider-ranging programme that was to be implemented in Portuguese Asia. For, where Asia was concerned, we see now, more clearly than ever before, the great difference between Gama and Afonso de Albuquerque, authoritarian though both were. Albuquerque had defended with vigour his conception of the western Indian Ocean as a chain of fortresses, from Kerala to East Africa, with Cochin, Cannanur, Goa and Hurmuz as key points, and Aden to be captured at a later date. In contrast Gama's position was recalled with nostalgia by the Duke of Bragança in a letter to D. João III in 1529, in which he wrote:

> The Count of Vedigeira in my view understood the affairs of India better than anyone else, and his vote was that Malaca should be sold to the king of Abitão [Bintang], and I do not recall whether with Ormuz too a similar bargain should be struck, and that these and all the other fortresses in India should be levelled, except Goa and Cochim, it being certain that if in the beginning of the decision to navigate [to India], this had been kept in mind, it would have been wonderful. If only the emperor [Charles V] had been left, by common consent, with the places [that depend] on the Algarve, that is Ceita, Alcácer, Tanger, Arzila, and only Azamor and Safym had remained with us, for which a means could be found to support them very easily and honourably, and they would cost very little money, and profit would come from them to the kingdom [Portugal].[35]

For once, it would seem, contemporary documentation backs up the extravagant claims made by Gaspar Correia in his *Lendas da*

João III na Junta de Badajoz-Elvas', in Teixeira da Mota (ed.), *A viagem de Fernão de Magalhães*, pp. 527–45.

[34] Archivo General de Simancas, Estado 367, Doc. 109, letter from Zúñiga at Évora to Charles V, 4 January 1524.

[35] AN/TT, Gavetas, XVIII/10–10, 'Carta e parecer do duque de Bragança sobre os lugares da Africa', Vila Viçosa, 12 February 1529, in Silva Rego (ed.), *Gavetas*, Vol. IX, p. 539.

Índia. In Correia's portrayal, Gama left Lisbon with a sizeable fleet of fifteen (rather than fourteen) ships, and powers and pomp which rivalled that of the King of Portugal himself (*'trouxe total poder de justiça e fazenda, como pessoa de El Rey'*). On board the fleet were some three thousand men, including a sizeable contingent of the upper and middling nobility as well as Gama's own sons Estêvão (appointed Captain-Major of the Indian seas) and Paulo.[36] Nevertheless, Correia stresses, Gama's purpose was mixed: some centralising initiatives, particularly in respect of finances and the soldiery, and some other conceptions of reform that were intended rather clearly to divest Portuguese Asia of what was perceived as deadweight, including a large number of unwanted fortresses.

It is of some importance to note, besides, that Gama received the title of not merely governor but viceroy, the first to have that dignity after D. Francisco de Almeida in 1505. As in the case of the fleet led by Gama in 1502, a fairly voluminous documentation exists concerning the supplies to the 1524 fleet, in the form of orders issued by the *vedor da fazenda*, Nuno da Cunha (later himself to be governor of Portuguese India), and in terms of receipts signed by the masters and intendants of ships on the eve of departure.[37] In the spring of 1524, just before leaving, Gama also made it a point to clarify the legal position of his varied and extensive fiscal privileges, both in Portugal and in India, receiving confirmations for several of these from D. João III. On the 28 February, Gama appeared formally before the king to accept the post of 'viceroy of the parts of India, and to remain there as captain-major and governor of the said parts', swearing an oath of fealty and homage (*preyto e menagem*) to the king in this capacity, three times over, in the presence among others of the Count of Vimioso.[38]

The unsavoury machinations revealed by Zúñiga's correspondence represent an important counterpoint to the official historiography, which sees in Vasco da Gama not a political actor, but an

[36] Correia, *Lendas da Índia*, Vol. II, pp. 815–16. For an analysis, based in large measure on Correia's narrative, see João Paulo Oliveira e Costa and Vítor Luís Gaspar Rodrigues, *Portugal y Oriente: El Proyecto Indiano del Rey Juan* (Madrid: Colecciones MAPFRE, 1992), pp. 183–7.

[37] Cf. for example, AN/TT, CC, II–113–113, II–113–164, II–113–165, II–114–42, orders from Nuno da Cunha to the *feitor e almoxarife da ribeira* in Lisbon, Sebastião Gonçalves, for supplies to the fleet of the Conde Almirante, respectively dated 27 February, 9 March, 22 February and 26 March 1524.

[38] AN/TT, CC, I–30–90, transcribed in Stanley, *The Three Voyages of Vasco da Gama*, Appendix, pp. viii–ix.

instrument of royal policy. This is a tradition harking back to the chronicles themselves, and we may cite, by way of example, the discussion in Francisco de Andrada's *Crónica de D. João III*. Andrada poses the sending of Gama to India as follows:

> The King Dom João, who at this time had held the sceptre of this, his kingdom, for no more than about two years, perceiving how important it was both for the honour and for the profit of the *Estado da Índia*, which his father the King Dom Manuel had left to him, having won it the cost of so much blood, so many lives, and so many valorous deeds on the part of his vassals, determined to send a man to govern there, through whom (since he would be the first that he was choosing for that post), India would perceive the great importance that was given to what was necessary for it, and with whom His Highness would be sure not only that he would preserve what had been gained, but increase it as far as possible.

Thus, the choice of successor for D. Duarte de Meneses is posed explicitly as a question both of consolidation and of further imperial expansion, an interesting distortion in view of the chronicle's composition in the early seventeenth century, and this chronicler's oft-remarked dependence on Gaspar Correia. Who then could be fit for this task?

> And it seemed to him [D. João III] that for this purpose, there could be no one more adequate than Dom Vasco da Gama, Count of Vidigueyra, and Admiral of the Indian Sea, who had discovered it, both on account of the experience that he had of her affairs, and because the Moors knew him from the time of the discovery, and from the other voyage that he had made afterwards, and on account of the great respect and reverence that they hence had for him.[39]

Gama is thus summoned to the court from Vidigueira, where he is 'resting from past travails' (*descansando já dos trabalhos passados*), and told why he has been called, and how important it is for him to go to India. Seeing that it is a matter of honour, Gama kisses the royal hand and agrees to go, so long as he is named viceroy (a title that he is to make use of, once he touches the first Portuguese fort in India), if all his sons are in turn given the captaincy of Melaka, and if his second son Estêvão is named *capitão-mor do*

[39] Francisco de Andrada, *Crónica de D. João III*, ed. M. Lopes de Almeida (Oporto: Lello e Irmão, 1976), pp. 151–2.

mar da Índia. The king agrees to these conditions, and the Count Admiral dutifully departs, in this bland official tale.

In point of fact, Gama went with an entire new 'team', as it were, to replace those in all the principal positions in India. Thus, accompanying him were new captains for Hurmuz, Goa, Cannanur, Cochin and Melaka, besides the complex and altogether enigmatic figure of a new *vedor da fazenda*, to replace the incumbent, Dr Pedro Nunes, whose powers had been whittled away, first by Diogo Lopes de Sequeira, and then by D. Duarte de Meneses.[40] This new *vedor* was Afonso Mexia (1477–1557), who had already served D. Manuel as *escrivão da fazenda*, and equally been factor at São Jorge da Mina, in west Africa.[41] A *cavaleiro* of the Order of Christ, and a rich, powerful and manipulative figure himself protected by his links to D. João da Silva de Meneses, Count of Portalegre (d. 1551), Mexia had managed at the end of some fifty years of Crown service to found a substantial property (*morgado*) in Campo Maior, despite his many powerful enemies.[42] On the outward-bound fleet of 1524, he appears as the right-hand man of the viceroy, a situation from which he was to draw considerable benefits. For once, unlike what had been the case with either Fernão de Alcáçova or Dr Pedro Nunes, it appeared that the executive and the fiscal authority in Portuguese India, though split, were in harmony. In some senses, Mexia was the first occupant of the position of *vedor da fazenda* in India to have any real political weight; some might argue, as we shall see below, that he in fact carried far too much weight.

It is clear from the documentation that besides the Castilians in the Moluccas, and the broad organisational reform, Gama had still another agenda, this one a more personal one. Preoccupied as he had been from early in his career with squabbles within the nobility, and between the military orders, the new viceroy intended to make the most of his powers, to humiliate and

[40] By way of example, in March 1524, D. Duarte de Meneses issued an *alvará*, giving powers to the captain of Goa, Francisco Pereira Pestana, to take charge of the *fazenda*, overruling the powers of the *vedor*; AN/TT, CC, II–114–26, dated 18 March 1524.

[41] For his *cartas de quitação*, see AN/TT, Chancelaria D. Manuel, Livro 3, fl. 29; Chancelaria D. João III, Livro 1, fls. 37, 38v, 39v, and Livro 4, fl. 18. His letter of nomination as *vedor da fazenda* in India is to be found in AN /TT, Chancelaria D. João III, Livro 45, fl. 132v.

[42] For biographical notes on Mexia, see Jorge Borges de Macedo, *Um caso de luta pelo poder e a sua interpretação n' 'Os Lusíadas'* (Lisbon: Academia Portuguesa de História, 1976), pp. 114–16, 127–8; also the earlier discussion in Anselmo Braamcamp Freire, 'Livro das tenças d'el Rei', *Arquivo Histórico Portuguez* 2, 6 (1904), 214–22.

disgrace the governor from whom he took charge, D. Duarte de Meneses. Meneses, unlike Gama, was a member of an old noble family; he was also still closely associated with the Order of Santiago, with which Gama's relations remained bitter. Besides, his comportment in his three years in India had been questionable on a number of fronts, as we have seen. True, one of the most swingeing and wide-ranging attacks on Meneses's government, that in a long and eloquent letter written to D. João III from India in October 1523, could not have been available to Gama until the very eve of his departure for India, but the Admiral had other agents and sources of information in place.[43]

We see this for example from an extensive document that was obviously prepared for his use, entitled 'Articles for the investigation of D. Duarte, Captain of India'.[44] These are accusations against the governor that are to be investigated, and questions using which potential witnesses are to be examined. By today's standards, many of these might be considered 'leading questions', blatantly designed to nudge the witnesses in a particular set of directions. They underline the gravity of the situation in which Meneses found himself with respect to the Crown by late 1524. To begin with, then: 'Did he trade with the King's money, and from which factories or revenues of the King did he take it, and by what means?' The list goes on. Did he trade in pepper? Did he trade in spices, or in other forbidden goods that formed a part of the Crown monopoly (such as sugar or iron, to Hurmuz)? Who were his agents in the trade, did he have his own ships for this purpose, and who were their captains, masters and pilots? Who were his business partners, if any, and did they include any Indians? The questions now become increasingly specific on financial matters. Did D. Duarte have the habit of dipping his hand into the money of deceased persons in Asia, which was notionally meant to be transmitted to their heirs in Europe? Had he asked his client (*seu criado*), Manuel de Frias, who was at that time captain on the Coromandel coast, to take hold of the money and goods of the deceased Florentine merchant, Piero Strozzi; and if so how much was it worth, and how had he invested it?

[43] AN/TT, CC, I–30–36, letter from António da Fonseca to D. João III, Goa, 18 October 1523, in *Documentos sobre os Portugueses em Moçambique e na África Central, 1497–1840*, Vol. VI (1519–37) (Lisbon: CEHU, 1969), pp. 180–237 (Portuguese text and English translation). Fonseca's letter, had it been written a half-century later, could easily have been read as another example of Portuguese *decadência*.

[44] AN/TT, CVR, Doc. 37 (5 folios), 'Artigos para por elles se tirar a residência a Dom Duarte Governador da Índia' (modern title).

How were the goods of Piero Strozzi evaluated and sold: was it by auction or by simple valuation, and in either event, who bought them?[45] The list goes on and on, accusing D. Duarte of having received bribes and gifts in matters of justice (and of thus having encouraged criminality), but equally in order to appoint certain men (who were not the king's men) to captaincies and other posts. It is suggested that the salaries of some public officials have been augmented without authorisation. He is accused, implicitly, of showing a lack of favour to Indian Christians, whom the Crown for its part wished to encourage. Then there is the question of prize-taking on the seas which obviously has as much to do with his brother, D. Luís de Meneses, as with D. Duarte himself. How had the captured slaves been divided, and had they (illegally) been given in place of salaries to soldiers and mariners? How had the division of the prizes been organised? Besides, did D. Duarte receive and pocket gifts and presents from Asian rulers that were in fact destined for the royal treasury?

It would be tedious to go on through the whole mass of accusations, and we may content ourselves by noting that they focus above all on financial matters (private trade, bribes, gifts), then on procedural matters (the overruling of royal regulations, the lack of cooperation with the *vedor da fazenda*), and thirdly on the governor's personal comportment. Was he given to attending extravagant dinners and banquets in his honour, sleeping with married women, and local Moorish and Gentile women (*mouras e gemtyas da terra*)? There is also the accusation that he had his eye on the riches possessed by the churches, their services, for example. Was it true, again, that several people had fled Portuguese India with Crown ships and become rebels (*alevantados*) because they were mistreated by him? There was also the question of payments to the soldiers. Procedures established late in D. Manuel's reign forbade payments in kind (say, copper or slaves), but D. Duarte was accused of doing just that. However, a deeper examination points to the fact that the accusations are often less ideological or procedural than personal or factional in nature. It was less a matter of what was done, than the fact that it was D. Duarte de Meneses who had done it. For instance, D. Duarte is accused (like Lopo Soares de Albergaria) of permitting a great many persons to build their own ships and

[45] For a brief biography of Piero di Andrea Strozzi (1483–1522), see Sanjay Subrahmanyam, *Improvising Empire: Portuguese Trade and Settlement in the Bay of Bengal, 1500–1700* (Delhi: Oxford University Press, 1990), pp. 1–15.

trade within Asia, and to have taken bribes from them for giving such permissions. Now, aside from the question of bribes, and whether wood, materials and craftsmen of the Crown were illegally used to build these private ships, the mere fact of encouraging such private trade was perfectly within the scheme of affairs envisaged by men like Vasco da Gama (and explicitly advocated by his brother, D. Aires da Gama, in his letter to D. Manuel of early 1519). The suspicions against D. Duarte have a far more malicious quality, ranging from the suggestion that he had received bribes from Malik Ayaz in order not to pursue matters to the hilt in Diu, to the accusation that he and his brother had mishandled the affairs of Hurmuz, both the dealings with Ra'is Sharafuddin and the cruises on the south Arabian coast (where it is alleged that D. Luís attacked ships carrying his own safe-conducts).

Now, the potential and actual informants using whom these charges were drawn up and could be investigated already in Lisbon in 1523–4, are listed in an appendix to the document. These included Garcia de Sá, who claimed to have bribed D. Duarte with a gold necklace and a gold cup, in order to be granted the captaincy of one of the return ships from India, Manuel Pacheco, António Coelho, Jerónimo Lobato, Aleixos de Sousa, Pêro Lopes de Sampaio, Jorge Botelho, Duarte Fernandes de Beja (brother of the by then deceased Diogo Fernandes) and a number of others, including a number of Gama's dependents (listed thus as *criados do almirante*) such as Domingos Fernandes, Tomé de Sousa and António Vaz (who had held office as bailiff in Chaul). We may recall that in order to pursue his own annual duty-free private trade to India (which he had received as a Crown grant), Gama was in the habit, from about 1515, of sending some of his chosen men to India; these men must equally have served him as informants and correspondents.

THE RETURN TO INDIA

The original intention appears to have been to have the fleet leave Lisbon by March, if not as early as February. However, as Juan de Zúñiga remarks in one of his letters (this one dated early May 1524), the fleet lifted anchor only on 9 April, and after having been forced to remain for eight days within sight of Lisbon (*ocho dias a vysta de Lisboa y de sus pontas*), on account of inclement weather in the Atlantic, finally left for the open sea on 16 April.

The ambassador attempted to ascertain the number of men on board, but came up against several and varied rumours. Some estimated the numbers at three thousand, including the merchants and everybody else aboard, others at two thousand, excluding the seamen on board (*sin los mariñeros*); whatever the case, the ambassador concluded that there were certainly over 2,500 men on board, and that the fleet (which he estimates at fourteen vessels, including three caravels and two galleons), was extremely well armed and supplied.[46]

A roughly similar picture emerges from the correspondence of the Cremonese merchant Affaitati, two of whose letters (respectively dated 8 and 16 April 1524) have come down to us. The first of these, more elaborate, and directed to a correspondent in Venice, runs:

> This fleet for India has not yet departed, and the only thing it awaits is good weather; they are 14 sail, amongst which are 8 large ships, and they are all very well in order: May God carry them. A small galleon came from there, which left in November; it brings news that this year a good quantity of spices would have arrived, if departures [viz. ships] were not lacking, there would have been no lack of these [spices]. The news that have come from there are of great moment. Some people have been lost in China, because those of the land have sent two ships to the bottom with their crews, so that it is now costing a great deal to keep India, and every year there is need of new people and gross capitals, and this fleet will cost 350,000 ducats, of which 100,000 ducats are being carried in cash.

He goes on to note that the Castilians have decided that year not to send ships to the Moluccas, in view of the negotiations under way, in which they believe they will succeed; thus, they do not wish to provoke the Portuguese any further. In his later letter, Affaitati notes that the fleet has departed at last, despite the lack of very good weather. He hopes that a good return cargo will be sent, with spices of every sort, but especially cloves and cinnamon.[47]

On this third voyage, Gama's fleet faced considerable difficulties. We may note in passing that some crucial details concerning

[46] Archivo General de Simancas, Estado 367, Doc. 87, letter from Juan de Zúñiga to Charles V, dated 9 May 1524.

[47] Letters from Francesco Affaitato to Sier Marco da Molin in Venice, dated 8 and 16 April 1524, in *I Diarii di Marino Sanuto*, vol. XXXVI, eds. F. Stefani, G. Berchet and N. Barozzi (Venice, 1892), pp. 352–3.

the makeup of this fleet are unclear; still, we are more or less certain that there were fourteen vessels (only Correia insists on the higher number of fifteen), and that a minority (four, or perhaps five) were caravels. We do know that Gama's own vessel was called the *Santa Catarina do Monte Sinai*, and we also have the names of seven others: *São Jorge* commanded by Francisco de Sá de Meneses, *São Sebastião* commanded by Pêro Mascarenhas, *São Roque* commanded by D. Henrique de Meneses, *Santo Espírito* commanded by D. António de Almeida, *Santa Elena*, *Espadarte*, and *Santa Clara*. The names of the captains are also roughly the same in most texts (with the exception of Correia); the one point of difference is that some of the chronicles and texts (like Castanheda) mention D. Jorge de Meneses as a captain, and others D. Simão de Meneses, captain-designate of Cannanur. We have already noted that ill-luck dogged this fleet, from its difficult departure in April from the banks of Tejo, onwards. Thus, once in the Indian Ocean, three of the ships – those captained by Francisco de Brito, Cristóvão Rosado and Mossem (or Misser) Gaspar Malhorquim (whose last name is an ethnic designation, to show he was from Majorca) – were separated from the main body of the fleet. The first two were apparently wrecked, while the ship of Gaspar Malhorquim seems to have witnessed a mutiny; the mariners, pilot and master killed him, and made off to the Red Sea, to engage in corsair activity. The galleon of D. Fernando de Monroy (captain-designate of Goa), for its part, hit a reef not far from Malindi, and lost sight of the fleet. Thus certainly four, and probably even five, of the original fourteen vessels did not make it to the Indian west coast. Arriving in Mozambique in August, after a rapid voyage, Gama for his part did not tarry, and made directly for Chaul, skirting the coast of Gujarat. Here, in the western Indian Ocean, on the night of 6 September, a massive and curious earthquake-like phenomenon shook the seas, making it appear to some of the more credulous among the Portuguese that they were witness to a celestial sign. After all, Vasco da Gama was on his way back to India.

Gama now had occasion to revisit the episode of the *Mîrî*. A few days after the earthquake, not far from Diu, the fleet encountered a Muslim-owned vessel returning to India from Aden, which was carrying goods and cash. This was taken by a single Portuguese vessel, in Castanheda's version that of D. Jorge de Meneses, on the open seas; Gama very scrupulously had all the goods and cash taken off, accounted for by a scribe, and kept apart; he wanted no scandals to circulate concerning his prize-taking. The ship apparently carried 60,000 *cruzados* in cash, and goods worth 200,000 *cruzados*, a rich enough prize by any measure. But this time, it would appear that Gama neither burnt the vessel, nor disdained its contents. Tomé Lopes might well have approved.

By mid-September, the fleet arrived at the Konkan port of Chaul, where a new fort had been built; this was Gama's first landfall in India, and here he could at last assume the title of viceroy, in keeping with the terms of his appointment. We may follow the succinct but revealing narrative of Castanheda here, setting out as it does the main elements of Gama's plan:

> And some days later he [Gama] arrived at the road in Chaul, and there he declared himself viceroy for it was thus stated in his instructions (*regimento*): and he remained there for three days without setting foot on land, or allowing anyone else to set out, save the magistrate (*licenciado*) João de Soiro [Osório] who was a judge of the court of appeals, and who accompanied him to India as *ouvidor geral*, and Bastião Luís who was named to the post of scrivener of salaries in Cochim, whom the viceroy ordered to inspect the fortress of Chaul on his behalf, and to make a public announcement in his name, that except for the settlers (*fronteiros e casados*), everyone else should embark immediately and leave with him, for otherwise they would be deleted from the list of salaries and maintenance-allowances (*soldo e mantimento*); and he also ordered them to tell Cristóvão de Sousa who was captain of the fortress, that if D. Duarte de Meneses, who was in Ormuz, arrived there on his return, he should not allow him to disembark, nor give him supplies for more than four days: all of which was done.[48]

Thus, on the one hand, a measure of fiscal economy, namely the desire to remove from the royal payroll all those who had no real place there; second, a form of strict discipline over his crew and even vigilance that is in marked contrast to Gama's comportment on earlier voyages. Thus, we hear that Gama was particularly insistent that no one should set foot on land, which occasioned some grumbling. Many on the fleet were unwell, and desperately eager to leave shipboard, but Gama would simply not allow it. Others were anxious to disembark to engage in some buying and selling: this too the viceroy saw fit to turn down. Instead, he departed for Goa, where he himself disembarked with some

[48] Castanheda, *História*, Vol. II, Livro VI, cap. 71, p. 265.

chosen companions, leaving D. Jorge de Meneses in charge of the fleet (according to the version of Castanheda).

The situation in Goa in 1523–4 has already been sketched out briefly, in the context of the conflicts between the captain, Francisco Pereira Pestana, and a group of settlers. We have also noted that D. Duarte for his part was largely inclined to favour Francisco Pereira in this quarrel. The captain was an ally of the governor, and allegedly had in place a series of associates, to aid him in his business ventures. On the other hand, there were others in Goa who were hostile to him, ranging from Rui Gonçalves de Caminha, at this time in charge of the finances of the horse-trade (but later to rise, in the 1540s, to the post of *vedor da fazenda*), to the factor Lançarote Fróis.[49] The *vedor da fazenda*, Dr Pedro Nunes, for his part, was kept at a safe distance, in Cochin. Gama, on his arrival in Goa, made substantial changes, putting in place as factor a man of his own choosing, Miguel do Vale, who appears to have already begun exercising this function by late September.[50] The events of the period are partly summed up by a letter from the Goa Municipal Council to Dom João III, dated 31 October 1524, and signed by four *casado*-representatives, Cristóvão Afonso, Diogo João (?), Paio Rodrigues and Pêro Gonçalves. After the usual expressions of loyalty and fealty, and noting the return to Goa of their scrivener, a certain Luís Fernandes Colaço, who had been sent to Portugal as agent of the Council, as well as the confirmation through him of the privileges granted to the city of Goa, the letter continues:

> The Count of Vidigueira arrived in this city on the 23rd of September of the present year of 1524 with nine vessels, it is said that five are missing from the number that left with him from there. It seems to us that he comes with good intentions and desirous of serving Your Highness and doing justice to all parties, which is much needed in this land, for we see that in the few days that he was in this city he made reparations to many persons, and

[49] The financial role of Rui Gonçalves de Caminha may be seen from a number of *mandados* and *conhecimentos* bearing his name, and dating from 1524: cf. AN/TT, CC, II–44–146, *conhecimento* dated 6 February 1524 signed by Pedro Cabreira; CC, II–113–76, another *conhecimento* dated 10 February 1524, signed Duarte Teixeira; CC, II–114–20, *mandado* from D. Duarte de Meneses to Rui Gonçalves, dated Goa, 15 March 1524; CC, II–116–132, *conhecimento* from Pedro Lopes de Tovar dated 4 July 1524; CC, II–119–31, *mandado* from Francisco Pereira Pestana to Rui Gonçalves, 11 September 1524. For his later career, see Luís de Albuquerque and Margarida Caeiro, *Cartas de Rui Gonçalves de Caminha* (Lisbon: Publicações Alfa, 1989).

[50] AN/TT, CC, II–119–118, *mandado* from Gama to Lançarote Fróis or Miguel do Vale, asking for flags for the galliot *Santa Clara*, Goa, 28 September 1524; AN/TT, CC, II–119–137, another *mandado* to Miguel do Vale, dated Goa, 1 October 1524.

> cleaned up many errors that had been committed in respect of Your Treasury. He was received by us in this city with the honour due to those who love justice and follow your orders. We presented him with our privileges and liberties, he said that he would respect them as they had been granted to us by Your Highness, and on account of the limited time that he was in this city, he was not able to take care of certain matters that we had requested, and since the time was short to make up the cargo of pepper, he did not even wish to look into many matters that [thus] remain for his return.[51]

A faint tone of complaint has begun to creep in here. Gama has not given Goa adequate attention, it would seem, from the viewpoint of its residents. However, they are also agreeably surprised with certain aspects of his comportment, and note that although – as with every new governor – several persons went to him with gifts, he 'did not wish to take anything from either Christian or Moor, not even from this City, which all of us here found strange, for it is the custom to take everything'. The City Council also goes on to note, at first rather blandly, that the captain, Francisco Pereira, has been substituted by D. Henrique de Meneses, 'since Dom Fernão [de Monroy] who it is said came to be captain, has not arrived'. On the subject of D. Henrique, the Municipal Council guarded its judgement: 'We can say nothing about him save all that one should say about a good *fidalgo*, and we are treated by him with reason and justice, and Your Highness has sent and commended him.' The letter then goes on to develop the theme of the importance of the Portuguese settlement at Goa, made up of some 450 or 500 *casados*, amongst whom they number a good sprinkling of '*fidalgos*, and cavaliers and squires [who are] your servitors, and other people of much merit'. God has so ordained that they should have left their native lands (*nosas naturezas*) to populate this distant land, but the least the Crown and its representatives can do is to protect their interests.

The actions of the former captain Francisco Pereira now come in for some guarded praise: after all, he has helped finish the construction of the Franciscan monastery, a hospital (near the Santa Catarina gate), and also had a stone jetty made for vessels.[52]

[51] Letter from the Goa Municipal Council to Dom Joao III, dated 31 October 1524, AN/TT, CC, I–31–83, reproduced in an appendix to Stanley, *The Three Voyages*, pp. x–xvi.

[52] On the Franciscan monastery in Goa, see the letter from Frey António Padram, O.F.M. to D. Manuel, dated Cochin, 27 October 1520, AN/TT, Fragmentos, Maço I, in Schurhammer, 'Carta inédita sobre a fundação do convento de S. Francisco de Goa', in *Orientalia*, pp. 207–12.

On the other hand, there were also negative aspects to his captaincy, to begin with the loss in April 1524 of the lands on the *terra firme* that had been conquered by Rui de Melo. Here, the tone of the letter becomes somewhat sharp.

> We cannot judge who should be blamed for the loss of this land. But Dom Duarte, who was governor in these parts, was in the road of this city with the fleet, ready to depart for Ormuz, and he was asked to help, and told that with a few people the Moors could be expelled from the land. He responded that he could not do it for he was on his way, and that even if Goa were lost, that would not prevent him from going to Ormuz, which was not really necessary for your service. His brother, Dom Luís too was in this city at that time, and went off to winter [the monsoon] in Cochym and he took all the people that he could.

Thus, the limited efforts of Francisco Pereira proved vain; the forces of the Bijapur commander Yusuf Lari (later to be far more celebrated as Asad Khan Lari of Belgaum) triumphed.[53] To close, the letter launches (after its earlier guarded tone) into a fully fledged attack on the ex-captain Francisco Pereira, who has systematically undermined the rights of the *casados*, imprisoned some, expelled others from their homes which he had then taken over, besides monopolising the grain trade to their disadvantage. So intense were the conflicts, that at best twenty of the *casados* in Goa now had no grudge against the former captain, amongst whose many sins were favouring the temporary residents (*homens ssolteyros*) over the *casados*. In this respect, we see something of a change in overall tone between this letter of the Municipal Council, and that of the previous year, where Francisco Pereira is not overtly attacked or criticised, and a judicious neutrality is maintained by the Council in respect of the rights and wrongs of the complaints made by other Portuguese.[54]

We are fortunate enough to have another perspective, in the form of a letter from almost exactly the same date, this one from D. Henrique de Meneses (who had just taken over as captain of Goa), to D. João III. His letter begins by noting his reluctance to

[53] We still lack a proper study of the fascinating career of Asad Khan Lari; but see Schurhammer, 'O tesoiro do Asad Khân', in *Varia I: Anhänge*, pp. 31–45, for some of the relevant Portuguese sources, as also the index entry in Schurhammer, *Die Zeitgenössischen Quellen zur Geschichte Portugiesisch-Asiens und Seiner Nachbarländer zur Zeit des Hl. Franz Xaver (1538–1552)*, reprint (Rome: Institutum Historicum S.I., 1962), pp. 529–30.

[54] AN/TT, CC, I-30-54, letter from the Goa Municipal Council to D. João III, dated 27 October 1523, cited above.

take over the captaincy of Goa, which had been imposed on him by Vasco da Gama, in the absence of D. Fernando de Monroy, and given the fact that Francisco Pereira had been removed from the post and imprisoned on the viceroy's orders ('until he has paid off his debts'). Meneses declares that he would much rather be giving battle to the 'Malavares, who go about in total rebellion, and who present great hindrances to your service'. But there is his own age and poor health to be considered, besides the fact that the viceroy has ordered that it should be thus.[55] D. Henrique, it emerges rather rapidly, had a poor opinion of the Goa *casados* and clergy, who it seemed to him were not interested in defending the town.

> Sire, as for the affairs of Guoa, I say that there are enough people there to defend it if they did what they should do, but I see it fuller of wolves in camel-hair (*lobos de chamalote*) and white hoods, than of arms, nor do I see the presumption of honour here, rather they are all or for the most part married with *negras* whom they take to church . . .[56]

Such people are, in the captain's view, ill brought-up and rebellious by nature; thus it is no surprise that both matters of justice and those of the treasury are in great disorder. Even if the walls of the city, facing the hinterland, have been reasonably repaired, there is still not enough artillery, for too much money has been spent on the hospital and the monastery of St Francis, both of which are out of all proportion with the importance of Goa itself.

> And if, as I have heard it said, the Rumes come, as is generally expected every year in India, and of which now there is much suspicion or [even] certainty that they will come within a year, if this is so, and I do not find out in advance, or whoever else is here, I believe that the city will see itself in great trouble, because even the towers of the four passes into this island are quite badly supplied and armed with artillery, and the hinterland (*a terra fyrme*) is in rebellion, and taken by Ydalcam [ᶜAdil Khan], who is rather powerful and even more so, desirous of this island and city.

In view of all this, and the lack of order and discipline amongst the inhabitants, D. Henrique makes it clear that he would rather

[55] There is rather good reason, on account of this statement alone, to reject the notion in some Portuguese encyclopaedia entries that D. Henrique de Meneses was born in 1496.

[56] AN/TT, CC, I-31-76, letter from D. Henrique de Meneses to D. João III, 27 October 1524.

give up the post as soon as D. Fernando, the rightful captain, arrives. He also makes it a point to state that, like the new viceroy, he too does not see much purpose to mollycoddling the settlers, and making contributions to the hospitals and Santa Casa da Misericórdia, and even less use in favouring the sizeable community of Indian Christians. The real priorities, both for him and for Gama, lie elsewhere: in defending Portuguese India against the Ottomans, and in breaking the Mappila threat on the west coast. There is also another purpose to the letter that gradually emerges: D. Henrique wishes rather insistently to have the captaincy of Hurmuz, which he understands, however, has been promised by D. João III to a certain Manuel de Lacerda. He points out his own considerable services, as well as his financial needs, including the pressing need to marry off his daughters before they grow too old; surely, Manuel de Lacerda can be given something else. He even suggests that he would be more than ready to give up the captaincy of Goa to some young *fidalgo*, so long as he were willing to 'take from me one of the many daughters I have'. However, it would be best if this marriage were arranged through his close allies in Portugal, Tristão da Cunha and his son, Nuno da Cunha.[57]

The letter closes by reiterating once more how ideal it would be, from various points of view, if he were to go to Hurmuz and settle the troubled affairs there. Of course, D. Henrique admits, were the viceroy himself to go there, he would take care of everything, and leave matters 'in the full peace and order of your service'; but Gama has already declared that he will not do so. D. Henrique declares that he himself is wasting his time in Goa, and would much rather be elsewhere. If he is refused Hurmuz, he will return to Portugal at the end of three years; in any event, he writes, 'I came here with the viceroy, which I would not have done with anyone else, as I told Your Highness many times.' Thus Gama's prestige had forced the hand of many a *fidalgo*, who might otherwise have been unwilling to go to India to serve under a lesser nobleman.

Gama appears to have remained in Goa for about a month; we have an order (*mandado*) from him to the factor, Miguel do Vale,

[57] Note that D. Henrique de Meneses, who was the illegitimate son of D. Fernando de Meneses *o Roxo*, and a certain Constança Vaz, was married to D. Guiomar da Cunha, daughter of Simão da Cunha, through whom he had three sons and three daughters. His wife was thus apparently a niece of Tristão da Cunha; cf. Banha de Andrade, *História de um fidalgo quinhentista*, pp. 20–1.

dated 20 October 1524, in Goa.[58] Besides the affair of Francisco Pereira, it would appear that Gama left an indelible impression in Goa as a disciplinarian, with a violence that was this time not directed so much at the opponents of the Portuguese, but at some of the Portuguese themselves. Castanheda, the most sober of the sixteenth-century chroniclers, reports that before leaving Belém, Gama had announced that no single women were to board the vessels, 'to avoid many sins that follow from bringing them [on shipboard]'. In Goa, it was found that two such women had secreted themselves aboard the vessels, since there were two men (presumably in India) who were to marry them; when Gama found this out, he had them publicly whipped on a yoke (*canga*), despite their pleas that if he did so, no one would be willing to marry them afterwards. Equally striking was the fact that the viceroy refused the many sick men on the fleet permission to have treatment in the new hospital at Goa, on the grounds that 'the king, his lord, had no need of hospitals in India, for if they were there, the men would always claim to be sick'.[59] This somewhat misanthropic judgement meant, writes Castanheda, that 'many died for want [of attention]'; it does appear though that later Gama did allow some men from his fleet to be treated at the hospital in Goa. Others, including incapacitated ex-soldiers, whose allowances and salaries were cut off as part of the new austerity measures, began to beg for alms, something that the chronicler claims had not been seen in Portuguese India till then. 'Everyone marvelled a great deal at this', he writes, and it is certain that the implications for Gama are not complimentary.[60]

During his stay in Goa, we are aware that Gama issued strict orders, as in Chaul, to strike off from the payroll of the Crown all those save the men of the garrison who presented themselves for a muster.[61] Equally, he ordered all the Portuguese living on the outskirts of the town to enter the walls, under pain of death; he also reduced the rations of those on board the ships in the road. Having put in place these draconian measures, he is reported to have gone to sea, with Cochin as his next destination. Off Goa, he

[58] AN/TT, CC, II–120–95, *mandado* dated 20 October 1524, ordering Miguel do Vale to pay the carpenter Pedro his salary, and the value of some wood for ship construction.

[59] Gaspar Correia, *Lendas*, Vol. II, p. 830, is for once more mild than Castanheda, claiming that Gama required the hospitals to treat only those 'who presented wounds and injuries', the others being classed as shirkers.

[60] Castanheda, *História*, Vol. II, Livro VI, ch. 71, p. 266.

[61] Historical Archives, Panaji, Goa, *Provisões, Alvarás e Regimentos* (Mss. 3027), fl. 221v, dated 1 October 1524, cited in Costa and Rodrigues, *El Proyecto Indiano del Rey Juan*, p. 185.

encountered the fleet of D. Luís de Meneses, who was on his way from Cochin, with the intention of meeting his brother D. Duarte, who was expected from Hurmuz. Gama forced D. Luís to return with him to Cochin, where the whole fleet arrived in late October 1524, after a halt in Cannanur.

Even this brief visit to Cannanur, late in the month of October, is not devoid of significance. We are not aware of what precisely transpired, though we do know that a change of regime was brought about in the fortress, with D. Simão de Meneses taking over as its captain. Enigmatic traces of a correspondence in Arabic exist from this time, in particular a letter from the *alguazil* (or minister) of the Kolattiri, a certain Kurup, directed to the viceroy, who is addressed in honorific terms not merely as Admiral, but as the *wazîr al-wuzarâ'*.[62] The letter complains against the Captain-Major (that is, D. Duarte de Meneses) who has shown hostility towards Cannanur and its ruler, and who has now been absent for some months from the region (a reference to D. Duarte's passage through the port, en route to Hurmuz). It is hoped by the writer that Gama's presence will help to sort matters out. This wish was not quite fulfilled, for Gama appears to have taken a hard line at Cannanur concerning Mappila commercial and raiding activities, apparently threatening the Kolattiri and his ministers with reprisals. In response to this, it appears that the Cannanur ruler decided to hand over one of the Mappila notables to him, who is identified variously as a certain Balia Hacem, as one of the brothers of Mammali Marikkar, or even (rather improbably) as Mammali himself.[63] This man was then imprisoned in the Portuguese fortress at Cannanur, only to be taken out and hanged summarily in late January 1525, after Gama's death.

By early November 1524, we find Gama in Cochin, embroiled in the conflict that was to occupy him through the last weeks of his life. Accompanying Gama was the new captain of the fort, Lopo Vaz de Sampaio, who took over from the incumbent D. Garcia de Noronha, himself a nephew of Afonso de Albuquerque. A series of increasingly complex factional fault-lines now devel-

[62] AN/TT, Casa Forte No. 189, Maço 1, Documentos Orientais, Doc. 47 (Arabic). This letter is reproduced with errors, and a false dating (27 May 1503, when Gama was not even in India), in Fr. João de Sousa, *Documentos Arábicos para a história portugueza copiados dos originaes da Torre do Tombo* (Lisbon: Academia Real das Sciencias, 1790), Doc. 2, pp. 4–5. For a discussion of this collection, see Jean Aubin, 'Les documents arabes, persans et turcs de la Torre do Tombo', *Mare Luso-Indicum* 2 (1973), 183–7, and Georg Schurhammer, 'Orientalische Briefe aus der Zeit des Hl. Franz Xaver (1500–1552)', *Euntes Docente* 21 (1968), 255–301.

[63] Bouchon, *Mamale de Cananor*, pp. 172–3, for a discussion.

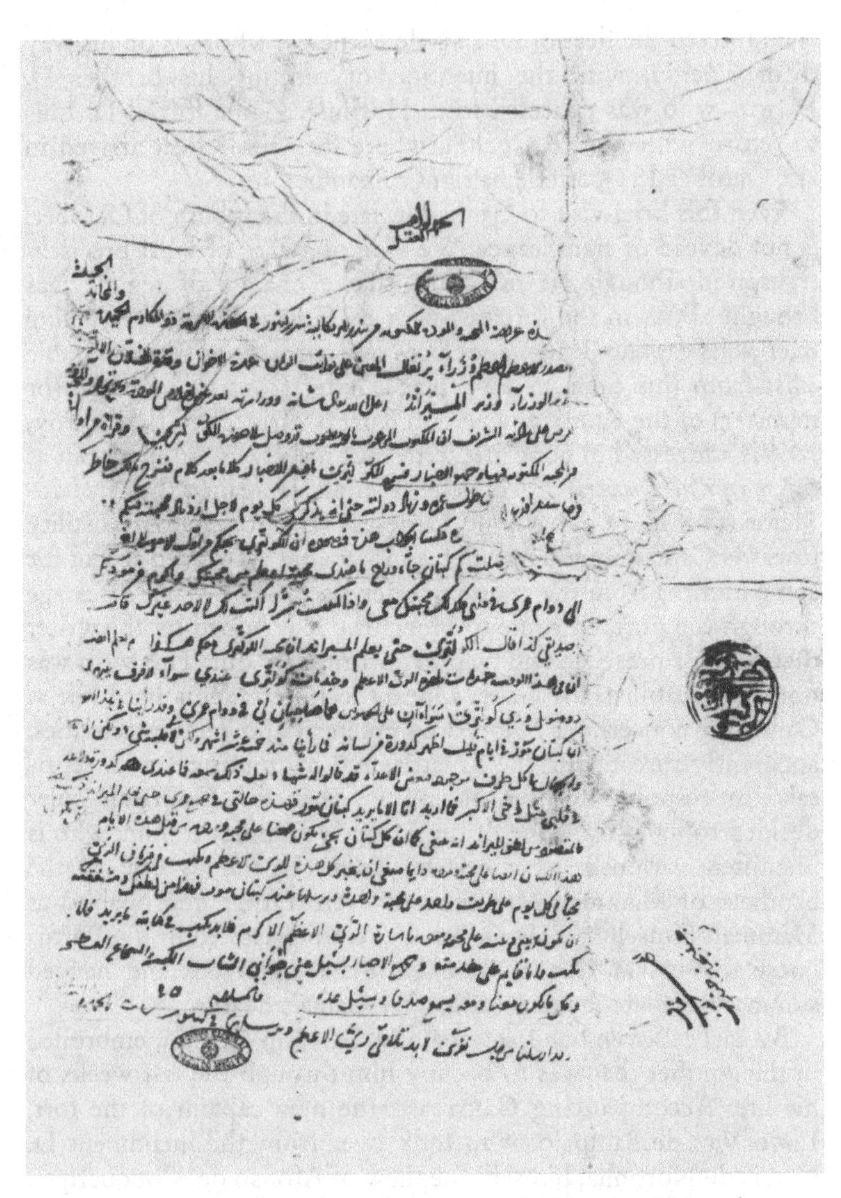

Plate 24 Arabic letter from the *wazîr* at Cannanur to Vasco da Gama, October 1524, complaining of the comportment of D. Duarte de Meneses.

oped, around the viceregal establishment. It is clear that to start with, Gama's own personal position was practically unassailable on account of a triple protection: first, his enormous personal prestige; second, the extensive powers that he had been given as viceroy by D. João III; and third, the brutally vigorous style in which he made use of those powers. The partisans of D. Duarte de Meneses, who was himself still awaited in early November at Cochin from his visit to Hurmuz, were thus forced to seek vulnerable points in those surrounding Gama (and who had accompanied him from Europe), or amongst those in Cochin who adhered to his line. For instance, it is clear that Gama sought out and chose to extend his protection to Dr Pedro Nunes, the *vedor da fazenda* who had received short shrift from both Diogo Lopes de Sequeira and D. Duarte de Meneses. Even if he was at the very end of his term, and replaced in December by Afonso Mexia, the viceroy made it clear that in his view, the *vedor* had acquitted himself very well, under difficult circumstances. In turn, this meant that the *vedor*'s enemies, notably the Portuguese factor in Cochin, Manuel Botelho, were piqued. A war of words arose over the question of responsibility for pepper procurement. Castanheda, here reflecting the line favoured by Gama himself, gives Pedro Nunes credit for improvements in the quality of pepper sent to Europe. The Mappila merchants, in his version, were in the habit of giving damp pepper (mixed with mud and sand) to the Portuguese, so that weight losses of the order of thirty to forty per cent occurred between delivery in Cochin, and the sale of the dried and cleaned pepper in Portugal. It is suggested that Pedro Nunes had hence decided to strike a deal with the Syrian (St Thomas) Christians of Cranganor who were in direct contact with the producers (thus circumventing the intermediary Mappila retailers), and by offering them monetary incentives, had managed to reduce the losses (*a quebra*) to a mere seven per cent.[64] In contrast, we have a long letter from Manuel Botelho, written in January 1525, in which he seeks to take credit for all matters related to pepper procurement, which he for his part claims to have negotiated through his own good relations with the Syrian Christians. The letter grumbles that after having been named to the Cochin factory for three years by D. Manuel (presumably to depart in the outward fleet of 1521), his powers and jurisdiction had been first reduced, then extended piecemeal, on account of

[64] Castanheda, *História*, Vol. II, Livro 6, ch. 72, p. 267.

which Botelho 'is in great discredit with the merchants of this land'. The factor believes he knows whose hand lies behind this, as he writes to D. João III:

> I was going to leave this year, but since the lading of the cargo had begun, I could not do so, and remain until next year, when I will leave come what may, and I will not complete the [additional] two years that Your Highness has granted me, for I cannot stay here all that time, harassed by your *veadores da fazenda*, who work as hard as they can to discredit me, and to remove from my charge here, some other things besides those which the King who is in Holy Glory had withdrawn, so that I was and am harassed by them in such a way that I cannot give an assurance to Your Highness that I will make up the cargo for the coming year, which I must begin in early March, in order to gather it together in the factory, for I have agreed upon it thus with the merchants.[65]

CHRISTIANS AND SPICES REVISITED

The increasingly hostile relations between the Mappilas and the Portuguese Crown, that we have described above, should undoubtedly be located in a rather more complex dialectic of local politics, in which the Syrian Christian communities of central Kerala must equally be included. As we have noted, in the early years of the sixteenth century, the Bishops of the community tended to make common cause with the Portuguese, as we see from their letter to the *Catholicos* of 1504. Two decades later, matters had evolved somewhat. After a period of what have been called 'tenuous relations', between about 1506 and 1516, a renewed interest was shown by some of the Portuguese in the Christian communities, particularly through the mediation of the curious figure of Padre Álvaro Penteado, who arrived in Cochin in 1516, after a peripatetic existence in various centres of Portuguese Asia in the preceding years.[66] Penteado's relations with the local clergy were mixed, but his actions certainly improved Portuguese knowledge concerning the character, and economic and political role, of the Syrian Christians. After a brief visit to Portugal, probably in 1519–20, he returned to India, with fairly extensive and independent powers of negotia-

[65] AN/TT, CVR, No. 16, Manuel Botelho to D. João III, Cochin, 21 January 1525, fl. 2v.

[66] The best discussion of this period is in João Paulo Oliveira e Costa, 'Os Portugueses e a cristandade siro-malabar (1498–1530)', *Studia* 52 (1994), 136–58. But also see A. Mathias Mundadan, *A History of Christianity in India*, Vol. I (*From the beginning up to the middle of the sixteenth century*) (Bangalore, 1984).

tion from D. Manuel (but which would lead him to a conflict with the Vicar-General); the discovery of the so-called tomb of the Apostle St Thomas in Mylapur some years earlier obviously helped him whip up enthusiasm in the late Manueline court for his projects.

On his return to Kerala in 1521, Penteado found his role as privileged intermediary with the Syrian Christians somewhat under threat, notably from a recently arrived Dominican friar of Castilian origin, Juan Caro. Caro for his part appears to have had particularly close relations with the Syrian bishop Mar Jacob (Ja^{c}qôb, whose original name was Raban Mas^{c}ud), and the two together clearly played a major role in strengthening commercial relations between Syrian Christian pepper wholesalers, and the Portuguese factory at Cochin. A letter from Mar Jacob to D. João III, apparently written in December 1524 (and possibly drafted by Caro himself, in view of its particular mixture of Castilian and Portuguese), clarifies matters. In it, Mar Jacob, who styles himself 'Bishop of India' (*episco de Hendu*), begins by noting that he rules over 'the Christians of India, who are said to be of Coulam', and that he has been sent to do so by 'the Patriarch of Babylonia'. He then goes on:

> It is about four years since there arrived in this land one Padre Mestre Joam Caro, from whom I have received many doctrines for my salvation and that of this people, and thus of things of your service. One of things of your service in which he instructed me, was that the Christians over whom I reign were receiving all the pepper from the hands of the farmers who harvest it, and that they had not dared to bring it to your factory out of fear that the Moors instilled in them, telling them falsehoods and deceit, that too much was weighed in the scales, and that it was very poorly paid for, and that they would be beaten and treated like slaves, and that when your Portuguese saw their pepper, they would rob them and make them prisoner, and send them to Portugal, and by way of proof the Moors produced some bad Christians who enjoyed advantages in the trade with them. And since they therefore did not bring it [to the factory], it was necessary for them to sell it to the Moors, who thus traded alone in this commerce, and thus the Moors had the power and reason enough to sell it [the pepper] dirty and full of water, and since your ships needed it for their cargo, they took what they gave, for there was no way of getting around this inconvenience, and there were also some Christians who did the same ill, learning it from the Moors . . .[67]

[67] AN/TT, CVR, No. 99, in Schurhammer, 'Three Letters of Mar Jacob', pp. 339–40. The

Mar Jacob goes on to note how under Juan Caro's influence, he decided to investigate the pepper trade, and found that what he said was in fact true: most of the Syrian Christians merchants kept their distance from the Portuguese out of fear, while others supplied them with poor quality, adulterated, pepper. He thus took the initiative and, visiting the Christian settlements (*aas povoações dos cristãos*), instructed them to deal with the Portuguese officials in Cochin. Their interlocutors in the pepper trade were three in number: the *vedor da fazenda* Pedro Nunes, the factor Manuel Botelho, and that altogether remarkable figure, 'Diogo Pereira fidalguo', the former opponent of Albuquerque, now a permanent fixture in Kerala. The Syrian Christian merchants, it is reported, were 'so greatly welcomed' by these three, and by some other Portuguese, that they decided to continue in this direct trade.

This *rapprochement* between the Syrian Christians (especially those of Cranganor) and the Portuguese establishment in turn brought forth serious political consequences. Let us recall once more that it was less the church hierarchy of the *padroado real* (Crown patronage of missions) that promoted this new relationship, than ambitious individual clergymen like Álvaro Penteado and Juan Caro. There was both political and economic mileage to be had out of the Syrian Christians, who represented a population of between twenty-five and thirty thousand, depending on which of several estimates we use. This fact was obviously equally perceived by the new Samudri Raja, who had succeeded in 1522, who thus took it upon himself to pressurise the Christian community in and around Cranganor. In late 1523, the principal church at Cranganor was attacked and burnt down, causing consternation amongst that rather well-established Christian community. This is referred to by a certain João Garcês, an interpreter (*língua*) with long experience in Kerala, in a letter of early 1529 to D. João III, but also finds mention in Mar Jacob's letter of late 1524.[68] The bishop writes, after recounting all his actions in favour of the Portuguese Crown:

letter is obviously written in November or December 1524, as demonstrated by its repeated references to the 'viceroy'. Only the last lines are in Syriac, by the Bishop himself.

[68] AN/TT, CC, I–24–3, João de Garcês at Cochin to D. João III, 2 January 1529, in Luís de Albuquerque and José Pereira da Costa, 'Cartas de "serviços" da Índia (1500–1550)', *Mare Liberum* 1 (1990), 330. The same letter appears under the name of João Carcere, in Silva Rego (ed.), *Documentação para a história das missões*, Vol. II, pp. 175–9.

> This, Sire, is the service that I have done in these parts, with the intention of moving you to help me in the expansion of these people [Syrian Christians] through this India in the faith of Jesus Christ, our Redeemer. And now there is more need of it [help] than ever, because since I have helped you as I have said, the Moors have robbed me and killed many of my people, and also burnt our houses and churches, by which we have been greatly distressed and dishonoured.

In the same context, he hence offered the aid of the Syrian Christians as an auxiliary military force, to aid the Portuguese, claiming that they represent 'over twenty-five thousand warriors' (*passante de vintecinquo mjll homens de guerra*).[69] The letter equally asks D. João III to order Vasco da Gama (*o Vissorey*) to intercede to recover a large piece of land that the Christians of Cranganor had long possessed (and which was confirmed by a grant on copper-plate), but which had been lost, 'usurped by many lords'; and still more generally that Gama be required to look after the interests of the community. In point of fact, direct contacts were apparently established between the Christians of Cranganor and Gama, sometime in November 1524; the viceroy is reported to have assured them all aid in reconstructing their church. Indeed, Mar Jacob's letter contains a none-too-subtle threat: 'If you do not do as I write [he states to D. João III], it may very well be that these Christians become dissatisfied with you, and then the only pepper you will have will be dirty and full of water'!

Mar Jacob also proposed the construction of a Portuguese fortress at Cranganor (a proposal that was put into effect a decade later, in 1536), and that the Crown should ensure that 'the pepper that goes to Calequu, should not go out from there'. As it happens, this latter idea was also one that found favour in Gama's eyes. The Calicut factory and fortress, let us recall, were the brainchild of Albuquerque, who had decided in 1513 to make peace with the Samudri, against the express advice of all those Portuguese who were gathered around the Cochin Raja. The idea was thus to participate in the economic life of Calicut, rather than to destroy the economic viability of the port. A decade later, many Portuguese (both those in India, and others who had returned to Europe) continued to see little sense in this policy, while at the same time – as we have noted above – the Samudri

[69] Schurhammer, 'Three Letters of Mar Jacob', p. 340.

himself was none too enthusiastic about the results of the Portuguese presence. In this context, it is now clear (despite some confusion in the literature) that Gama came to India with orders to dismantle the Calicut fortress and factory, and then to proceed to make war on the Samudri himself, and on the Mappilas of Ponnani, Tanur and other coastal centres that fell under the Samudri's tutelage, from a position of security at Cochin. These orders were only put into effect, however, in late 1525, when the Calicut fort was eventually given up. Ironically, by that time, D. João III had changed his mind, partly on account of the offer (however dubious) of military aid from the Syrian Christians; he now formulated orders, in 1525 (but which reached India too late, that is only in 1526), to strengthen the fort at Calicut, and to use it to make war on the Samudri.[70]

It should be noted that underlying these twists and turns of policy was a particular conception on the part of the Portuguese establishment, concerning the character of the pepper trade from Kerala. This may not be the place to enter into this issue in all its intricacies, but a brief statement is nevertheless in order. Let us note that through the first three decades of the sixteenth century, Portuguese exports of pepper to Europe by the Cape route very rarely exceeded 25,000 *quintais* (roughly the cargo taken back by Gama in 1503), and were usually below 20,000 *quintais*.[71] To obtain cargoes exceeding 25,000 *quintais* was extremely difficult, not because they were unavailable but because the Portuguese state simply did not furnish adequate capital to its factors in Cochin and elsewhere to do so. Indeed, even if we take the somewhat low estimates of authors like V.M. Godinho into consideration, it is clear that Portuguese purchases could not have accounted for the greater part of pepper production in Kerala.[72]

In this situation of capital scarcity, it was common for Portuguese factors from as early as the 1520s to raise advances, often from their own private funds, or from those of other officials and private traders, to tide over the gap. But, at the same time, there was every effort to keep purchase prices of pepper low, either by

[70] Cf. João Paulo Costa, 'Os Portugueses e a cristandade siro-malabar', pp. 160–1.

[71] In 1518, a truly exceptional year, the returning fleet carried back a cargo of some 40,000 *quintais*; cf. Geneviève Bouchon, *Navires et cargaisons retour de l'Inde en 1518* (Paris: Société de l'Histoire de l'Orient, 1977).

[72] One of the clearer discussions to date on pepper production and market-shares in sixteenth-century Kerala is that in Jan Kieniewicz, 'Pepper gardens and market in precolonial Malabar', *Moyen Orient et Océan Indien* 3 (1986), 1–36, which contains, *inter alia*, a critique of the unreliable account in K.S. Mathew, *Portuguese Trade with India in the Sixteenth Century* (New Delhi: Manohar, 1983).

circumventing intermediaries (thus, the alliance with the Syrian Christians), or by driving as hard a bargain as possible. The unwillingness of Portuguese factors to accept the validity of market prices for pepper is at one and the same time the result of their relative penury from early on, and the cause of their portrayal of the market as fundamentally organised around political rivalries rather than economic factors. To hear their version, the market was segmented: while others could obtain pepper cheaply, pepper prices were kept artificially high for the Portuguese. Their solution to this was equally draconian and political: namely to blockade as many alternative outlets as possible, to ensure that local merchants had to offer pepper to them on the cheap.

We catch echoes of this in Manuel Botelho's letter of early 1525, cited above. The Cochin factor is inclined to congratulate himself for getting together the following cargo in the Cochin factory for the return fleet of that year: 17,802 *quintais* of pepper, 872 *quintais* of cinnamon, 334 of cloves, and small quantities of nutmeg, mace, rhubarb and other goods. The total pepper cargo, including that from Kollam and Cannanur, is, he states, of the order of 30,000 *quintais*. He then goes on to lament the state of trade, in precisely the terms that we have outlined above. Thus:

> In the letters I wrote Your Highness last year [1524], I said that I was hoping to make up a large cargo this year, and that I greatly feared that I would not have the pepper, since the Moors were taking as much as they could to Cambaya. It was so because in this land, there was no one to forbid it, and Your Highness knows already who is to be blamed for this. So, from the crop of last year there remained very little old pepper in this land, for as I say above, the Moors took as much as they could. That which I could have for this cargo I received, but it was little until the new crop came in, which I did not wish to receive or take, but I was forced to do so for the ships that were ready to load up are very large, and have need of much pepper, and this new pepper that they brought me was not properly dried, and its purchase gave a great deal of trouble, as it had to be dried in this weighing-house in the sunny places (*soalheiras*), that were always full of it [pepper] until the lading of the ships was completed.[73]

In sum, of the 17,800 *quintais*, only 7,000 was from the previous year's stock, the rest being green and partly damp, requiring

[73] AN/TT, CVR, No. 16, Manuel Botelho to D. João III, Cochin, 21 January 1525, fl. 1v.

drying at the last moment. But Botelho has determined that it was better thus, than to receive poor adulterated pepper. He gives his reasons: first, because if freight goods of private parties are carried instead of pepper, the Crown will nevertheless have some revenue from them; second, because by this means the local merchants will see that the Portuguese are not desperate to have pepper, and are even willing to let their ships return partly empty rather than compromise on quality. Nevertheless he complains: 'The people of this land are very difficult to trade with, because there is no truth in them', the only exception being the Syrian Christians, who have appeared at the Cochin factory that year in far larger numbers than before. They at least should be favoured in his view. Botelho's emphasis is thus thoroughly traditional, but he does betray another set of concerns at one point, insisting that he be sent minted gold (*dinheiro em ouro amoedado*) from Portugal, in the form of *portugueses* and *cruzados*, 'because it is with good payment that one gets good pepper'.

The letter (to which we shall return below briefly) then concludes with a traditional refrain of the 'Cochin group', and indeed of D. Aires da Gama: the importance of Kerala in the Portuguese scheme, very nearly to the exclusion of all else. Thus:

> In the guarding of this coast from Coulaão to Calecut, it is necessary now, more than ever, to take great care, and Your Highness should order your Captains-Major that it be thus, because it is from this piece of land (*deste pedaço de terra*) that you receive all the profit, and not from the fleets that go to the Straits [of the Red Sea].

In sum, Botelho's conception was not wholly different from that to which Gama himself (as well as Duarte Barbosa and others) appears to have subscribed: a concentration on Kerala (Kollam, and Cannanur, but above all Cochin), the deployment of patrolling fleets against their rival 'Moors', a relative tolerance for private Portuguese trade, both within Asia and even on the Cape route. Indeed, one of Botelho's purposes in writing his letter was to renew a demand for the right to send a hundred *quintais* of pepper on his own account to Portugal, for every year that he remained in Cochin.

The Syrian Christians were seen in all of these conceptions in singularly pragmatic terms, as potential allies for what were essentially commercial ends, thus in marked contrast to the Manueline conception. We see this, for example, in D. João III's reply to Mar Jacob's letter, which was probably drafted in late 1525 or early 1526. Rather than enter into flights of mystical fancy, the letter runs, soberly enough:

> Abuna [Bishop], I Dom Joam etc. send you hearty greetings. I have seen the letter that you wrote me, in which you informed me of what you had done in order to serve me, on account of the information that you were given by Mestre Frey Joam Caro, concerning how the Moors harass the Christians of the land in order that they do not come with their pepper to my factory, and that on account of what you had worked in this matter, it had already started coming, and it would always do so, and I thank you very much for all that you have done in this matter, and after their conversion to our Holy Faith, you could not have served me more in any other matter, and I commend you strongly that now that you have begun it so well, you should complete it, and leave no place for the malice of those enemies of our faith, rather the Christians should do so because it is to their own profit that the Moors withdraw from the trade and they come to have it wholly in their hands, for thus I too shall be greatly pleased. And I am writing to my Captain-Major and Governor in those parts and to my *veeador da fazenda*, that the Christians should be very well treated and welcomed, favoured and honoured, and very well paid for the pepper that is purchased, and they will do so.[74]

The letter concludes by promising Mar Jacob himself the aid and favour of the governor, the continuation of an annual pension (*tença*) that had been granted him by D. Manuel, and a piece of blue satin for a garment. It is thus a thoroughly businesslike letter, in which – beyond the token genuflection in the direction of matters of faith – centre-stage is occupied by practical matters. As has been remarked, D. João III was for the most part willing to support the authority of the Syrian bishops, rather than insist on the points of difference between their religious observances and those of the Catholic clergy. A gradual process of 'acculturation' may have been envisaged in religious matters, an idea to which Mar Jacob appears not to have been entirely hostile, although other Syrian Christians (including the other rather less malleable bishop, Mar Denha) saw matters rather differently. In the latter half of the 1520s, these divisions amongst the Syrian Christians manifested themselves in larger alliances: as Mar Jacob wrote in a

[74] AN/TT, Núcleo Antigo, No. 875, 'Sumário das cartas que vieram da Índia no ano de 1525 e respostas d'el-rei a elas', fls. 155v–56, in João Paulo Costa, 'Os Portugueses e a cristandade siro-malabar', pp. 176–7.

letter to D. João III in December 1530, those rulers in Kerala who were opposed to the Portuguese (and the Samudri is obviously meant here) took to favouring the elements among the Syrian Christians who were 'hard of heart' (*duros de coraçam*), that is strongly opposed to the Catholics.[75] The pepper trade for its part, continued to pose problems; the Portuguese apparently expected the Syrian Christian merchants to bring them cheap pepper at the Cochin factory, even when better prices were to be had elsewhere, in brief, quite literally to pay a price for an imagined religious solidarity.[76]

A HOUSE DIVIDED

In sum then, the affairs of Calicut, the Syrian Christians and the Mappilas, all of these while high up on Vasco da Gama's list of priorities, did not find a resolution in the last months of 1524. By November, it was known very widely in Cochin that Gama was seriously ill, and unable to embark on an attack on Calicut, as he wished. Instead, in order to aid the hard-pressed D. João de Lima, captain of the Calicut fortress, he sent out a fleet of small galleys under a certain Jerónimo de Sousa, which had a certain success against a flotilla of Mappila vessels under the command of Kutti ᶜAli of Kappatt. But the overall policy lacked coherence. Castanheda cites the case of the coastal fleet of Luís Machado, whose function it was to patrol the coast near Goa; Gama decided to withdraw this fleet, and to take it with him to Cochin in October, doubtless considering it to be wasteful expenditure. The next month, since D. Henrique de Meneses found that a substantial number of vessels were passing Goa en route to Gujarat from Kerala, with pepper and spices on board, he was obliged to purchase a merchant vessel, and mount an improvised patrol under the command of his nephew, D. Jorge Telo, around Goa. This patrol had a certain success: by late December 1524, D. Henrique was thus able to send a pepper cargo to Afonso Mexia in Cochin 'from the prizes that were taken by Dom Jorge and António'.[77] The lack of success in preventing departures from

[75] Letter from the Bishop Mar Jacob ('Jaᶜqôb dabsem episco') to D. João III, Cochin, 16 December 1530, AN/TT, Gavetas, XV/19–36, in Schurhammer, 'Three Letters of Mar Jacob', pp. 343–4.

[76] Cf. João Paulo Costa, 'Os Portugueses e a cristandade siro-malabar', pp. 166–7.

[77] AN/TT, CC, II–122–120, *mandado* dated 29 December 1524 from D. Henrique de Meneses, captain of Goa, to the factor Miguel do Vale, ordering him to hand over the pepper from the *presas* to Parigi Corbinelli, captain of the ship *São Jorge* that is en route

Kerala for the Red Sea is also remarked in the Venetian diaries of Marino Sanuto, where it is reported in a letter from Cairo, that in winter 1524–5, a fair number of *sambuqs* had arrived from Calicut (and elsewhere) in India at Jiddah. The cargoes that are reported are mostly of ginger, not pepper, but it is equally noted that the prices of most spices are low in Cairo in expectation, since a good number of other ships are known to have left Calicut. The Portuguese, it is noted, boast that they have had a notable success at Calicut, killing over 400 people there, but it is well known that 'they are not very powerful'.[78]

As Gama's health worsened, more and more of his powers were delegated to the new captain of Cochin, Lopo Vaz de Sampaio, and to the *vedor da fazenda*, Afonso Mexia. Meanwhile, D. Luís de Meneses, who had been brought back to Cochin against his will, expressed his rancour in no uncertain terms, particularly in respect of the viceroy's young and inexperienced son, D. Estêvão da Gama, who was to replace him as Captain-Major of the Seas. As before, at Chaul and Goa, Gama announced rather severe austerity measures in Cochin. Castanheda, whose view is rather critical, states that the viceroy wished to discontinue the system of clientship, by which soldiers were supported 'at the tables' (*à mesa*) of prominent captains and *fidalgos*, who subsequently took them along on their fleets, or to their posts, as dependents. Indeed, fighting crews and garrisons in Portuguese India were usually constituted in this manner, since there was a marked reluctance – as we have noted – to accept the 'Swiss system' of paid professional soldiery.[79] The chronicler goes on to note that as a result of these measures, many of the Portuguese in Cochin were unhappy with the viceroy, and made common cause with D. Luís de Meneses, who was already rather well liked; the resultant public gatherings and disturbances reportedly had the captain Lopo Vaz on his toes 'day and night', to ensure that no violence broke out. Some disgruntled Portuguese also reputedly left Cochin for the usual resorts, namely the ports of the Coromandel coast, 'and other parts where they went about

to Cochin; for the context, see Castanheda, *História*, Vol. II, Livro 6, Ch. 74, pp. 269–70.

[78] 'Copia di capitolo di lettere di sier Beneto Bernardo qu. sier Francesco, scritte a sier Mafio, suo fratello, date al Cayro a di 14 Marzo 1525, ricevute a di 9 Zugno', in *I Diarii di Marino Sanuto*, Vol. XXXIX, ed. F. Stefani *et al.*, (Venice: The Editors, 1894), pp. 43–5.

[79] For a discussion of the conflict between these alternative systems, see Vítor Luís Gaspar Rodrigues, 'A organização militar do Estado Português da Índia (1500–1580)', unpublished thesis (Lisbon: Instituto de Investigação Científica Tropical, 1990).

outside the service of the King'. Besides, the Mappila merchants of Cochin, 'who trembled when they took his [Gama's] name', are equally reported to have discreetly left the town.[80] Manuel Botelho, for his part, mentions a part of Gama's austerity measures, in his letter to D. João III written a few weeks after the viceroy's death.

> The viceroy, who is with God, had told me that he held it was greatly to your service to bring this factory back to the state in which it was earlier, and not to have in this Cochym, so many unnecessary officials, and who account for so many salaries, and he was going to do that, and Our Lord chose to carry him off, by which Your Highness suffered a great loss in matters of your service.[81]

On the other hand, if Gaspar Correia is to be believed, this austerity was applied by Gama to others, rather than to his own household and person. The large sum of money spent on his fleet (which some historians estimate at as much as 200,000 *cruzados*) has already been noted.[82] Jean Aubin has succintly noted Gama's extravagant style on his third voyage, drawing on the information of the chroniclers: he portrays Gama as surrounding himself with royal splendour, with ushers in silver livery, pages in gold collars, making use of silver tableware in a decor dominated by Flemish tapestries, with royal etiquette at his table, and a ceremonial according to which the secretary addressed him while on bended knee.[83] This contrast between viceregal splendour, and general frugality, may have been intended to impose Gama's authority and make his presence all the more awe-inspiring; but its effects on the Portuguese in Cochin were rather more ambiguous.

It was to this divided Cochin, full of resentment against a celebrated, powerful and authoritarian viceroy, that D. Duarte de Meneses returned, in early December 1524. On his way back from Hurmuz, he had put in at Chaul, where he was given a cold reception by the captain Cristóvão de Sousa, then at Goa, where D. Henrique de Meneses equally showed little enthusiasm towards him, before making his way to Cochin. Though the viceroy was by now gravely ill, and apparently unable to act

[80] Castanheda, *História*, Vol. II, Livro 6, Chs. 73 and 75, pp. 267–70.
[81] AN/TT, CVR, No. 16, Manuel Botelho to D. João III, Cochin, 21 January 1525, fl. 2v.
[82] Borges de Macedo, *Um caso de luta pelo poder*, p. 40.
[83] Aubin, 'Préface', in *Voyages de Vasco de Gama*, p. 26. For example, Gama had two specially designated 'doorkeepers to his chamber' (*porteiros da Câmara*), Francisco Mendes and Pedro de Palma. For salary payments to them, see AN/TT, CC, II–122–30, *mandado* dated Cochin, 14 December 1524.

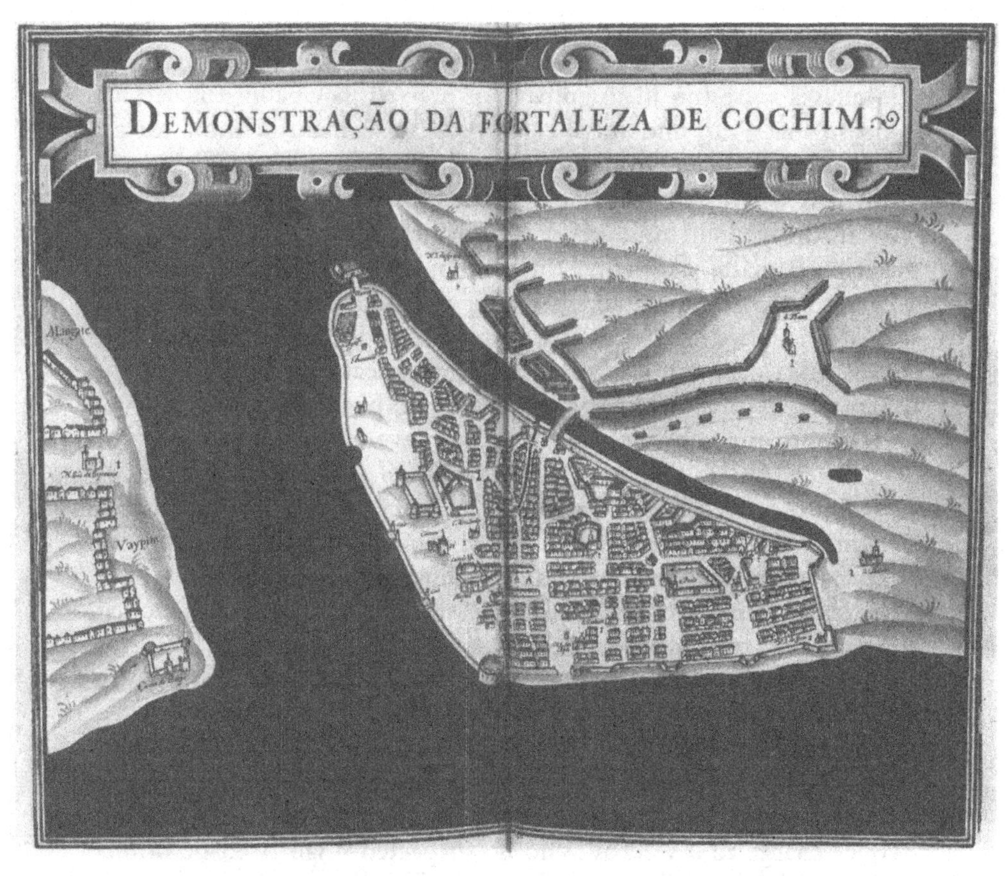

Plate 25 Seventeenth-century plan of the Portuguese quarter of Cochin, Bocarro-Resende, *Livro das Plantas* (Paris text).

directly, the reception committee had been prepared: it comprised Afonso Mexia, Lopo Vaz de Sampaio, and the *ouvidor-geral* João de Osório. At first, Gama reputedly forbade D. Duarte even to step on land, so that all transactions had to be carried on while he was still on board ship. The committee presented him with a long royal letter, composed at Évora on 25 February 1524, which Castanheda reproduces in its entirety. The letter is a puzzling one. It begins by asking D. Duarte to return immediately to Portugal, handing over all power and jurisdiction to Vasco da Gama as soon as he arrives. However, certain conditional clauses are then set out. D. Duarte is to remain in Cochin or Cannanur during the time that the ships for Portugal are not ready to leave; further, his own servitors (*criados*), those of his father, the Count-Prior D. João de Meneses, and his brother D. Luís, and his other close relatives, are all exempted from the civil and criminal jurisdiction of the viceroy and the new *ouvidor-geral*. Further, if D. Duarte de Meneses were too ill to embark or arrived in Cochin too late to do so that year, and was hence forced to remain in India for an extended period of time, he was asked to spend it in Cannanur fortress where, however, he could exercise extraordinary powers over the captain and other authorities, albeit not in fiscal or financial matters. Finally, on handing over charge to Gama, he was to provide a statement listing the ships, arms, artillery and soldiery under his charge that his successor inherited.[84]

If this document is accurate, it shows great care and tact on the Crown's part in dealing with D. Duarte, perhaps motivated by a desire to prevent an outbreak of severe factional hostilities between the followers of the house of Tarouca and that of Vidigueira. D. Duarte was clearly not one to take his own status lightly. Together with this letter, Lopo Vaz handed him another far more severe one from Gama, telling him (as noted above) not to leave shipboard; at this, he is reported to have reprimanded Lopo Vaz, reminding him that he had been raised to the status of cavalier in North Africa, by his father D. João de Meneses. D. Duarte is equally reported by Castanheda to have had a sharp exchange with João de Osório, whom he saw as a mere *letrado*, (and hence addressed slightingly as 'bacharel'), and thus of far lower social status than himself. Despite a great reluctance to hand over charge to a viceroy who he suspected might well be on his deathbed (in which case D. Duarte and his brother appear to have believed that the post of governor would revert to one of them), he is reported to have done so eventually, on 4 December 1524. Thereafter, Mexia, Osório and Sampaio are reported to have forced D. Luís de Meneses, who for his part was still in the town of Cochin, to get on board his ship in order to avoid further conflicts.

All evidences point to a situation therefore in which, by early December 1524, effective power was being exercised in Cochin by Afonso Mexia, in a fragile alliance with a group of which the

[84] Castanheda, *História*, Vol. II, Livro 6, Ch. 76, pp. 271–3. For a fairly clear idea of the level of resources left by D. Duarte de Meneses to Vasco da Gama, see the anonymous text dated August 1525, in AN/TT, Colecção São Vicente, Vol. XI, fls. 1–36, 'Lembrança d'algumas cousas que sam passadas em Malaqua e asy nas outras partes da Imdea', published in Rodrigo José de Lima Felner (ed.), *Subsídios para a história da Índia Portugueza* (Lisbon: Academia Real das Sciências, 1868), Part III; however, it is necessary to distinguish additions and changes made by D. Henrique de Meneses in the first half of 1525.

leading element was Lopo Vaz de Sampaio (who for his part made no secret at this time of his contempt for the *vedor da fazenda*). It was only later that their alliance became firm, on the occasion of a major conflict, which we shall have occasion to discuss below (albeit briefly), that broke out in 1526–7, in Cochin and elsewhere in Portuguese India, concerning the succession to the post of governor. It is worth remarking that both Mexia and Sampaio were men who, like Vasco da Gama himself, were relatively closely associated with D. Jaime, Duke of Bragança; indeed, some years later, when Sampaio was in disgrace, the Duke vigorously interceded with D. João III in his favour.[85] But this was not enough to guarantee an immediate alliance in India, where other factors – including personal ambitions – had precedence.

As his malady grew more grave, Vasco da Gama decided to remove to the house in the outskirts of Cochin, of a figure who is by now familiar enough to us, Diogo Pereira, *o Malabar*. This wily 'old India hand', and opponent of Albuquerque, had gradually built up a formidable power base in Cochin; so much so that, besides the *vedor da fazenda* and the factor, it is to him that Mar Jacob refers in his letter of 1524 as an authority in matters concerning the pepper trade. The Cochin Raja handed over the island of Vendurutti in the Cochin harbour to him as a sort of 'fief', and he would later serve as intermediary in the 1530s, in the foundation of the Portuguese fortresses of Chaliyam and Cranganor.[86] Gama is also reputed at this time to have instructed the principal *fidalgos* in Cochin that they were to follow the instructions of Lopo Vaz de Sampaio if he were incapacitated, and also in the event that he died, until the succession papers were opened. Gama, unlike any of his predecessors, carried a set of sealed letters from the king, designating a series of successors by priority, in the event he died. This practice was later to become commonplace, but in 1524 it was still unusual; and no one knew who in fact had been designated to succeed the viceroy in the first place. D. Luís de Meneses, on board his ship in Cochin harbour, hence still held out hopes, as indeed did Lopo Vaz de Sampaio. Another powerful *fidalgo*, the captain-designate of Melaka Pêro Mascarenhas, did not leave Cochin for his post until 1525; it is unclear

[85] AN/TT, CC, I–45–103, letter from D. Jaime to D. João III, cited in Borges de Macedo, *Um caso de luta pelo poder*, p. 128.

[86] Cf. the detailed biographical study of this figure by Luís Filipe F.R. Thomaz, 'Diogo Pereira, *o Malabar*', *Mare Liberum* 5 (1993), 49–64.

whether he too was waiting to see how the succession turned out.[87]

The atmosphere in Cochin in these last months of Gama's life is brought out rather well not only by the chroniclers (including Gaspar Correia, who may himself have been present in the port), but by the letters of the period, of which we have already made mention of a handful.[88] Unfortunately, the bundle of 'India letters' sent out to Portugal in early 1525 has not survived in its entirety; what has come down to us is an invaluable volume entitled, 'Summary of the letters that came from India in the fleet that came in the past year of 1525'.[89] In the case of some of these letters (those of D. Henrique de Meneses of late October 1524, of Mar Jacob, and of Manuel Botelho of January 1525) we have both the full text and the summary; a comparison of the two suggests in each of these cases that the summary represents a rather close paraphrase of the original.

Of the eighty-odd letters summarised in the volume, some naturally command our attention more than others. The volume begins with the summary of a letter by Gama himself, that he had begun in December 1524; the original was apparently accompanied by a detailed account (in two copies) in the form of a diary of all that transpired between the fleet's departure from Belém in April to December.[90] Gama's letter itself is rather flat and uninteresting, at least in its summary form. It begins with practical shipboard problems, including the fact that captains on the *Carreira da Índia* routinely suffered from a lack of water, since they did not take enough on board on leaving from Portugal. The capture of the ship from Aden is next mentioned, as also the fact

[87] AN/TT, CC, I–32–106, letter from Pêro Mascarenhas at Melaka to D. João III, 1 September 1525, published in Borges de Macedo, *Um caso de luta pelo poder*, pp. 139–44. Mascarenhas arrived in Melaka only on 16 July 1525, and claims that his departure from Cochin was delayed by D. Henrique de Meneses and Afonso Mexia.

[88] Cf. AN/TT, CC, II–116–108, *conhecimento* dated 2 July 1524, by which a certain Gaspar Correia, *almoxarife* at Cochin acknowledges receipt of some livestock fom Lançarote Fróis, factor at Goa. For his account of affairs in Cochin, see Correia, *Lendas da Índia*, Vol. II, pp. 843–5.

[89] Compare the summary of the 1506 letters in AN/TT, Gavetas, XX/4–15, 'Sumários das cartas que vieram da Índia para el-Rei de Portugal', in Silva Rego (ed.), *Gavetas*, Vol. X, pp. 356–72; and of the 1533–4 letters in AN/TT, Gavetas, XX/1–53, 'Sumário das cartas que vieram este anno de 1534 na armada da Índia e no caravelam em que veyo por capitam', published in Georg Schurhammer, 'The India Letters of 1533', in Schurhammer, *Orientalia*, pp. 153–84. Indeed, a surprisingly large number of letter-writers from 1524–5 can be found in India in 1533–4, as a comparison of the documents shows.

[90] AN/TT, Núcleo Antigo, No. 875, 'Sumário das cartas que vieram da Índia na Armada que veio o anno passado de mil b^{c} xxb', fls. 2–3, 'Huma carta que tinha começado o viso rey para vosa alteza antes que falecese'.

that those on board had told Gama that Ra'is Salman had a fleet of twenty galleys ready in Jiddah, and another twenty-two vessels under another Ottoman captain at Suez. The letter now passes on to the affairs of Goa, beginning with how Gama's fleet had seized a Crown vessel captained by a certain Bastião Fernandes, who stated that he was a *criado* of D. Duarte de Meneses; the vessel carried a cargo of ivory, which it was claimed belonged to Vasco Fernandes Coutinho. The debts and misdemeanours of the captain of Goa, Francisco Pereira, are the next subject, as also his relations with merchants from Safavid Iran and Vijayanagara (*mercadores de narcyngua e xeque ysmael*). Gama reports how the captain had paid back over 25,000 *pardaos*, as well as a substantial quantity of copper that belonged to the Crown. In all these respects then, Gama's letter confirms the broad accuracy of the account to be found in Castanheda, even if it glosses over the fleet's arrival in Chaul, the affair of the women who were publicly scourged, or the draconian measures to cut down expenses. Instead, Gama focuses largely on the strength of the Muslim opponents of the Portuguese, and his own efforts to raise revenues, noting for example that he has given out the land-revenues in Goa in farm, for a sum that was 8,000 *pardaos* more than what was taken earlier. In this context, it is noteworthy that he writes approvingly of the activities of the celebrated Brahmin entrepreneur from Goa, Krishna, who apparently provided him valuable help and information; the Goa Municipal Council for its part took a rather more negative view of Krishna's extensive fiscal powers.[91]

The summary equally mentions Gama's concern with the 'robberies and rebellions of Malabar, and how the Moors are very strong and the Christians very weak', before passing on to issues concerning the transition in government. The Act by which charge was handed over (*auto da emtregua da governamça*) by D. Duarte to Gama in early December was apparently reproduced in the original, but is merely mentioned in the summary. The summary does not provide us any further details, passing on to the extensive letters and papers sent by Afonso Mexia, and by other persons. Thus, the trace of Gama's last letter is altogether

[91] On the career of Krishna, see P.S.S. Pissurlencar, *Agentes da diplomacia portuguesa na Índia* (Bastorá-Goa, 1952), pp. 1–21; for his later conflicts with Nuno da Cunha in the early 1530s, see Schurhammer, 'The India Letters of 1533', pp. 175–7. Cf. also the letter from Krishna to D. João III, in AN/TT, Núcleo Antigo, No. 875, fl. 45v, recounting his meeting with Gama, and the letter of complaint from the Goa Municipal Council dated 10 January 1525 (fls. 45v–46v).

disappointing, providing little more than a confirmation of matters of which we are already cognisant through other sources. The Admiral, to the last, remains a man of few words.

The rather more garrulous Afonso Mexia takes it upon himself to say far more. The summary of one of his letters, dated 30 December 1524, opens with the usual mention of the voyage, and the ship that was taken off Diu on its voyage from Aden; mention is made of a cargo worth 60,000 *cruzados*. The usual statement of the affair of Francisco Pereira may be found once more. Interestingly, Mexia notes that Gama had already sent an earlier vessel (between October and December) back to Portugal, with a small cargo of 600 *quintais* of pepper, and his first impressions of the situation in India; little other trace survives in the archives of this vessel. For the main fleet, it is noted, Gama had decided to use the oldest and largest ships possible, keeping the newer and sleeker ones for military use in the Indian Ocean. This had apparently created conflicts with some of the captains, who considered the older vessels unsafe, as we see from other letters. Mexia then adds details concerning the pepper weighing-house in Cochin, where he claims smuggling and embezzlement were common, so that Gama had decided to shift its location to a place where it could be supervised more closely. This, for its part, had occasioned conflicts with the Cochin ruler.

The letter shows clearly enough that Gama had made himself unpopular rather rapidly, with a number of interests and in a variety of ways. He had found that artillery was greviously short in the fortresses, and on demanding records for their upkeep from the *almoxarifes*, found that the latter had burnt the books. Mexia alleges that the bulk of these guns had been taken by individuals for their own use, often on mercantile voyages; he also suggests rather clearly that D. Duarte de Meneses was complicit in the matter. Mexia's own predecessor, that pillar of honesty in a sea of corruption, Dr Pedro Nunes, comes in for praise in this and other matters relating to the pepper trade. But Mexia's account also adds considerable nuances to the account of the chroniclers. He insists, for example, that Gama himself patronised various *fidalgos* and Crown servants at his table every day, an interesting contrast to Castanheda's complaint that the viceroy prevented other *fidalgos* from doing the same! The *vedor da fazenda* does not seek to gloss over the problems of transition, noting that there were great conflicts with D. Duarte and D. Luís de Meneses, though principally on account of persons who had 'bad thoughts'.

Fortunately, he notes, he together with Lopo Vaz, Pêro Mascarenhas and the secretary Vicente Pegado (whose letters we shall discuss below), had ensured that these rumblings were kept from going out of hand.

On account of Gama's illness, he goes on to state, pressing affairs had to be decided by others such as the men mentioned above, in consultation with officials like Dr Pedro Nunes and Francisco de Sá. Thus, the ships and fleets that were sent out in late 1524 to different parts of Asia, were decided and organised by these men in a notional collective. In some instances, royal orders from Portugal were already in effect, and were simply continued with. This was the case, for example, with Lopo de Azevedo, who in early 1524 had already replaced the captain of the Coromandel fleet (and D. Duarte de Meneses's nominee) Manuel de Frias, and who thus returned to Cochin after his round trip at this very time.[92] Mexia returns time and again to the social conflicts that surrounded Gama's brief government. Even relatively trivial conflicts engage his attention: Gama, he notes, had forbidden the captains on board his fleet to load casks of wine on board without express royal permission. On arriving in India, he found several who had done so, and promptly confiscated the wine and had it stored away. Nobody protested overtly, but there was resentment on this front once more.

Mexia's posthumous image of Gama (and we may note that this letter was written a mere week after the viceroy's death), is thus of a stern disciplinarian, a veritable Hercules in the business of cleaning out the Augean stables. He reports Gama's dying statement, and it comes down to us in the following summary:

> And he [Mexia] says that the viceroy before his death told him that he should write to Your Highness that his wife and sons should be looked after, for they would be forsaken after his death, and that it should be remembered that he need not have made that voyage had it not been for the fact that matters of his service in those parts pained him a great deal, and that he had spent a good deal on that voyage without any profit, and he asked Your Highness as a favour to take on his servitors, and that Your Highness should do so since on a number of occasions he [Gama] had been told that in

[92] AN/TT, Núcleo Antigo, No. 875, fls. 28v–29, 'Huma carta de Lopo d'Azevedo sprita em Cochy a b de Janeiro de mil xxb'. Also the brief discussion in Jorge Manuel Flores, '"Cael Velho", "Calepatanão" and "Punicale": The Portuguese and the Tambraparni Ports in the Sixteenth Century', *Bulletin de l'Ecole Française d'Extrême-Orient* 82 (1995), 17–19.

> the provision that he had to take along some people every year, he should not take any of his own servitors (*criados*).[93]

Here then is an image of the dying Gama, somewhat at variance with the usual one derived from Gaspar Correia, in which he discreetly has a sum of money paid to the women whom he had had scourged in Goa. If Mexia spoke for Gama in this letter, his view of Portuguese India was altogether an interesting one: what was lacking were not men (in Mexia's view there were, if anything, too many Portuguese in India!), but artillery and discipline. The project, we may recall, was to reduce not only the size of each establishment, but the number of establishments as such. Since Colombo and Pasai had been abandoned by September 1524 (the latter rather more ignominously than the former), this project was already partially being implemented by the time the viceroy's fleet arrived in India.

But Gama's 'last wishes' as recorded by Mexia also point in a different direction than the application of abstract principles – namely his concrete concern with the protection of his own family and clients. This protection extended to Diogo Pereira, in whose house in Cochin Gama passed his last days, and whom D. João III had ordered back to Portugal. Gama, and as a consequence Mexia and others, decided to suspend this royal order, and the *vedor* thus wrote rather blandly to the king that Pereira 'was needed there for some matters concerning the service of Your Highness', the more so since they did not know why he had been ordered to come back!

The secretary of Portuguese India, Vicente Pegado, adds further details to Mexia's account in a letter of 1 January 1525, of which an extensive summary equally exists. Pegado, we may note, was named to this post by Gama in place of the official nominee, António Rico, who had failed to complete the voyage to India (and who eventually took the post over only in 1526). He thus had every reason to be grateful to the viceroy, and his letter is perhaps even more positive than that of Mexia in respect of Gama's memory. But the narrative is nevertheless a complex one. Pegado's letter (or rather, the summary available to us) begins squarely with the 'great past disorders from the time of Dom Duarte' that Gama found in India, and then goes on to speak of officials helping themselves freely to the king's goods in the

[93] AN/TT, Núcleo Antigo, No. 875, fls. 4v–7v, 'Huma carta de Afonso Mexia sprita em Cochy a xxx de dezembro de mil b^c xxb [*sic*] por duas vias'.

factories: 'And on the viceroy asking why this was done, the officials replied that they did not wish to be martyrs for Your Highness.' The scandals were many, from Crown ships that were used to raise money through freight-charges, to Crown ships (such as one commanded by Lopo de Azevedo) that made the lucrative voyage from Bhatkal to Hurmuz without Crown goods, or the large-scale export of spices to the same Hurmuz by private merchants who had dispensations from D. Duarte. The source of all of this was easy to see: as in Mexia's letter, Pegado's account too claims that there are too many Portuguese in India. In Cochin alone, he claims, there were 4,500 to 5,000 of them at the time of the dispute between Gama and Meneses, who thus constituted a considerable source of instability. These men, he reports, 'are ill-advised in all the evil usages and customs, and this cannot be amended save with a great effort'. Gama, for his part, had made a start, concentrating in particular on the war with the Mappilas, to which end he had had some smaller vessels (brigantines and others) constructed. The imminent Ottoman threat is mentioned, as is also the fact that an embassy had been sent to Süleyman the Magnificent from Diu to solicit his aid.

Pegado also takes some care to mention the details of the conflicts between the viceroy and the brothers Meneses, here adding valuable new details to other accounts. For example, he notes that Gama had punished some of D. Luís de Meneses's servants, a fact that the latter greatly resented. He also notes that when D. Duarte de Meneses arrived from Hurmuz, he reputedly carried over 100,000 *pardaos* in tribute from there, besides another 30,000 from a ship that had been captured in the Persian Gulf, and still another sum of 15,000 *cruzados*, sent from Portugal in the ship *Salvador*, to be employed in spices. Gama apparently demanded this money, but was refused it by D. Duarte. Here, Pegado's letter suggests that the two actually met and had an altercation, noting that 'the viceroy had had words with Dom Duarte about this money, and about a ship that it was said belonged to Francisco Pereira'; in contrast, several of the chronicles (notably Castanheda) suggest that the two did not ever meet, and that Meneses remained on board ship while Gama lay sick on land.[94] In an earlier letter, of 29 December 1524, Pegado has further details on this conflict, noting for example 'an inquiry that

[94] AN/TT, Núcleo Antigo, No. 875, fls. 7v–10v, 'Huma carta de Vicente Pegado secretario scripta em Cochim a primeiro de Janeiro de [1]:525'. João de Barros's version appears to concur though with that of Pegado.

the viceroy secretly conducted on Dom Duarte, in which only five witnesses could be examined on account of his indisposition'; the two copies of the testimonies that accompanied the letter have unfortunately not come down to us. Pegado paints a very tense situation in Cochin in early December, even noting that 'a mutiny (*alvoroço*) had begun to arise concerning what had passed between the viceroy and Dom Duarte'. It is obvious that besides the *fidalgos* who had accompanied Gama to India (and who were probably suspect in the eyes of the several thousand Portuguese already in Cochin), an important role was played in calming the atmosphere by a number of 'old India hands' whom Pegado mentions with some approval. These were men like Pêro de Faria, who had already served some sixteen years in India, and who is reported to have offered his services to Gama immediately after the latter's arrival in Cochin, or António de Miranda de Azevedo, Francisco de Sousa, Manuel da Gama, and Manuel Sodré – in sum, a handful of influential *fidalgos* with their own clients in place among the soldiery.

The four letters from Pêro de Faria that are summarised in the same collection suggest, equally, that among the 'old India hands' it was not Diogo Pereira alone whom Gama had chosen to favour. Faria was later to attain a certain notoriety, twice being named captain of Melaka (in the late 1520s and again in the late 1530s); he died in 1546, after nearly forty years in Portuguese Asia. One of these letters, written before Gama's arrival, appears to set out Faria's vision of both political and commercial prospects in Portuguese Asia, and suggests a rather cautious line, by which 'the Kings and Lords are to be favoured and permitted to live by the customs under which they were born, without changing them or proposing novelties, for it is thus that goodwill and strange lands are won over'. Faria thus emerges, like Diogo Pereira, Duarte Barbosa and others, as a proponent of a policy of minimal interference with local rulers, but equally a fervent supporter of a particular engagement between the Crown and old India hands. Time and again, his letters return to the theme of how the Crown has neglected men like himself (he claims twenty years' service in India), instead granting plum captaincies for as long as nine years to others who had freshly arrived from Portugal. He portrays himself as a simple soldier who had never learnt letters or the fine manners of a courtier, and with only the Count of Vimioso to argue his case in Portugal. But it is on account of men like him that Portuguese India is safe from the Turk, for when push comes to shove, they will 'pull out their beards in shame', and do whatever is necessary. Indeed, in his view, no *fidalgo* or cavalier should be allowed to return to Portugal until the Ottoman threat has passed.[95]

It is one of the letters dated 28 December 1524 that is of particular interest for describing the relations between Faria and Gama. The letter begins as usual, by complaining that he has not been granted the captaincy of a fortress that year, then passes quickly to lament the great loss that the death of the viceroy represents. Indeed, to hear Faria tell it, Gama 'already had the affairs of India placed in good order'; it is only a question of not retreating from the stands that he had taken, particularly with respect to the Mappilas. But, as Faria brings out thereafter, Gama had more than his share of opponents. His letter thus declares that, 'the captains, factors, scriveners and other officials would be very satisfied at the death of the viceroy, for they do not wish to have the justice in their own house that he brought'. Singled out for ironical attack is D. Duarte de Meneses:

> And he [Faria] says that if any man served Your Highness badly, it was Dom Duarte, who committed many thefts and took a great deal of money, and that on account of this he would live long, and the viceroy as he was a most loyal vassal, and had served so loyally and with such truth and justice, had been carried off by God, and that Your Highness should pardon him if in so saying he appeared to be a bad Christian.

Faria now goes on to stress that Gama and he had been very closely associated in the last weeks of the former's life; thus, of all the *fidalgos* who had asked the viceroy for a captaincy, he had said he would give one to Faria alone. As a measure of this closeness, Faria had been asked to accompany D. Estêvão da Gama whenever he took the fleet out on the high seas. Faria praises young D. Estêvão inordinately, claiming that he was modest and courteous, not at all arrogant on account of being the viceroy's son. D. Estêvão and D. Paulo, the viceroy's two sons, were – we may note – promptly sent back to Portugal in the fleet of early 1525. This was in keeping with royal orders of what was

[95] AN/TT, Núcleo Antigo, No. 875, fls.16v–19, summaries of letters from Pêro de Faria dated 15 November and 28 December 1524, and 11 January 1525. An extensive later correspondence exists between Faria and D. João III; cf. for example AN/TT, CC, I–59–105 and I–60–17, respectively dated Goa, 20 October and 18 November 1537; CC, I–66–37, letter dated Melaka, 25 November 1539, etc. For further details, see Schurhammer, *Die Zeitgenössischen Quellen*, entries 612–14, 1582, 1670, 1936.

to transpire if Gama were to die in India. Thus, the Admiral's intention to send out a strong *armada* to the Red Sea under the command of D. Estêvao was not realised. It was only much later, as Governor of Portuguese India in 1540–1, that Dom Estêvão da Gama could lead a celebrated expedition up the Red Sea, as far as Suez and Mount Sinai.[96]

A CONTESTED HERITAGE

Vasco da Gama died, then, in Cochin, on Christmas Eve 1524, some three months after his arrival in Goa, and about eight weeks after setting foot for the third time in his career on the Kerala coast. He was buried with full honours, and considerable ceremony, in the Franciscan church of Santo António, one of the earliest to be built by the Portuguese in Asia.[97] After Mass had been said for his soul, the *fidalgos* and notables are reputed to have gathered at the See church, where the sealed royal letter of succession was produced by the *vedor da fazenda*, Afonso Mexia. This was the first time that such a procedure was being followed (indeed, no earlier governor had died in office before the naming of a successor). The letter, dated 10 February 1524 at Évora, was hence read aloud by Vicente Pegado, the secretary, and the successor's name came as something of a surprise: for it was D. Henrique de Meneses. The expectations of Lopo Vaz, Pêro Mascarenhas and D. Luís de Meneses were thus dashed, and two of them made their displeasure known to the Crown in no uncertain terms. The volume of summary letters we possess laconically notes 'a letter from Dom Luís de Meneses in which he complains greatly of the harassment to which the viceroy put him, and even more of the fact that Your Highness did not grant him the government of India, that he even now requests most strongly, and that he expects to serve in this to the complete satisfaction of Your Highness'.[98] There is also the summary of a letter from Lopo Vaz de Sampaio, 'in which he recounts everything that transpired in the voyage until the death of the viceroy, and after

[96] The expedition is very extensively documented; eg. AN/TT, CC, I–70–109, letter from D. Estêvão da Gama at Goa to the King, dated 16 October 1541. On D. Estêvão's desire to carry out what had been denied to him in 1524–25, cf. Castanheda, *História*, Vol. II, Book 9, ch. 39, pp. 940–1.

[97] Cf. Schurhammer, 'Carta inédita sobre a fundação do convento de S. Francisco', pp. 210–11.

[98] AN/TT, Núcleo Antigo, No. 875, fl. 23v. Apparently, at the time he wrote this letter, D. Luís de Meneses still intended to remain in India.

that he complains greatly that D. Anrique was given the succession to the government of India, since he [Lopo Vaz] has more qualifications for this than the said Dom Amrique'.[99] Nevertheless, word was sent to D. Henrique de Meneses in Goa by some galleys, and the galleon *São Jerónimo* (commanded by his cousin D. Jorge de Meneses), and in the meanwhile Lopo Vaz took charge of affairs, sending one fleet under Simão Sodré to take prizes in the Maldives, another under António de Miranda de Azevedo to do the same off Cape Gardafui, several merchant vessels to Hurmuz (as noted above), and a single ship to bring back pitch from Malindi.

By about 21 January 1525, the fleet to Portugal was ready to depart: on board were Gama's two sons, and also his bitter rivals, the brothers D. Duarte and D. Luís de Meneses. At least one letter from the period, that by Simão Sodré, notes that Gama's sons were rather badly treated in these weeks, now that they were stripped of their father's protection.[100] The fleet does not appear to have awaited the arrival from Goa of D. Henrique de Meneses, who left behind Francisco de Sá as his successor in the captaincy there; it did however carry back several letters from D. Henrique, including one from Goa dated 9th January 1525, in which he thanks the king for the honour done to him, and recounts at some length the exploits of his nephew D. Jorge Telo, against the Mappila *paraos* off Goa.[101] As an anonymous account from the epoch informs us, the new governor assumed office on 8 January 1525 (all of fourteen days after Gama's death), then set off for the south a week later, on 16 January, with a fleet of five vessels, determined to take the attack to the Mappila settlements on the Kanara and Malabar coasts.[102] In a series of brutal attacks, he first mounted a raid on Mappila vessels of Bhatkal, and then did the same off Mt Eli, near Cannanur. Arriving in Cannanur on 26

[99] AN/TT, Núcleo Antigo, No. 875, fls. 13v–14. In another letter dated 15 January 1525 (fls. 14–14v), Lopo Vaz complains bitterly against the *vedor da fazenda*'s office, which reduces that of the captain of Cochin in importance, and mounts a thinly veiled attack on Afonso Mexia!

[100] AN/TT, Núcleo Antigo, No. 875, fls. 31v–32. The summary of text notes: 'diz como faleçeo o Viso Rey e a perda de sua pesoa e como a seus filhos foram feitos grandes agravos por os mandar Vosa Alteza tam secamente embarcar'.

[101] AN/TT, Núcleo Antigo, No. 875, fls. 15v–16. The first letter that I have been able to trace from Cochin to Goa that mentions Gama's death is AN/TT, CC, II–122–108, letter from Afonso Mexia to Miguel do Vale, 28 December 1524. But as this letter mentions, the galleon *São Jerónimo* had already been despatched before this date.

[102] AN/TT, Colecção São Vicente, Vol. XI, fls. 37–46v, 'Notícia do governo de D. Henrique de Meneses que começou a governar a Índia a 8 de Janeirro de 1525, por successão de Dom Vasco da Gama'.

January, the very same day he had 'the uncle of Baleacem', a Mappila notable who was held captive in the fort, hanged secretly. Later raids followed on Dharmapatam and other settlements of northern Kerala, before the new governor arrived in Cochin on 4 February 1525. Here he remained for two weeks, before setting out again with a formidable force of 1,500 Portuguese, to attack Ponnani and Calicut itself. Throughout 1525, D. Henrique appears to have concentrated his attention on the Kerala coast, leading a number of attacks and raids, in defence of Calicut, and against Mappila settlements like Quilandi. In brief, what Gama had been unable because of his ill health to implement, his successor managed to do within two months of his taking over as governor. This did not mean that the Portuguese fortress at Calicut was secured: rather, by the end of 1525, it had been abandoned like Colombo and Pasai (albeit with more premeditation than the latter).[103]

But, as so often in the first half of the sixteenth century, Portuguese policy with respect to the Indian Ocean at the court had meanwhile taken another of those curious turns. The letters sent in the fleet that left Cochin in the last days of January 1525 (we are uncertain exactly when) probably reached Lisbon in the autumn of that year.[104] D. Duarte de Meneses, who carried some of them back on his ship, was treated with a harshness in proportion with the strength of the criticisms to which he was subjected, being imprisoned at the castle of Torres Vedras and elsewhere, in which situation he remained for all of seven years before eventually regaining a measure of royal favour through the intercession of the Count of Castanheira (and becoming captain of Tangiers once more in 1531). As for his brother D. Luís de Meneses, who made the return voyage on the *Santa Catarina do Monte Sinai* (on which Gama had gone out to India), he never returned to Portugal; some rumours had it that he had decided to turn rebel on the coast of East Africa, others that he had been boarded off the Portuguese coast, reputedly by one of a number of French corsairs active in those years, and killed.[105] If the

[103] For the events leading up to the construction of the Pasai fortress (and rather more briefly, its abandonment), see Jorge Manuel dos Santos Alves, 'Princes contre marchands au crépuscule de Pasai (c. 1494–1521)', *Archipel* 47 (1994), 125–45.

[104] Thus, D. João III in one of his letters to D. Henrique de Meneses mentions a letter that the latter had written at Cannanur on 27 January 1525, but which arrived 'na naao em que vynha Dom Duarte de Meneses'; AN/TT, Núcleo Antigo, No. 875, fl. 112v.

[105] For the latter version, see Correia, *Lendas*, Vol. II, pp. 854–5. The attacks on Portuguese shipping by French corsairs from Brittany and Normandy, like the celebrated Jean Ango, were a major concern in the early and middle years of D. João

laissez-faire politics that D. Duarte had implemented in the Indian Ocean thus received a visible reprimand, the policies that Vasco da Gama had wished to implement – namely the concentration on a few centres, the abandonment of a number of outlying fortresses – were set aside as well. The Admiral's death meant that his political weight and prestige were no longer behind the case that he had argued, and the mid-1520s see a return to a phase of projects of expansion, that eventually find concrete expression during the long governorship of Nuno da Cunha (1529–38), when new fortresses were set up in Chaliyam, Cranganor, Bassein and Diu.

As has been noted above, by September 1525, the decision to abandon the Calicut fortress had already been reversed by D. João III, albeit too late for it to be communicated in time to the governor in India.[106] An extremely detailed royal letter, addressed to D. Henrique de Meneses, notes this reversal of policy: whereas earlier, it had been thought that the best policy would be to withdraw the fortress and attack the Samudri Raja from the exterior, it had now been found more appropriate to strengthen the fortress with as many men, munitions and guns as possible. Indeed, these letters of late 1525 posit two priorities for the governor, namely the war against Calicut and that against 'Cambaya' (here, effectively, Diu). 'And since it seems to us that both these wars of Cambaya and Calecut cannot be managed at the same time, we find it well that principally and first you proceed in making this war against Calecut, and in cleaning up the entire coast of Malabar of the *paraos* that the Moors possess there in such a number and as well-armed as you know.'[107]

In late 1525 or early 1526, D. Henrique sent the Cochin factor Manuel Botelho back to Portugal with details of his anti-Mappila exploits at Ponnani, Calicut and Quilandi. To add to his earlier tensions with Lopo Vaz de Sampaio, the governor's relations with Afonso Mexia, the *vedor da fazenda*, had soured within months of his takeover of office, and he had instead come to rely more

III's reign, and precede the expedition of Jean and Raoul Parmentier to the Indian Ocean in the late 1520s; cf. Charles Schefer (ed.), *Le discours de la navigation de Jean et Raoul Parmentier de Dieppe* (Paris: Ernest Leroux, 1883). For an overview, see Vitorino Magalhães Godinho, *Mito e mercadoria, utopia e prática de navegar, séculos XIII–XVIII* (Lisbon: Difel, 1990), Ch. 13, pp. 459–75.

[106] Cf. the discussion in João Paulo Costa, 'Os Portugueses e a cristandade siro-malabar', pp. 160–1, citing the nomination on 26 September 1525 of Francisco de Sousa Tavares to the post of captain of Calicut.

[107] AN/TT, Núcleo Antigo, No. 875, fl. 120v, undated letter from D. João III to D. Henrique de Meneses (fls. 117v–22).

and more on his own cousin D. Jorge de Meneses, D. Simão de Meneses (captain of Cannanur, and now named Captain-Major of the Indian Seas), his nephew D. Jorge Telo, and Manuel Botelho. In letters that the Cochin factor carried back, D. Henrique appears to have announced that the anti-Mappila campaigns had enjoyed some success, and that the attack on Diu (leading to its capture) could now be pursued whole-heartedly. He thus sent a drawing of the town, including the section known as 'a villa dos Rumes', to Portugal, proposing the landing of a substantial force against the town. The Crown responded enthusiastically, and in a series of letters dated early September 1526, sent through a single vessel that year, promised the governor major reinforcements to be despatched in February 1527: thirteen or fourteen ships, and a force of 3,000 men.[108]

But, as had happened earlier, events in India overtook plans in Portugal. In late 1525, D. Henrique de Meneses set out briefly for Goa, to prepare for his campaign against Diu, and having carefully spread a rumour that he was in fact on his way to Massawa in the Red Sea, returned to gather his forces on the Kerala coast. In early February 1526, the governor, who had been suffering from a problem in one leg as a result of exertions in various campaigns, died at Cannanur after an illness of roughly a month. This was a piquant situation indeed for, as has been remarked above, between the foundation of the state of Portuguese India in 1505 and 1524, no governor or viceroy had died in office before the arrival of his successor. Now, in the space of fifteen months, there were not one but two successions.

We may rapidly summarise what followed. The succession letter being opened by Afonso Mexia, it was found that the captain of Melaka, Pêro Mascarenhas, had been named to succeed D. Henrique de Meneses. But the *vedor da fazenda*, and a faction amongst the *fidalgos*, had a marked preference for the captain of Cochin, Lopo Vaz de Sampaio. Thus, on the pretext that Mascarenhas was too far away, the next letter of succession was opened, and using it, Lopo Vaz was declared governor. It is clear that the complex figure of Afonso Mexia, whose ambitions and inflated sense of the centrality of his post often led him to such sharp practices and abuses, lies at the centre of the

[108] Cf. the valuable collection of letters dated September 1526, and written in Tomar, in AN/TT, CVR, No. 17, letters to D. Henrique de Meneses and Afonso Mexia (28 ff.). This force corresponds roughly to that sent not in 1527, but a year later, in 1528, under Nuno da Cunha.

problem.[109] The analysis of the problem of institutions and personalities must await another study; suffice it to note here that this breach of procedure opened up a celebrated intra-élite conflict in Portuguese India that could be resolved only in 1529, with the long-delayed arrival of a new governor, Nuno da Cunha.[110] Mascarenhas, who returned to India from Melaka to claim his due, was set aside, and returned to Portugal to complain. Later, the comportments of both Lopo Vaz and, to a more limited extent, Afonso Mexia came under judicial scrutiny, and the former received exemplary punishment, after being taken back as a prisoner to Portugal to answer charges.

This conflict, which in a certain sense froze new lines of development in Portuguese Asia at the level of policy for a period of three years, nevertheless did not interrupt, in the medium term, the embryonic changes that we have noted for 1525–6. In sum, D. João III had already moved by these years from a simple reaction against his father's megalomaniac imperialism to a more complex policy, particularly bearing in mind the interests and pressures of 'old India hands'. The implementation of this knee-jerk anti-Manueline reaction had, in a manner of speaking, been Vasco da Gama's last project: to dismantle large parts of the *Estado da Índia*, reduce extravagant expenditure, limit proselytisation, and promote the private trade of the Portuguese nobility in the Indian Ocean. Indeed, many of these elements merely develop ideas set out by D. Aires da Gama, in his letter written from Cannanur to D. Manuel in 1519. Naturally, by 1524, this was tempered by three additional considerations: defending the core of the *Estado* against the Ottomans, the anti-Castilian programme in the Moluccas, and the anti-Mappila campaign, which Gama doubtless saw as a prestige issue. By 1525, even Joanine pragmatism had begun to dream more grandiose thoughts, as resources shrank in a plague-ridden Portugal, and the possibilities of major revenues from expansion into the trade of Gujarat appeared too tempting to be ignored. Had Vasco da Gama been alive in 1525 or 1526, he might have found himself therefore once more at odds with the policies of the Portuguese Crown.

[109] Cf. AN/TT, CC, I-33-36, letter from Afonso Mexia to D. João III dated Cochin, 10 December 1526, concerning his interference in the correspondence with the Cochin ruler; also the exchange of correspondence between Nuno da Cunha and Afonso Mexia in early 1530 in AN/TT, CC, I-44-58.

[110] The usual references on this conflict are Borges de Macedo, *Um caso de luta pelo poder*, and António Coimbra Martins, 'Correia, Castanheda e as Diferenças da Índia', *Revista da Universidade de Coimbra* 30 (1984), 1–86.

It is thus tempting to see Gama in the end of his life as a man already out of touch with circumstances, linked to the (literally) reactionary politics of an earlier reign, and soon to be overtaken by events. But that would be to underestimate the difficulties that even intelligent contemporaries had in seeing the shape of developments to come. For let us not forget that already in 1525, some were writing the obituary of the Portuguese Asian enterprise. To gain a flavour of these, we could probably do little better than to quote the returning Venetian ambassador to Charles V, Gasparo Contarini, in his address to the Senate on 16 November 1525:

> Having had information concerning the affairs of Portugal, I believe first of all, as has been affirmed to me by men most familiar with that kingdom, that that king has a far smaller sum of money than is commonly believed, for he spends a very large sum in maintaining that voyage to India, and the needs of various fortresses and diverse fleets, which cost him a considerable amount of money: so that I believe that this being so, these voyages of his to the Indies will diminish, rather than grow; and since, as has been said, it represents a great expense, and the Portuguese are much detested in almost all the parts of the Indies, as the countrymen (*li paesani*) have seen that they go about fortifying themselves little by little, and making themselves the lords of those lands. From trustworthy persons, I have come to know that two years ago, five Portuguese ships were robbed, and captured by those of China, who like them had made up a fleet. So that, since they are detested, and since the natives of those lands are continuously becoming expert in navigation and making war, I think that the difficulties will increase each day. Besides, this new young king does not apply the diligence that his father did, and already some of his captains who are in India have begun to compete and have conflicts between them. And that suffices for the East Indies.[111]

But perspectives change with time. In the mid-1560s, a Portuguese *fidalgo*, who had accompanied Gama on his last voyage, could ruminate on the good old days when the great Vasco had presided over a prosperous Cochin, so unlike the 'decline' that he claimed to witness forty years later. Thus, D. Jorge de Castro, former captain of the fortress of Chaliyam, in a letter to the Cardinal D. Henrique:

[111] Cf. 'Relazione di Gasparo Contarini, ritornato ambasciatore da Carlo V, letta in Senato a di 16 Novembre 1525', in Eugenio Alberi (ed.), *Relazioni degli Ambasciatori Veneti al Senato*, 1st series, Vol. II (Florence: Clio, 1840), p. 49.

> I confess to Your Highness that it hurts my heart these past years to see the destruction and neglect of this famous river of Cochin, and the warehouses which at another time were warehouses, and the treasure-houses and storehouses for goods and pepper and the factory and fortress of this city, all of which I saw in their prosperity about forty years ago, when Dom Vasco da Guama came to these parts as viceroy, of all of which I can barely say now where they have gone . . .[112]

But underlying this is a familiar, indeed recurrent, sixteenth-century theme: the unjust neglect of Cochin, which was once the centre of Portuguese trade in India, in favour of other centres of far less interest. A theme that, appropriately enough, would have found an echo in the sentiments and priorities of Vasco da Gama, even if his bones no longer rested, by 1563, in that city.

[112] AN/TT, CC, I–106–52, letter from the *vedor da fazenda* D. Jorge de Castro to the Cardeal Infante, Cochin, 24 January 1563.

CHAPTER SEVEN

FINALE: THE JUDGEMENTS OF POSTERITY

The commemoration of the fifth centenary of the birth of Vasco da Gama is not only the act of a country that celebrates the memory of one of its most illustrious children, but equally of all of humanity, which sees in Vasco da Gama one of the great argonauts of Modern Times.

Joaquim Veríssimo Serrão, 'Le Voyage de Vasco da Gama' (1970)[1]

THE EXPLORER'S VISAGE

LET US OPEN, MORE AT LESS AT RANDOM, A BOOK ON THE history of Portugal and search its index for Vasco da Gama. Consider Durand's recent *Histoire du Portugal* (1992), a book that forms part of a series entitled *Nations d'Europe*, which in itself is a project designed to promote a sense of European 'unity-in-diversity' in the context of the closing of the ranks in the European Union of the late twentieth century.[2] A sober and sensibly put together work by a specialist in the rural history of medieval Portugal, this introduction to Lusitanian history contains in its index seven references to Gama, fewer than those to the Infante D. Henrique ('dit le Navigateur') or to Camões, but far more than to Pedro Álvares Cabral. The importance of the 'extraordinary adventure' of the first voyage is stressed, the type of ship used in it is mentioned, and even the second voyage (in which Gama 'sowed terror') is described in summary. The links between Gama and the *Lusíadas* are brought out succinctly. For

[1] Fondation Calouste Gulbenkian, *Cinquième Centenaire de la Naissance de Vasco de Gama (1469–1969): Exposition bibliographique et iconographique* (Paris: Centre Culturel Portugais, 1970), p. 19.

[2] Robert Durand, *Histoire du Portugal* (Paris: Hatier, 1992), pp. 75, 79, 82, 88, 93, 98, 106. The portrait is reproduced on p. 83. The same portrait appears on the cover of Carmen M. Radulet, *Vasco da Gama: La prima circumnavigazione dell'Africa, 1497–1499* (Reggio Emilia: Edizioni Diabasis, 1994).

Plate 26 Anonymous portrait of Vasco da Gama, Museu de Arte Antiga, Lisbon, school of Gregório Lopes.

good measure, one even finds the black-and-white reproduction of a portrait in oils on wood of Vasco da Gama from the Museu Nacional de Arte Antiga in Lisbon, where the white-bearded Argonaut, slightly favouring his right profile, with a humble cap on his head in the style of portraits of Columbus, and the cross of the Order of Christ on his chest, gazes calmly at the reader. In his left hand, he holds a piece of paper with some indecipherable lines on it. The portrait belonged initially to Gama's descendants, the Counts of Vidigueira and the Marquises of Niza, was sold in 1845 to the Count of Farrobo, then acquired by the king D. Fernando II, who in the 1860s presented it to the museum.

A late copy of this same portrait (by Joaquim Pedro de Sousa) had appeared as the frontspiece to Teixeira de Aragão's late nineteenth-century volume, *Vasco da Gama e a Vidigueira*,

Plate 27 Portrait of a standing Vasco da Gama in the winter of his life, *O Livro de Lisuarte de Abreu*, late sixteenth century, anonymous artist.

ironically with the signature of the factor of Hurmuz in the 1520s, Cristóvão da Gama, added to it. Other copies, notably by the French painter Armand Dumaresq, and in lithograph by Mauricio José Sendim, also exist, in minor variations on what had by the early twentieth century become the accepted 'official portrait' of Gama. The conservator of the Museu Nacional de Arte Antiga, Luís Keil, who in 1934 published an entire booklet, devoted with the subtlety of a sledgehammer to demonstrating that the signature of Cristóvão da Gama attached to his hero's portrait by a number of painters and authors was false, nevertheless insisted that the portrait itself, of the 'school of Gregório Lopes, [was] possibly painted in Lisbon or in Évora in the last months of 1523 or in the first months of 1524'. Elsewhere in the same work, commenting on the water-colour portrait of Gama made by Gaspar Correia (in the mid-sixteenth century) as part of his *Lendas da Índia*, Keil comments that 'this drawing, and the painting from the Museu de Arte Antiga are the oldest portraits, that are presently known, of Vasco da Gama'.[3] Correia's portrait has a rather well-dressed full figure, favouring his left profile, leaning on a long cane, with one hand on the hilt of a sword that peeps out of his cloak. In turn, this portrait bears a rather close resemblance to (but is not an exact copy of) the Vasco da Gama of the Codex Lisuarte de Abreu, a sixteenth-century collection of documents and drawings.[4] Gama's features in both, but especially in the latter, are fine and regular, his eyes large and reflective. In Correia's portrait, he appears more vigorous than in the other, where the subtitle might well be 'The Admiral in his Autumn'.

Keil, elsewhere in the same work, goes even further in defence of his chosen representation, the portrait of the Museu de Arte Antiga: 'As for me, I would judge that the portrait of the Museu de Arte Antiga, executed by a painter of the school of Gregório Lopes, by a real artist of the epoch, should be considered as the only contemporary iconographic document of the great Admiral.' From 'portrait', we have thus moved to the rather more authoritative 'document'. Correia's portrait he judged inadmissible in the final analysis, not least of all because Correia himself admitted he had been aided by a 'native painter' (*pintor homem da terra*),

[3] Luís Keil, *As assinaturas de Vasco da Gama: Uma falsa assinatura do navegador português. Crítica, comentários e documentos* (Lisbon, 1934), Plates 1–4, and 26. Keil's particular sacrificial victim on the occasion was a certain Jean-Paul Alaux, author of *Vasco de Gama ou l'Epopée des Portugais aux Indes* (Paris: Editions Duchartre, 1931).

[4] The Pierpont Morgan Library, New York, 'O Livro de Lisuarte de Abreu', M. 525.

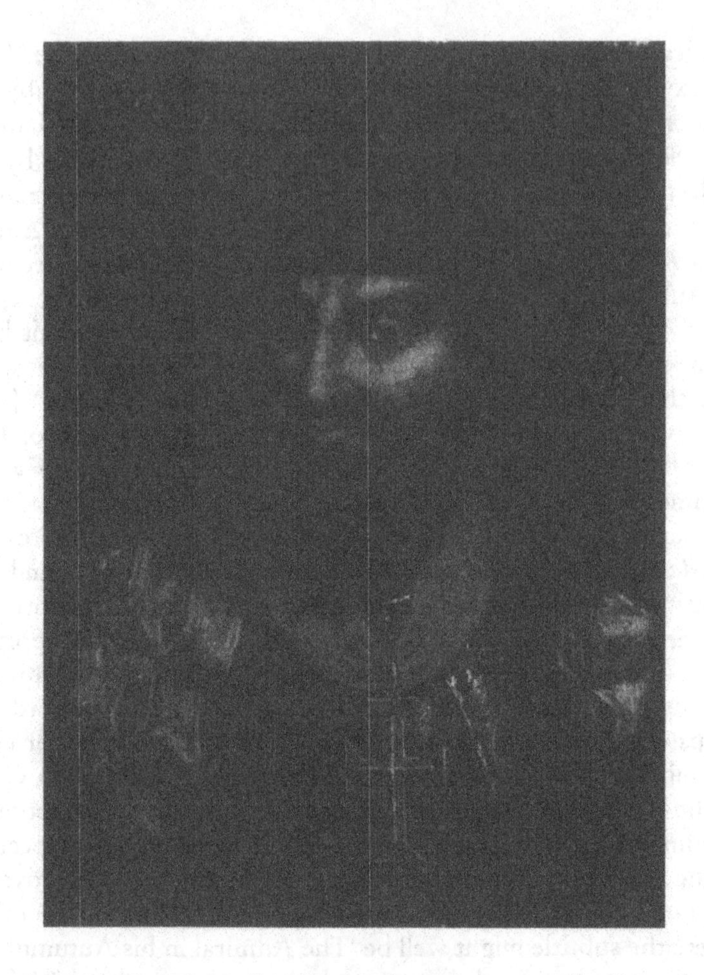

Plate 28 Portrait in oils of Vasco da Gama, reputedly from a lost original, Sociedade de Geografia, Lisbon.

which 'immediately implies the borrowing of processes with a clearly oriental character, which would alter the structure of European figures'. Gama could obviously only be drawn or painted accurately by a European! It would be churlish to suggest that the conservator was interested above all in promoting the portrait in the museum under his own charge. But let us note that academic and artistic judgements have evolved since the time of Keil. Rather than the portrait proposed by him as authentic and contemporary, more recent books favour another painting, in the possession of the Sociedade de Geografia in Lisbon, which equally belonged to the descendants of Gama, the Counts of

Vidigueira and Marquises of Niza. This portrait in oils on canvas, judged by Keil as belonging to 'the last years of the seventeenth century, or the beginning of the eighteenth century', was nevertheless chosen in the early 1970s by the fervently (and even conspiratorially) nationalist historian Armando Cortesão to illustrate his book *The Mystery of Vasco da Gama*. Cortesão describes the portrait simply as an 'oil painting according to a contemporary prototype', and the same portrait has been reproduced in recent times in the Portuguese press, and in the French edition (by Paul Teyssier and others) of Gama's first two voyages. Gama appears here, as elsewhere, favouring his left side, and sporting the cross of the Order of Christ on his chest. He is portrayed as black-bearded, and in his youth or early middle-age. Above all, Gama in this portrait can hardly be termed a model of masculine beauty. With a long nose dominating his face, thin and cruel lips, and an expression between determination and malevolence, the sombre aspect of the portrait is aided by the dark background, which lends to Gama's face, which is half in the shadow, a particularly sinister aspect. This Gama could well be used to illustrate Tomé Lopes's text, or that of Saradindu Bandhyopadhyaya: one can imagine him looking out with this very expression on the ill-fated *Mîrî*, in early October 1502, except for the fact that he was not a member of the Order of Christ at the time.

The point we are making need not be laboured: the iconography of men like Gama and Columbus is an essential part of their legend. In the context of the latter figure, the vast iconographic corpus, ranging from Sebastiano del Piombo's portrait (often dated to 1519), to the amusing late nineteenth-century lithograph of the legend of Columbus and his egg, has been studied with great care by a number of writers, perhaps most comprehensively by Rosanna Pavoni.[5] Columbus appears bearded, emaciated and towering in some versions, and the plump and slightly grumpy epitome of a clean-shaven medieval Italian burgher in others. And, as Pavoni puts it, 'For each objective, whether or not of propaganda, for each set of declared national values, one attributes to Christopher Columbus, to the Messenger of Christ, to the Colonialist, a face, a sentiment, a gesture, an attitude.' Columbus, like Gama (or like Joan of Arc, or William Shakespeare, in Pavoni's terms), comes rapidly to be emptied of

[5] Rosanna Pavoni, *Christophe Colomb: Images d'un visage inconnu*, tr. Federica Bevilacqua-Mariat (Paris: Vilo, 1990).

all specific historical content, thus gradually becoming no more than a symbolic universal shell that all of later humanity can then fill as it wishes.

Put this way, matters may appear rather too deliberate and conspiratorial, from a number of viewpoints. Iconographic images and the represented intentions, after all, need not have one-to-one correspondences, for otherwise it would be difficult to explain how the Gama of the nationalist Armando Cortesão could as well serve anti-Portuguese propaganda. The explanation here is simple and based on a shared aesthetic judgement: between two portraits, one rather more ugly than the other, it would be assumed by most historians today that the one with more regular features, and a more benevolent expression, is less authentic and more of an idealisation. This would appear to underlie the judgements concerning relative 'authenticity' of both Armando Cortesão and, say, the French translator of the travel-accounts, Paul Teyssier, who equally chose the portrait of the Sociedade de Geografia for the cover of his work.

A recent comic-book produced in Portugal, and which enjoyed a certain commercial success in the 1994 Lisbon Book Fair, serves as another example to demonstrate the complexity of the situation, particularly when the iconography is accompanied by a written text. Written and drawn by Fernando Santos and Carlos Santos, twin writer-artists, born in Mozambique (and who returned to Portugal after decolonisation, in 1976), the comic-book entitled *Vasco da Gama e a Índia* combines drawings that are a very close reproduction of the style of the celebrated Asterix series of Goscinny and Uderzo (even down to individual incidents), with a text that is nevertheless almost totally solemn in intention, the very occasional humour being slapstick in nature.[6] Borrowing in good measure for the text and the 'cast of characters' from the *Lusíadas*, the giant Adamastor and the villainous 'Catual' are both present here; as for the drawings, they are often marked by a curious, rather naive, pseudo-exoticism, with the interior of the Samudri Raja's palace being shown, for example, as a vaulted gothic church. Despite his somewhat ridiculous appearance (that draws in part, nonetheless, on Keil's designated 'authentic' portrait, even reproducing the cross of the Order of Christ), the received nationalist version of Gama's legend, as it

[6] Carlos Santos and Fernando Santos, *Vasco da Gama e a Índia* (Odivelas: Europress, 1992).

appears in the textbooks of Salazar's *Estado Nôvo*, is left substantially intact in the comic-book. It is, of course, obvious that in the 1950s or 1960s, even the measure of iconographic irreverence shown here would have been subjected to censorship in Portugal.

One is tempted to compare the vision of the brothers Santos to that of the official Indian school-level book. We do dispose of a slim volume published in 1993 in Hindi, by the National Council of Educational Research and Training, entitled 'Vasco da Gama: The first European to seek the maritime route to India'.[7] This is not a textbook, but rather a part of a series termed 'Read and Learn', designed to provide low-priced books of general interest to a school-level audience. Its author begins by asking the rhetorical question of why a railway station in India is named Vasco da Gama. After all, 'Vasco da Gama does not sound like an Indian name'! So who was he? Where did he live? Why did he come to India? What did he do to make his name so famous?

In order to answer these questions, the author produces a rather detailed summary in eleven chapters, and roughly eighty pages of text (with a handful of illustrations) of the celebrated anonymous journal of the first voyage, obviously using the English translation by E.G. Ravenstein. Certain obvious errors committed by other later authors (for instance, the identification of the Gujarati pilot with Ahmad ibn Majid) are thus avoided, for the simple reason that Ravenstein did not commit them either. On the other hand, the perspective of the European text is followed closely to the extent of calling the Muslims *mûr* rather than the usual *musalmân*. Of particular interest are the introduction and conclusion within which this rather flat narrative is framed. The introductory chapters reproduce late nineteenth-century European notions concerning the heroic and fearless character of Gama, and attribute to him a considerable knowledge of navigation. Thus, after stating that Gama was born in Sines in 1460, the text goes on:

> No description is available of Vasco da Gama's youth. But one thing is certain: that he was an extraordinary young man. He was learned, courageous, bold, and a skilful manager. He had served in Portugal's navy. He had an extensive knowledge of maritime matters. The king had a good deal of confidence in him.

Elsewhere, in the preface, the reasons for the exploration of the

[7] Rajendra Singh Vatsa, *Vasko da Gâmâ: Bhârat ke liye samudrî mârg ḍhûnḍhnevâlâ pahlâ yûropîya* (New Delhi: NCERT, 1993).

sea-route by the Portuguese are summed up as follows. In ancient times, India's culture was world famous; foreigners used to call India 'the Golden Bird'. Scents, handicrafts, spices and textiles used to go from this highly developed India to a Europe whose countries were 'in a backward state'. Indian black pepper being prized above all in Europe, it used to pass through west Asia, in order to reach those markets. However, west Asia was in the grips of the 'huge Turkish empire', and once a long war began between Turkey and Europe, the trade route through west Asia was closed. In search of black pepper then, a number of courageous European mariners began the search for a maritime route to India. 'But only Vasco da Gama was successful.'

The last few pages of the book once more place Gama's career in a longer-term perspective. Cabral's expedition, and the fleets that followed one after the other, are noted; the Portuguese are reported to have abandoned Calicut for Goa. After the Portuguese, other Europeans followed. The English for their part saw that Indian monarchs were weak and divided, and prone to a feckless and pleasure-loving life. Seizing the occasion, they conquered India, ruling over it for two hundred years. But even when they left, in 1947, the Portuguese 'who were the first to come to India were also the last to leave'. Gama's name is still remembered through the 'railway station' of that name, 'perhaps in the hope that now an independent India and Portugal may be able to establish friendly relations, and forget the old sorrow-laden memories'.

It is to be stressed that the work summarised above, though clearly produced as part of a project of mature nationalist history-writing (as witness the conclusion), represents a conception of Gama's role that is rather different from that presented both by nationalists of the 1940s and 1950s, the generation of Sardar K.M. Panikkar, inventor of the notion of 'the Vasco da Gama epoch', and by certain present-day Third World nationalists.[8] In a certain measure, it represents the success of the European myth-building enterprise around Gama, when even Indian nationalists accept their premises and conclusions; at the same time, official Portuguese historians of the National Commission for the Commemoration of the Portuguese Discoveries, even in the 1990s, have at times given a meaning to the 'Vasco da Gama epoch' (*a era de*

[8] K.M. Panikkar, *Asia and Western Dominance* (1st edn, London, 1959; reprint Kuala Lumpur: The Other Press, 1993). Also see Ashis Nandy *et al.*, *The Blinded Eye: 500 Years of Christopher Columbus* (Goa: The Other India Press, 1993).

Vasco da Gama) that is the polar opposite of Panikkar's intention.[9] In another context, that of the celebrated figure of Captain James Cook, the Sri Lankan anthropologist Gananath Obeyesekere has pointed to the considerable investment of western imperial (and even post-imperial) historiographies in emblematic figures of explorers and/or conquerors of the early modern era. Thus, he writes, 'One of my basic assumptions is that mythmaking, which scholars assume to be primarily an activity of non-Western societies, is equally prolific in European thought'.[10] The long debate between Obeyesekere and his anthropologist adversary, Marshall Sahlins, is not our primary concern here; after all, in no version of the narrative of Gama's voyage does he ever appeared to have been deified, save in one produced by his own countryman Camoẽs. But this may not be so irrelevant, after all, to the 'apotheosis of Captain Cook': Gama, even more than Cook, is a sort of 'secular God', representing the notion of the Promethean Superman, who stole the fire of modernity from the Gods.

The argument running through this book has been that the creation of the legend of Vasco da Gama began in his own lifetime, and that he participated in it. In part, this was because the symbolic capital of the legend, if one will, could be made to yield actual financial, fiscal and material returns, in terms of the construction of a career trajectory that took Gama from a relatively marginal position in the lower nobility to a major landed title and position of grandee. But the political position of Dom Manuel in his own realm, and that of Portugal in Europe, were also responsible for the creation of Gama's legend, which was built after all in competition with that of Columbus, the only other emblematic figure of similar dimensions from his generation, who ironically was located across the border in rival Spain. If in his lifetime, Gama had been created Admiral because Columbus had had that honour, in retrospect Gama remained the key Portuguese national treasure, to be trotted out on all occasions. This was a process that was renewed generation after generation in Portugal with varying meanings and significations,

[9] For a sample of the polemic on the question, see the characteristic essay by Teótonio R. De Souza, '"North–South" in the *Estado da Índia?*', *Mare Liberum* 9 (1995), 453–60.

[10] Gananath Obeyesekere, *The Apotheosis of Captain Cook: European Mythmaking in the Pacific* (Princeton: Princeton University Press, 1992), p. 10. This book is largely a critique of two earlier works by Marshall Sahlins, who has since replied with another book; the debate continues.

Plate 29 Stylised portrait of Vasco da Gama from the *Tratado dos Vice-Reis* (c. 1635) (Paris text).

from Camões in the late sixteenth century, to Francisco Toscano's comparison in the 1620s between Gama and Aeneas, to Júlio de Mello's literary parallel in the 1770s between the founder-king of Portugal D. Afonso de Henriques and the 'incomparable discoverer of India', and José Agostinho de Macedo's narrative poem of 1811, *Gama*. After a notable high point in the 1880s and 1890s, in the first half of the twentieth century examples of myth-making around Gama are still legion: for do we really need to look farther than Fernando Pessoa's *Mensagem*, wherein 'the heavens open the chasm to the soul of the Argonaut'? As late as 1988, when the Instituto de Ciências Sociais conducted a study on 'Nationalism and Patriotism in Contemporary Portuguese Society', it was found that the most admired personage in Portuguese history was Vasco da Gama (58.8%), followed by the Infante D. Henrique (45.2%).[11]

In the closing years of the twentieth century, five hundred years from the voyage of the *São Gabriel*, the myth of Vasco da Gama has been successfully exported from Portugal through the entire world. We could do worse than to look back once more, in closing, at the first of these grand celebrations, that was organised at Goa in the closing years of the sixteenth century, to mark the first centenary of Gama's voyage. Let us recall that the moving force behind that celebration was Vasco da Gama's own great-grandson, the Count of Vidigueira, D. Francisco da Gama, who became viceroy of Portuguese India twice, from 1597 to 1600, and then again from 1622 to 1628, before dying in disgrace in Oropesa (Spain) in 1632. It would appear that unlike his ancestor, Dom Francisco had no well-defined political programme in Asia, save to enrich himself while playing Asian politics by the rules of Machiavelli, and to destroy as comprehensively as possible the reputation of his predecessor as viceroy, Matias de Albuquerque (1591–7) – for his part a distant relative of Afonso de Albuquerque.[12] There are some who would see in this late sixteenth-century struggle the revival of tensions between the proponents and the opponents of Manueline imperialism, but this is surely

[11] As cited in Maria Isabel João, 'O regresso do infante', *Público*, (Special supplement, 4 March 1994), 26. One may read this otherwise, and presume that this means that over 40% of the Portuguese did not admire Vasco da Gama!

[12] I am preparing a series of studies on Dom Francisco da Gama, fourth Count of Vidigueira. At present we dispose of only somewhat hagiographic accounts, such as António da Silva Rego, 'O início do segundo governo do vice-rei da Índia D. Francisco da Gama, 1622–1623', *Memórias da Academia das Ciências de Lisboa, Classe de Letras* 19 (1978), 323–45.

Plate 30 Stylised portrait of D. Francisco da Gama, in the same attitude as his great-grandfather, from the *Tratado dos Vice-Reis* (c. 1635).

too simplistic. The dialectic was not an eternal one; Francisco da Gama was not Vasco da Gama, any more than his predecessor as viceroy, Matias de Albuquerque, was the Terrible Afonso (to whom he was, despite claims by his panegyrists, not closely related in any case)![13] But nor was this the mere expression of the animal spirits of factional rivalry, the usual framework utilised to analyse earlier quarrels like those between Pêro Mascarenhas and Lopo Vaz de Sampaio in 1526–8, between Martim Afonso de Sousa and Dom João de Castro in the 1540s, and so on.[14] Rather, the context was a Portuguese Asia in which the events of the years between 1498 and 1524 had already become overlain with a thick crust of pseudo-Hellenistic mythology, to which Camões himself had contributed in no small measure. The struggle was thus not only between two clans and the social networks in which they were located, but between two competing mythological constructions of the establishment of Portuguese power in India, and thus a concrete expression of differences over legendary 'property rights'.

In 1557, Afonso de Albuquerque's son, Brás de Albuquerque, had published the *Comentários do Grande Afonso de Albuquerque*, a text that served in the dying moments of the Joanine reign to remind the Portuguese public of the great Crusade that his father had carried out against the Moors in the Orient. Arriving as viceroy in Goa, Dom Francisco da Gama sought to produce a chronicle that would surpass this, celebrating not merely the founder-figure of his lineage, the first Count of Vidigueira, but the three generations of descendants who had followed. This task was given to that able but cynical chronicler Diogo do Couto, and the product was the so-called *Tratado dos Gama* ('Treatise of the Gama'), a text for which Couto admits to borrowing freely from Damião de Góis and João de Barros, 'in many matters word for word'.

Even if, at first glance, it would seem that Gama and his men were no less than Jason and the Argonauts to Couto, and the influence of the *Lusíadas* on the chronicler need hardly be stressed, the end-product is not devoid of the occasional surprise. Couto agrees, for example, with the view that Gama in 1524

[13] Biblioteca Pública e Arquivo Distrital, Évora, Codex CXV/1–13, 'Vida de Mathias de Albuquerque', ch. 1, fos. 2–2^{v}, 215–21. Also see the discussion in Sanjay Subrahmanyam, '"The Life and Actions of Mathias de Albuquerque" (1547–1609): A Portuguese Source for Deccan History', *Portuguese Studies* 11 (1995), 62–77.

[14] For a more general discussion of this question, largely in the context of the seventeenth century, see Subrahmanyam, *The Portuguese Empire in Asia*, pp. 235–8.

wished to do away with the fortresses at Kollam, Calicut and Colombo, 'for they served no purpose than to create additional expenses', replacing them with simple factories.[15] This, of course, runs counter to the rather bellicose character of his other writings, wherein it appears that fortresses are all to the good, indeed, the more the merrier. But in his closing view of Gama, at the moment he lies freshly buried under the supervision of his son D. Estêvão in the monastery of St Francis (in fact, at the time still Santo António), Couto does little more than hark back to the obituary judgement of João de Barros:

> This Count, D. Vasco da Gama, son of Estêvão da Gama, was a man of middling stature, broad-shouldered, and fleshy, a great cavalier in his person, very resolute in his counsel, and daring in the enterprise of any action; he was unbending in his orders, and greatly to be feared in his fury, he suffered many travails and was very fair in his justice, a great executor of punishments.

It is somewhat astonishing that this was the best (or worst) that a panegyrical work could offer. For all his wiles as a courtier then, Couto knew where, as a chronicler, he had to draw the line. The rest he left to Camões, or to others who sang the glories of expanding Lusitania in various major or (in the late sixteenth century) minor keys.

We are aware that Couto, whose warts were many in number, nevertheless strove manfully in his *Décadas* at least to collect and synthesise a series of rather diverse materials, in a number of European and Asian languages. From Mughal ambassadors in Goa, he sought to gather materials on the origins of the Timurids, even if he went on to mix this up with materials of a rather different provenance concerning the presence of the Apostle St Thomas amongst the Mongols in Central Asia. Elsewhere, he spoke with renegades and mercenaries, gathered together insights into the Taung-ngu dynasty in Burma, myths concerning the origins of Vijayanagara, rumours of the existence in ancient times of Indo-Greek kingdoms in Gujarat. All of this, though embedded in a work primarily designed to celebrate Portuguese conquests and victories in Asia, represents a sort of openness of spirit from which many far later historians, and especially those working within the paradigm of 'European expansion', have frankly regressed. The doyen of historians of Portuguese expan-

[15] Biblioteca Nacional de Lisboa, Reservados, Códice 462, 'Tratado de todas as cousas socedidas ao valeroso capitão D. Vasco da Gama . . .', ch. 24, fl. 87.

sion in the post-World War II period could thus have the temerity, in the late 1980s, to produce a programmatic text on Portuguese expansion, which wholly excluded both Asian and African source-materials, and even Asian and African historians. All this was in the name of proving that the Portuguese too were true Europeans, and that 'one of the fundamental traits of the European spirit is to anticipate others in the denunciation of our own errors and crimes'.[16]

A perhaps apocryphal story, circulating in Portugal in the late 1980s, may help set matters in perspective. It was said that in 1987, when the Portuguese government wished to commemorate the voyage to the Cape of Bartolomeu Dias, a sailing-ship that was a replica of the caravels of the late fifteenth century was rigged up, to make a landing at the very spot where Dias had been, in 1487, in South Africa. The Portuguese on board were to be met by half-naked bushmen, in the interest naturally of historical veracity. There was a problem however, namely that the area that was identified as the one where Dias had been was now a 'Whites Only' beach (this being during the last years of apartheid). A solution was eventually found: whites in black body-paint greeted the modern-day Dias. Apartheid was maintained; historical 'authenticity' was satisfied. The point is not whether the story is true, but that it circulated amongst modern-day Portuguese, who thus took delight in mocking the myth of the Discoveries, and of the Portuguese Superman on the poop of his caravel or carrack.[17] The myth of Vasco da Gama has been quietly resisted in Portugal too.

Since that time, other, rather more strident voices have been heard in Portugal. A 'black book' on Portuguese expansion has appeared, alleging that Portuguese expansion is no different from Nazi Germany's genocide of Jews and Gypsies. The deaths caused by bombardments in East Africa in 1505 by D. Francisco de Almeida, are equated to the napalm massacres of the Baixa de

[16] Vitorino Magalhães Godinho, 'Portugal e os descobrimentos', in *Mito e mercadoria, utopia e prática de navegar, séculos XIII-XVIII* (Lisbon: Difel, 1990), p. 55.

[17] It is perfectly possible that this story is true; after all, the Portuguese Discoveries Commission sought, on 10 June 1994 (as well as earlier in the Universal Exposition at Seville), to 'recreate' in Oporto the embassy of Tristão da Cunha to Rome, replete with the elephant and cheetah. This spectacle was announced in *O Independente*, No. 311 (29 April 1994), Supplement 'Comemorações'. Such spectacles can undoubtedly be seen as contributing to the development of a 'historical consciousness', rather like other attempts to recreate medieval life in historical sites. But the Disneyland aspect of the exercise ought not to be neglected either, as well as the ethnocentric readings that are likely to be given to such 'exotica' by spectators.

Cassenge in 1962: for the 'stench of the dead' in the two cases is the same.[18] This is surely no solution to the problems of the historiography, gratifying though it may be to certain polemicists of a Third World persuasion. In adding one more to the many contested portraits of Vasco da Gama, this book has intended to make an implicit plea for a rather more nuanced, indeed ironical, look at the history of 'European expansion', both the inflation (thus, expansion) of European claims and pretensions, and the outward-looking processes by which Europeans sought to redefine the commercial and political networks of the early sixteenth century. No dramatic shifts of paradigm have been proposed, merely the careful sifting of a mass of tangled materials, to which the continuing research of other historians will doubtless add over the years. For the task admits of no easy solutions, with perhaps the greatest of our burdens being the historian's classic sin: anachronism.

In a recent essay, a well-known Portuguese historian, examining the vicissitudes of Portuguese corsair activity in the fifteenth and sixteenth centuries, begins by quoting Spinoza: 'Concerning human actions, I have tried not to laugh, not to weep, not to detest them, but to understand them.'[19] To the reader of this volume, I hope the message that I have conveyed has been somewhat different, and rather less 'Christian': concerning past human actions, to laugh when they are ridiculous, to weep when they are tragic, to detest them as they were often detested by those who were their victims, for how else would we ever come close to understanding them?

[18] Ana Barradas, *Ministros da Noite: Livro Negro da Expansão Portuguesa*, 2nd edn (Lisbon: Edições Antigona, 1992), pp. 10–12. This volume appears to represent the work of an organisation ironically calling itself MAR.

[19] Luís Filipe F.R. Thomaz, 'Do Cabo Espichel a Macau: Vicissitudes do corso português', in Artur Teodoro de Matos and Luís Filipe F. Reis Thomaz, eds., *As Relações entre a Índia Portuguesa, a Ásia do Sueste e o Extremo Oriente: Actas do VI Seminário Internacional de História Indo-Portuguesa* (Macau/Lisbon: The Editors, 1993), p. 537. It is of course typical of Spinoza that he would produce such a Christian-sounding dictum.

Bibliography

I. CONTEMPORARY EUROPEAN SOURCES

Alberi, Eugenio (ed.), *Relazioni degli Ambasciatori Veneti al Senato*, 1st series, Vol. II (Florence: Clio, 1840).

Albuquerque, Luís de (ed.), *Crónica do Descobrimento e primeiras conquistas da Índia pelos Portugueses* (Lisbon: Imprensa Nacional-Casa da Moeda, 1986).

Albuquerque, Luís de, and Rodrigues, Vítor Luís Gaspar (eds.), *Bartolomeu Dias: Corpo Documental, Bibliografia* (Lisbon: Comissão Nacional para as Comemorações dos Descobrimentos Portugueses, 1988).

Albuquerque, Luís de, and Caeiro, Margarida (eds.), *Cartas de Rui Gonçalves de Caminha* (Lisbon: Publicações Alfa, 1989).

Albuquerque, Luís de, and Costa, José Pereira da, 'Cartas de "serviços" da Índia (1500–1550)', *Mare Liberum*, No. 1 (1990), pp. 309–96.

Andrada, Francisco de, *Crónica de D. João III*, ed. M. Lopes de Almeida (Oporto: Lello e Irmão, 1976).

Baião, António, Magalhães Basto, A. de, and Peres, Damião (eds.), *Diário da Viagem de Vasco da Gama*, 2 vols. (Lisbon, 1945).

Barbosa, Duarte, *Livro em que dá relação do que viu e ouviu no Oriente Duarte Barbosa*, ed. Augusto Reis Machado (Lisbon: Agência Geral das Colónias, 1946).

Barros, João de, *Da Ásia*, Décadas I–IV (Lisbon: Livraria Sam Carlos, 1973), facsimile of the 1777–8 Régia Oficina Tipográfica edition.

Cà Masser, 'Relazione de Lunardo da cha Masser', published in Prospero Peragallo, 'Carta de El-Rei D. Manuel ao Rei Cathólico narrando as viagens portuguezas à Índia desde 1500 até 1505 reimpressa sobre o prototypo romano de 1505, vertida em linguagem e anotada. Seguem em appêndice a Relação análoga de Lunardo cha Masser e dois documentos de Cantino e Pasqualigo', *Memórias da Academia Real das Sciências*, 2ª Classe, Lisbon, Vol. VI, No. 2 (1892), pp. 67–98.

Camões, Luís de, *Os Lusíadas*, introduction and notes by Maria Letícia Dionísio (Lisbon: Publicações Europa-América, 1985).

Castanheda, Fernão Lopes de, *História do Descobrimento e Conquista da Índia pelos Portugueses*, ed. M. Lopes de Almeida, 9 books in 2 vols. (Oporto: Lello e Irmão, 1979).

Coelho, Ramos (ed.), *Alguns Documentos da Torre do Tombo acerca das navegações e conquistas Portuguesas* (Lisbon, 1892).

Correia, Gaspar, *Lendas da Índia*, ed. M. Lopes de Almeida, 4 vols. (Oporto: Lello e Irmão, 1975).

Cortesão, Armando (ed.), *A Suma Oriental de Tomé Pires e o Livro de Francisco Rodrigues* (Coimbra: Universidade de Coimbra 1978).

Cortesão, Armando, and Thomas, Henry (eds.), *Carta das novas que vieram a el Rei nosso senhor do descobrimento do Preste João (Lisboa 1521): Texto original e estudo crítico, com vários documentos originais* (Lisbon, 1938).

Costa, Abel Fontoura da (ed.), *Roteiro da Primeira Viagem de Vasco da Gama (1497–1499) por Álvaro Velho* (Lisbon: Agência Geral das Colónias, 1940).

Costa, Abel Fontoura da (ed.), *Os sete únicos documentos de 1500 conservados em Lisboa referentes à viagem de Pedro Álvares Cabral*, 2nd edn (Lisbon, 1968; first edn 1940).

Costa, Leonor Freire, and Pinto, João Rocha, 'Relação anónima da Segunda Viagem de Vasco da Gama à Índia', in *Cidadania e História: Em homenagem a Jaime Cortesão* (Lisbon: Livraria Sá da Costa, 1986), pp. 141–99.

Couto, Diogo do, *Da Ásia*, Décadas IV–XII, facsimile of the Régia Oficina Tipográfica edition, 1788 (Lisbon: Livraria Sam Carlos, 1974).

Denucé, Jean, *Calcoen: Recit flamand du second voyage de Vasco de Gama vers l'Inde en 1502–1503* (Antwerp, 1939).

Diarii di Girolamo Priuli [AA. 1494–1512], Vol. I, ed. Arturo Segre (Città di Castello, 1921).

Diarii di Marino Sanuto, Vols. IV, V and XXXVI, eds. Nicolò Barozzi, Federico Stefani and Guglielmo Berchet (Venice, 1880–92).

Dinis, António Joaquim Dias (ed.), *Monumenta Henricina*, Vols. V and XIV (Coimbra: CMIH, 1963–73).

Documentos sobre os Portugueses em Moçambique e na África Central, 1497–1840, eds. António da Silva Rego and T.W. Baxter, 7 vols. to date (Lisbon: CEHU, 1962–).

Felner, Rodrigo José de Lima, (ed.), *Subsídios para a história da Índia Portugueza*, in 3 parts (Lisbon: Academia Real das Sciências, 1868).

Ficalho, Conde de, *Viagens de Pedro da Covilhan* (Lisbon: Imprensa Nacional, 1898).

Figueroa, Martín Fernández de, *Conquista de las Indias de Persia y Arabia*, ed. J. Augur (Salamanca, 1512), facsimile in James B. McKenna, *A Spaniard in the Portuguese Indies: The narrative of Martín Fernández de Figueroa* (Cambridge, Mass.: Harvard University Press, 1967).

Formisano, Luciano (ed.), *Letters from a New World: Amerigo Vespucci's Discovery of America*, tr. David Jacobson (New York: Marsilio, 1992).

Galvão, Duarte, *Crónica de D. Afonso Henriques*, ed. Castro Guimarães (Cascais, 1918).

Góis, Damião de, *Crónica do felicíssimo Rei Dom Manuel*, 4 vols. (Coimbra: Universidade de Coimbra, 1949–55).

Gonneville, Binot Paulmier de, *Campagne du Navire 'L'Espoir' de Honfleur (1503–1505): Relation Authentique du voyage du Capitaine de Gonneville ès Nouvelles Terres des Indes publiée integralement pour la premiere fois avec une introduction et des éclaircissements*, ed. M. D'Avezac (Geneva: Slatkine Reprints, 1971).

Harrisse, Henry, *Document inédit concernant Vasco de Gama: Relation adressée à Hercule d'Este, Duc de Ferrare par son Ambassadeur à la Cour de Portugal* (Paris, 1889).

Krása, Miloslav, and Polisensky, Josef, and Ratkos, Peter (eds.), *European Expansion (1494–1519): The Voyages of Discovery in the Bratislava Manuscript Lyc. 515/8 (Codex Bratislavensis)* (Prague: Charles University, 1986).

Longhena, Mario (ed.), *Viaggi in Persia, India e Giava di Nicolò di Conti, Girolamo Adorno e Girolamo di Santo Stefano* (Milan: Alpina, 1962).

Major, R.H. (ed.), *India in the Fifteenth Century: Being a Collection of Narratives of Voyages to India, in the Century preceding the Portuguese Discovery of the Cape of Good Hope; from Latin, Persian, Russian and Italian Sources* (London: The Hakluyt Society, 1857).

Marques, J.M. da Silva (ed.) *Descobrimentos Portugueses*, Vol. III (Lisbon: Instituto de Alta Cultura, 1971).

Nikitine, Athanase, *Voyage au-delà des trois mers*, trans. Charles Malamoud (Paris: F. Maspéro, 1982) (Russian edn, *Khozenie za tri morja*, Moscow, 1958).

Ordenações do Senhor Rey D. Manuel, 5 vols. in 4 (Coimbra, 1797).

Pato, R.A. de Bulhão, and Mendonça, H. Lopes de (eds.), *Cartas de Afonso de Albuquerque seguidas de documentos que as elucidam*, 7 vols. (Lisbon: Academia das Ciências, 1884–1935).

Peragallo, Prospero, 'Viaggio di Matteo da Bergamo in India sulla flotta di Vasco da Gama, 1502–1503', *Bollettino della Società Geografica Italiana*, Series IV, Vol. III (1902), reprinted from *Studi bibliografici e biografici sulla storia della geografia in Italia* (Rome, 1875), pp. 92–129.

Pina, Rui de, *Crónicas de Rui de Pina*, ed. M. Lopes de Almeida, (Oporto: Lello e Irmão, 1977).

Radulet, Carmen M., *Vasco da Gama: La prima circumnavigazione dell'Africa, 1497–1499* (Reggio Emilia: Edizioni Diabasis, 1994).

Ramusio, Giovanni Battista, *Navigazioni e viaggi*, ed. Marica Milanesi (Turin: Giulio Einaudi, 1978).

Ravenstein, E.G., *A journal of the first voyage of Vasco da Gama, 1497–1499* (London: The Hakluyt Society, 1898).

Rebello, Joaquim Inácio de Brito, 'Navegadores e exploradores portuguezes até o XVI século (Documentos para a sua história): Vasco da Gama, sua família, suas viagens, seus companheiros', *Revista de Educação e Ensino* (Lisbon), Vol. XIII (1898), pp. 49–70, 124–36, 145–67, 217–30, 274–85, 296–313, 366–70, 473–5, 508–22; Vol. XIV (1899), pp. 560–5; Vol. XV (1900), pp. 28–32, 90–2.

Rego, António da Silva (ed.), *As Gavetas da Torre do Tombo*, 12 vols. (Lisbon: CEHU, 1960–77).

Rego, António da Silva (ed.), *Documentação para a história das missões do padroado português do Oriente: Índia*, 12 vols., (Lisbon: Agência Geral do Ultramar, 1947–58).

Resende, Garcia de, *Crónica de Dom João II e Miscelânea*, ed. Joaquim Veríssimo Serrão (Lisbon: Imprensa Nacional-Casa da Moeda, 1973).

Rivara, J.H. da Cunha, *Archivo Portuguez-Oriental*, 6 Fascicules in 9 Parts (Goa, 1857–76).

Rohr, Christine von, *Neue Quellen zur Zweiten Indienfahrt Vasco da Gamas* (Leipzig, 1939).

Schefer, Charles (ed.), *Le discours de la navigation de Jean et Raoul Parmentier de Dieppe* (Paris: Ernest Leroux, 1883).

Schefer, Charles (ed.), *Le Voyage d'Outremer de Jean Thenaud (Egypte, Mont Sinay, Palestine) suivi de La Relation de l'Ambassade de Domenico Trevisan auprès du Soudan d'Egypte, 1512* (Paris: Ernest Leroux, 1884).

Sousa, Frei Luís de, *Anais de D. João III*, ed. M. Rodrigues Lapa, 2nd edn (Lisbon: Sá da Costa, 1951).

Teyssier, Paul, and Valentin, Paul, *Voyages de Vasco de Gama: Relations des expéditions de 1497–1499 & 1502–1503* (Paris: Editions Chandeigne, 1995).

Valière, Pierre, *Le voyage de Magellan raconté par un homme qui fut en sa compagnie* (Paris: Centre Culturel Portugais, 1976).

Velho, Álvaro, *Roteiro da Primeira Viagem de Vasco da Gama*, ed. Neves Águas (Lisbon, 1987).

II. SOURCES AND TRANSLATIONS FROM NON-EUROPEAN LANGUAGES

ᶜAbd al-Razzaq ibn Ishaq Samarqandi, *Matlaᶜ al-Saᶜdain wa Majmaᶜ al-Bahrain*, Vol. I, ed. ᶜAbd al-Husain Nawa'i (Teheran: Tahuri, 1353); Vol. II, ed. Muhammad Shafiᶜ (Lahore: Gilani, 1946–9); Indian section translated by Wheeler M. Thackston, *A Century of Princes: Sources on Timurid History and Art* (Cambridge, Mass.: The Aga Khan Program for Islamic Architecture, 1989), pp. 299–321.

Abu Makhrama, Muhammad al-Tayyib ibn ᶜAbd Allah, *Political History of the Yemen at the beginning of the 16th century, Abu Makhrama's account of the years 906–927 H. (1500–1521 A.D.)*, ed. and trans. Lein Oebele Schuman (Groningen: Druk V.R.B., 1960).

Ahmad ibn Majid, *Arab navigation in the Indian Ocean before the coming of the Portuguese, being a translation of the 'Kitâb al-Fawâ'id fî usûl al-bahr wa'l-qawâ'id' of Ahmad b. Mâjid al-Najdî*, ed. and trans. G.R. Tibbetts (London: Oriental Translation Fund New Series, 1971).

Al-Maqqari, *The History of the Mohammedan Dynasties in Spain*, tr. Pascual de Gayangos, 2 vols. (London: W.H. Allen and Co, 1840–3).

Ibn Battuta, Shams al-Din ᶜAbdullah bin Muhammad, *The Rehla of Ibn Battuta (India, Maldive Islands and Ceylon)*. ed. and trans. Mahdi Husain (Baroda: Oriental Institute, 1976).

Ibn Iyas al-Hanafi al-Misri, *Badâ'iᶜ al-Zuhûr fî Waqâ'iᶜ al-Duhûr*, ed. M. Sobernheim, Paul Kahke and Muhammed Mustafa, as *Die Chronik des Ibn Ijâs*, 5 vols. (Wiesbaden 1960–74). Translated into French by Gaston Wiet: *Histoire des Mamlouks Circassiens* 2 vols. (Cairo: Institut Français d'Archéologie, 1945), and *Journal d'un bourgeois du Caire: Chronique d'Ibn Iyâs*, 2 vols. (Paris: Librairie Armand Colin, 1955–60).

Keralolpatti, ed. Hermann Gundert (Mangalore, 1881); translated as *Keralolpatti: A legendary history of Malabar, beginning with the mythical Parasu-rama, in Malayalam*, (Kottayam, 1863).

Ma Huan, *Ma Huan's Ying-yai Sheng-lan, the overall survey of the Ocean's shores (1433)*, ed. and trans. J.V.G. Mills (Cambridge: Cambridge University Press, 1970).

Muhammad Qasim Hindushah Astarabadi Firishta, *Gulshan-i Ibrâhîmî*, 2 vols. (Lucknow: Kali Prashad, 1864–5). Translated by John Briggs, *History of the Rise of Mahomedan Power in India*, 4 vols., reprint (Calcutta: R. Cambray, 1908–10).

Qutb al-Din Muhammad ibn Ahmad al-Nahrawali, *Ilam al-ᶜalam ba Akhbâr al-Masjid al-Haram*, abridged ᶜAllama ᶜAbd al-Karim, ed. A.M. Jamal and ᶜAbd al-ᶜAziz Rafai (Basra, 1950). Edited and translated into German as *Geschichte der Stadt Mekka und ihres Tempels von Cutb ed-Din Muhammed Ben Ahmed el-Nahrawali*, by Ferdinand Wüstenfeld (Leipzig: F.A. Brockhaus, 1857).

Qutb al-Din Muhammad ibn Ahmad al-Nahrawali, *Al-Barq al-Yamânî fî al-Fath al-ᶜUsmânî*, ed. Ashraf ᶜAli Tabye and Hamd al-Jasir (Riyadh, 1967). Translated David Lopes, *Extractos da história da conquista do Yemen pelos Othomanos* (Lisbon: Imprensa Nacional, 1898).

Serjeant, R.B. trans. *The Portuguese off the South Arabian Coast: Hadrami Chronicles*, (Oxford: Clarendon Press, 1963).

Sikandar ibn Muhammad urf Manjhu ibn Akbar, *Mirât-i Sikandarî*, ed. S.C. Misra and M.L. Rahman (Baroda: Department of History M.S. University, 1961). Translated in *The Local Muhammadan Dynasties: Gujarat*, by E.C. Bayley, (London: W.H. Allen, 1886).

Sousa, Fr. João de, *Documentos Arábicos para a história portugueza copiados dos originaes da Torre do Tombo* (Lisbon: Academia Real das Sciências, 1790).

Zain al-Din ibn ^{c}Abd al-^{c}Aziz Ma^{c}bari, *Tuhfat al-mujâhidîn fî ba^{c}z Ahwâl al-Burtukâliyyîn*, text ed. and trans. by David Lopes as *História dos Portugueses no Malavar por Zinadim*, (Lisbon: Imprensa Nacional, 1899).

III. SECONDARY LITERATURE

Alaux, Jean-Paul, *Vasco de Gama ou l'Epopée des Portugais aux Indes* (Paris: Editions Duchartre, 1931).

Albuquerque, Luís de, 'A viagem de Vasco da Gama entre Moçambique e Melinde, segundo *Os Lusíadas* e segundo as crónicas', *Garcia da Orta* (1972), 11–35.

Albuquerque, Luís de, and Feijó, Rui Graça, 'Os pontos de vista de D. João III na Junta de Badajoz-Elvas', in Avelino Teixeira da Mota (ed.), *A viagem de Fernão de Magalhães e a questão das Molucas* (Lisbon: IICT, 1975), pp. 527–45.

Albuquerque, Luís de, 'Sur quelques textes que Camões consulta pour écrire *Os Lusíadas*', *Arquivos do Centro Cultural Português* 16 (1981), 35–50.

Albuquerque, Luís de, *Navegadores, Viajantes e Aventureiros Portugueses: Séculos XV e XVI*, Vol. I (Lisbon, 1987).

Albuquerque, Luís de, and Domingues, Francisco Contente (eds.), *Dicionário de História dos Descobrimentos Portugueses*, 2 vols. (Lisbon: Círculo dos Leitores, 1994).

Albuquerque, Luís de, *Introdução à história dos descobrimentos portugueses*, 3rd edn (Lisbon: Publicações Europa-América, n.d.).

Albuquerque, Martim de, *O poder político no renascimento português* (Lisbon: Instituto Superior de Ciências Sociais e Política Ultramarina, 1968).

Almeida, Fortunato de, *História da Igreja em Portugal*, rev. edn by Damião Peres (Oporto-Lisbon: Livraria Civilização, 1968).

Alves, Ana Maria, *Iconologia do Poder Real no Período Manuelino* (Lisbon, 1985).

Alves, Jorge Manuel dos Santos, 'A Cruz, os Diamantes e os Cavalos: Frei Luís do Salvador, primeiro missionário e embaixador português em Vijayanagar (1500–1510)', *Mare Liberum* 5 (1993), 9–20.

Alves, Jorge Manuel dos Santos, 'Princes contre marchands au crépuscule de Pasai (c. 1494–1521)', *Archipel* 47 (1994), 125–45.

Andrade, António Alberto Banha de, *Mundos Novos do Mundo: Panórama da difusão pela Europa de notícias dos Descobrimentos Geográficos Portugueses*, 2 vols. (Lisbon: Junta de Investigações do Ultramar, 1972).

Andrade, António Alberto Banha de, *História de um fidalgo quinhentista português: Tristão da Cunha* (Lisbon: Instituto Histórico Infante D. Henrique, 1974).

Aragão, A.C. Teixeira de, *Vasco da Gama e a Vidigueira: Estudo histórico*, 2nd edn (Lisbon: Imprensa Nacional, 1898).

Assayag, Jackie, 'L'Aventurier divin et la Bayadere immolée: L'Inde dans l'Opera', in Catherine Weinberger-Thomas, (ed.), *L'Inde et l'Imaginaire* (Paris: Editions EHESS, Collection Puruṣârtha 11), pp. 197–227.

Aubin, Jean, 'Albuquerque et les négociations de Cambaye', *Mare Luso-Indicum* 1 (1971), 3–63.

Aubin, Jean, 'Les documents arabes, persans et turcs de la Torre do Tombo', *Mare Luso-Indicum* 2 (1973), 183–7.

Aubin, Jean, 'Duarte Galvão', *Arquivos do Centro Cultural Português* 9 (1975), 43–85.

Aubin, Jean, 'L'ambassade du Prêtre Jean à D. Manuel', *Mare Luso-Indicum* 3 (1976), 1–56.

Aubin, Jean, 'Le Prêtre Jean devant la censure portugaise', *Bulletin des Etudes Portugaises et Brésiliennes* 41 (1980), 33–57.

Aubin, Jean, 'Le capitaine Leitão: Un sujet insatisfait de D. João III', *Revista da Universidade de Coimbra* 29 (1983), 87–152.

Aubin, Jean, 'Le Portugal dans l'Europe des années 1500', in *L'Humanisme Portugais et l'Europe* (Paris: Centre Culturel Portugais, 1984), pp. 219–27.

Aubin, Jean, 'L'apprentissage de l'Inde: Cochin, 1503–1504', *Moyen Orient et Océan Indien* 4 (1987), 1–96.

Aubin, Jean, 'D. João II et Henry VII', in *Congresso Internacional, Bartolomeu Dias e a sua época – Actas*, Vol. I (Oporto, 1989), pp. 171–80.

Aubin, Jean, 'La noblesse titrée sous D. João III: Inflation ou fermeture?', *Arquivos do Centro Cultural Português* 26 (1989), 417–32.

Aubin, Jean, 'La crise égyptienne de 1510–1512: Venise, Louis XII et le Sultan', *Moyen Orient et Océan Indien* 6 (1989), 123–50.

Aubin, Jean, 'D. João II devant sa succession', *Arquivos do Centro Cultural Português* 27 (1991), 101–40.

Aubin, Jean, 'Les frustrations de Duarte Pacheco Pereira', *Revista da Universidade de Coimbra* 36 (1991), 183–204.

Aubin, Jean, 'Como trabalha Damião de Góis, narrador da segunda viagem de Vasco da Gama', in Helder Macedo, (ed.), *Studies in Portuguese Literature and History in Honour of Luís de Sousa Rebelo* (London: Tamesis Books, 1992), pp. 103–13.

Aubin, Jean, 'Ormuz au jour le jour à travers un registre de Luís Figueira, 1516–1518', *Arquivos do Centro Cultural Português* 32 (1993), 15–42.

Aubin, Jean, 'La politique orientale de Selim Ier', in *Res Orientales VI: Itinéraires d'Orient, Hommages à Claude Cahen* (1994), 197–216.

Ayalon, David, *Gunpowder and Firearms in the Mamluk Kingdom: A Challenge to a Medieval Society* (2nd edn London, 1978).

Azevedo, Sílvia Maria, 'O Gama da História e o Gama d'*Os Lusíadas*', *Revista Camoniana*, New Series, 1 (1978), 105–44.

Bacqué-Grammont, Jean-Louis, *Les Ottomans, les Safavides et leur voisins: Contribution à l'histoire des relations internationales dans l'Orient islamique de 1514 à 1524* (Istanbul, 1987).

Bacqué-Grammont, Jean-Louis, and Anne Kroell, *Mamlouks, Ottomans et Portugais en Mer Rouge: L'Affaire de Djedda en 1517* (Cairo: Annales Islamologiques, Supplément, 1988).

Baião, António, *et al.*, *Duas conferências no Paço Ducal de Vila Viçosa* (Lisbon: Fundação da Casa de Bragança, 1956).

Bandyopadhyay, Saradindu, 'Rakta Sandhya', in *Saradindu Amnibasa*, Vol. VI (Calcutta: Ananda Publishers, 1976).

Barata, António Francisco, *Vasco da Gama em Évora, com várias notícias inéditas* (Lisbon: Typographia B. Dias, 1898).

Barber, Malcolm, *The new knighthood: A history of the order of the Temple* (Cambridge: Cambridge University Press, 1994).

Barbosa, Isabel Maria Gomes Fernandes de Carvalho Lago, 'A Ordem de Santiago em Portugal na Baixa Idade Média', Mestrado dissertation in Medieval History (University of Oporto, 1989).

Barradas, Ana, *Ministros da Noite: Livro Negro da Expansão Portuguesa*, 2nd edn (Lisbon: Edições Antigona, 1992).

Barreto, Luís Filipe, *Descobrimentos e Renascimento: Formas de ser e pensar nos séculos XV e XVI*, 2nd edn (Lisbon: Imprensa Nacional, 1983).

Barros, Henrique da Gama, *História da administração pública em Portugal*, 11 Vols. (Lisbon, 1945–54).

Beazley, Charles Raymond, *Prince Henry the Navigator: The hero of Portugal and of modern discovery, 1394–1460 A.D.* (New York: G.P. Putnam, 1895).

Bismut, Roger, *'Les Lusiades' de Camões, confession d'un poète* (Paris: Centre Culturel Portugais, 1974).

Bouchon, Geneviève, 'Les rois de Kôṭṭê au début du XVI siècle', *Mare Luso-Indicum* 1 (1971), 65–96.

Bouchon, Geneviève, 'Les Musulmans du Kerala à l'époque de la découverte portugaise', *Mare Luso-Indicum* 2 (1973), 3–59.

Bouchon, Geneviève, *Mamale de Cananor: Un adversaire de l'Inde portugaise (1507–1528)* (Geneva-Paris: Librairie Droz, 1975).

Bouchon, Geneviève, 'Le premier voyage de Lopo Soares en Inde (1504–1505)', *Mare Luso-Indicum* 3 (1976), 57–84.

Bouchon, Geneviève, 'L'inventaire de la cargaison rapportée en Inde en 1505', *Mare Luso-Indicum* 3 (1976), 101–36.

Bouchon, Geneviève, *Navires et cargaisons retour de l'Inde en 1518* (Paris: Société de l'Histoire de l'Orient, 1977).

Bouchon, Geneviève, 'A propos de l'Inscription de Colombo (1501): Quelques observations sur le premier voyage de João da Nova dans l'Océan Indien', *Revista da Universidade de Coimbra* 28 (1980), 233–70.

Bouchon, Geneviève, 'Glimpses of the beginnings of the Carreira da Índia (1500–1518)', in Teotónio R. De Souza (ed.), *Indo-Portuguese History: Old Issues, New Questions* (New Delhi: Concept Publishing Co. 1984), pp. 40–55.

Bouchon, Geneviève, 'L'interprète portugais en Inde au début du XVIe siècle', in *Dimensões da alteridade na culturas da língua portuguesa – O Outro* (1° Simpósio interdisciplinar de Estudos Portugeses – Actas) (Lisbon: Universidade Nova de Lisboa, 1985), Vol. II, pp. 203–13.

Bouchon, Geneviève, *L'Asie du Sud à l'époque des Grandes Découvertes* (London: Variorum Reprints, 1987).

Bouchon, Geneviève, 'Un microcosme: Calicut au 16ème siècle', in Denys Lombard and Jean Aubin (eds.), *Marchands et hommes d'affaires asiatiques dans l'Océan Indien et la Mer de Chine (13e–20e siècles)* (Paris: EHESS, 1988), pp. 49–59.

Bouchon, Geneviève, *Albuquerque, le lion des mers d'Asie* (Paris: Editions Desjonquères, 1992).

Bouchon, Geneviève, 'Timoji, un corsaire indien au service de Portugal (1498–1512)', in *Ciclo de Conferências, Portugal e o Oriente* (Lisbon: Fundação Oriente-Edições Quetzal, 1994), pp. 7–25.

Bouchon, Geneviève and Thomaz, Luís Filipe, *Voyage dans les Deltas du Gange et de l'Irraouaddy. Relation Portugaise Anonyme (1521)* (Paris: Centre Culturel Portugais, 1988).

Boxer, Charles R., 'Diogo do Couto (1543–1616), Controversial Chronicler of Portuguese Asia', in R.O.W. Goertz (ed.), *Iberia – Literary and Historical Issues: Studies in Honour of Harold V. Livermore* (Calgary, 1985), pp. 49–66.

Braga, Teófilo, 'O Centenário de Camões em 1880', *Revista de Philosophia e Positivismo* (Oporto) 2 (1880), 1–9.

Braga, Teófilo, *Os Centenários como Síntese Afectiva nas Sociedades Modernas* (Oporto: A.T. da Silva Teixeira, 1884).

Brummett, Palmira, *Ottoman Seapower and Levantine Diplomacy in the Age of Discovery* (Albany: SUNY Press, 1994).

Caetano, José A. Palma, *Vidigueira e o seu concelho: Ensaio monográfico* (Vidigueira: Edição da Câmara Municipal, 1986).

Cahen, Claude, and R.B. Serjeant, 'A Fiscal Survey of the Mediaeval Yemen', *Arabica* 4, 1 (1957), 22–33.

Campbell, Eila M.J., 'Gama, Vasco da, 1^{eiro} Conde (1st Count) da Vidigueira (b. c. 1460, Sines, Port. – d. Dec. 24, 1524, Cochin, India)', *The New Encyclopaedia Britannica: Micropaedia*, 15th edn (Chicago, 1989), Vol. V, pp. 100–1.
Chagas, Manuel Pinheiro, 'A trasladação dos ossos de Vasco da Gama em 1880', *A Ilustração Portuguesa* Vol. II, nos. 49–51 (1885–6) (in three parts).
Chagas, Manuel Pinheiro, 'O Naufrágio de Vicente Sodré', in *Brinde aos Senhores Assignantes do Diário de Notícias em 1892* (Lisbon: Typographia Universal, 1892), pp. 5–80.
Chandeigne, Michel (ed.), *Lisbonne hors les murs, 1415–1580: L'invention du monde par les navigateurs portugais* (Paris: Editions Autrement, Séries Memoires no. 1, 1992).
Chelhod, Joseph, 'Les Portugais au Yémen d'après les sources arabes', *Journal Asiatique* 283, 1 (1995), 1–18.
Chittick, H. Neville, *Kilwa: An Islamic Trading City on the East African Coast* (Nairobi: British Institute in East Africa, 1984), 2 vols.
Chorão, Maria José Mexia Bigotte, *Os Forais de D. Manuel, 1496–1520* (Lisbon: Arquivo Nacional da Torre do Tombo, 1990).
Chumovsky, T.A., *Três Roteiros Desconhecidos de Ahmad ibn-Mâdjid, o piloto árabe de Vasco da Gama*, tr. Myron Malkiel-Jirmounsky (Lisbon: Comissão Executiva das Comemorações do V Centenário da Morte do Infante D. Henrique, 1960).
Cinquième Centenaire de la Naissance de Vasco de Gama (1469–1969): Exposition bibliographique et iconographique (Paris: Centre Culturel Portugais, 1970).
Cirurgião, António, 'A divinização do Gama de *Os Lusíadas*', *Arquivos do Centro Cultural Português* 26 (1989), 513–38.
Ciutiis, Salvatore de, *Une ambassade portugaise à Rome au XVIe siècle* (Naples, 1899).
Coelho, José Maria Latino, *Vasco da Gama* (Oporto: Lello e Irmão, 1985; reprint of the 1882 text).
Cordeiro, Luciano, *De como e quando foi feito Conde Vasco da Gama* (Lisbon: Sociedade de Geografia, 1892).
Cordeiro, Luciano, 'Os Restos de Vasco da Gama', *Boletim da Sociedade de Geografia de Lisboa*, 15ª Série 4 (1896), 191–200.
Cordeiro, Luciano, *Questões Histórico-Coloniais*, Vols. II and III (Lisbon: Agência Geral das Colónias, 1936).
Cortesão, Armando, 'The mystery of Columbus', *The Contemporary Review* 151 (1937), 322–30.
Cortesão, Armando, *The Mystery of Vasco da Gama* (Coimbra: Junta de Investigações do Ultramar, 1973). (Portuguese version, *O mistério de Vasco da Gama*, in the same year).
Cortesão, Jaime, *A expedição de Pedro Álvares Cabral* (Lisbon, 1922).
Cortesão, Jaime, 'Do Sigilo Nacional sobre os Descobrimentos: Cro-

nicas desaparecidas, mutiladas e falseadas – Alguns dos feitos que se calaram', *Lusitánia* 1 (1924), 45–81.

Cortesão, Jaime, *A política de Sigilo nos Descobrimentos* (Lisbon, 1960).

Cosme, João Ramalho, and Maria de Deus Manso, 'A Ordem de Santiago e a Expansão Portuguesa no Século XV', in *As Ordens Militares em Portugal: Actas do I.° Encontro sobre Ordens Militares* (Palmela: Câmara Municipal de Palmela, 1991), pp. 43–56.

Costa, C.D.N. (ed.), *Medea (by Seneca)* (Oxford: Clarendon Press, 1973).

Costa, João Paulo Oliveira e, 'Do sonho manuelino ao realismo joanino: Novos documentos sobre as relações luso-chinesas na terceira década do século XVI', *Studia* 50 (1991), 121–56.

Costa, João Paulo Oliveira e, 'Os Portugueses e a cristandade siro-malabar (1498–1530)', *Studia* 52 (1994), 121–78.

Costa, João Paulo Oliveira e, 'Simão de Andrade, fidalgo da Índia e capitão de Chaul', *Mare Liberum* 9 (1995), 99–116.

Costa, João Paulo Oliveira e, and Rodrigues, Vítor Luís Gaspar, *Portugal y Oriente: El Proyecto Indiano del Rey Juan* (Madrid: Colecciones MAPFRE, 1992).

Coutinho, Carlos Viegas Gago, 'Discussão sobre a rota seguida por Vasco da Gama entre Santiago e S. Brás', *Anais da Academia Portuguesa da História*, 2ª Série, 2 (1949), 99–131.

Cruz, Maria Augusta Lima, *Diogo do Couto e a Década Oitava da Ásia* (Lisbon: Imprensa Nacional, 1992–5), 2 vols.

Cruz, Maria Augusta Lima, 'As andanças de um degredado em terras perdidas – João Machado', *Mare Liberum* 5 (1993), 39–47.

Cunha, Mário Raul de Sousa, 'A Ordem Militar de Santiago (Das Origens a 1327)', Mestrado dissertation in Medieval History (University of Oporto, 1991).

Cunha, Rosalina Silva, 'Sociedade de Geografia: A Colecção Vidigueira', *Boletim Internacional de Bibliografia Luso-Brasileira* 1, 1 (1960), 65–99.

Curtin, Philip D., *Cross-cultural trade in World History* (New York: Cambridge University Press, 1984).

Darrag, Ahmad, *L'Egypte sous le règne de Barsbay, 825–841/1422–1438* (Damascus: Institut Français de Damas, 1961).

Denucé, Jean, *Magalhães, la question des Moluques et la première circumnavigation du globe* (Brussels, 1911).

Desai, Ziyauddin A., 'Relations of India with Middle Eastern Countries during the 16th–17th centuries', *Journal of the Oriental Institute* (Baroda) 23 (1973–4), 75–106.

Deswarte, Sylvie, *Les enluminures de la "Leitura Nova", 1504–1522: Etude sur la culture artistique au Portugal au temps de l'humanisme* (Paris, 1977).

De Souza, Teotónio R., '"North–South" in the *Estado da Índia*?', *Mare Liberum* 9 (1995), 453–60.

Dias, José Sebastião da Silva, *Os Descobrimentos e a problemática cultural do século XVI*, 3rd edn (Lisbon: Editorial Presença, 1988).

Dias, Luís Fernando de Carvalho, 'As Ordenações da Índia', *Garcia da Orta* (Special No.) (1956), 229–45.

Dias, Manuel Nunes, *O capitalismo monárquico português (1415–1549): Contribuição para o estudo das origens do capitalismo moderno*, 2 vols. (Coimbra, 1963–4).

Diffie, B.W., and G.D. Winius, *Foundations of the Portuguese Empire, 1415–1580* (Minneapolis: University of Minnesota Press, 1977).

Disney, Anthony, 'Vasco da Gama's reputation for violence: The alleged atrocities at Calicut in 1502', *Indica* 32, 2 (1995), 11–28.

Domingues, Francisco Contente, 'Colombo e a política de sigilo na historiografia portuguesa', *Mare Liberum* (1990), 105–16.

Dornellas, Affonso de, 'Bases genealógicas dos Ataídes', in Dornellas, *História e Genealogia*, Vol. I (Lisbon, 1913), pp. 107–42.

Duby, Georges, *The Three Orders: Feudal Society Imagined*, tr. Arthur Goldhammer (Chicago and London: The University of Chicago Press, 1980).

Duchac, René Virgile, *Vasco de Gama: L'orgeuil et la blessure* (Paris: L'Harmattan, 1995).

Durand, Robert, *Histoire du Portugal* (Paris: Hatier, 1992).

Dutra, Francis A., 'Membership in the Order of Christ in the Sixteenth Century: Problems and Perspectives', *Santa Barbara Portuguese Studies* 1 (1994), 228–39.

Earle, T.F., and Villiers, John, *Albuquerque, Caesar of the East: Selected Texts by Afonso de Albuquerque and His Son* (Warminster, 1990).

Eça, Vicente Almeida d', 'Almirante da Índia', *Revista Portugueza Colonial e Marítima*, (special commemorative number 20 May 1898), 26–31.

Ehrhardt, Marion, *A Alemanha e os Descobrimentos Portugueses* (Lisbon: Texto Editora, 1989).

Faria, António Machado de, 'Cavaleiros da Ordem de Cristo no século XVI', *Arqueologia e História* 6 (1955), 13–73.

Faria, Francisco Leite de, 'Pensou-se em Vasco da Gama para comandar a armada que descobriu o Brasil', *Revista da Universidade de Coimbra* 26 (1978), 145–85.

Faria, Francisco Leite de, 'O mais antigos documentos que se conservam, escritos pelos Portugueses na Índia', in Luís de Albuquerque and Inácio Guerreiro, (eds.), *Actas do II Seminário Internacional de História Indo-Portuguesa* (Lisbon: IICT, 1985).

Farinha, António Dias, 'A dupla conquista de Ormuz por Afonso de Albuquerque', *Studia* 48 (1989), 445–72.

Farinha, António Dias, 'Os Portugueses no Golfo Pérsico (1507–1538): Contribução Documental e Crítica para a sua História', *Mare Liberum* 3 (1991), 1–159.

Fernández, Luis Suárez, *Política Internacional de Isabel la Católica: Estudio y Documentos*, Vol. II (1482–8) (Valladolid: Universidad de Valladolid, 1966).

Ferrand, Gabriel, 'Le pilote arabe de Vasco da Gama et les instructions nautiques arabes au XVe siècle', *Annales de Géographie* 172 (1922), 289–307.

Ferrand, Gabriel, 'Les Sultâns de Kilwa', in *Mémorial Henri Basset: Nouvelles Etudes Nord-Africaines et Orientales* (Paris: Librarie Orientaliste Paul Geuthner, 1928), pp. 239–60.

Finlay, Robert, 'Portuguese and Chinese Maritime Imperialism: Camões's "Lusiads" and Luo Maodeng's "Voyage of the San Bao Eunuch"', *Comparative Studies in Society and History* 34, 2 (1992), 225–41.

Fischel, Walter J., 'The spice trade in Mamluk Egypt', *Journal of the Economic and Social History of the Orient* 1, 2 (1958), 157–74.

Flores, Jorge Manuel, 'The Straits of Ceylon and the Maritime Trade in Early Sixteenth-Century India: Commodities, Merchants and Trading Networks', *Moyen Orient et Océan Indien* 7 (1990), 27–58.

Flores, Jorge Manuel, 'Os Portugueses e o Mar de Ceilão, 1498–1543: Trato, Diplomacia e Guerra', Mestrado dissertation in History (Universidade Nova de Lisboa, 1991).

Flores, Jorge Manuel, '"Cael Velho", "Calepatanão" and "Punicale": The Portuguese and the Tambraparni Ports in the Sixteenth Century', *Bulletin de l'Ecole Française d'Extrême-Orient* 82 (1995), 9–26.

França, José-Augusto, *Le Romantisme au Portugal: Etude de faits socio-culturels* (Paris: Editions Klincksieck, 1975).

Freire, Anselmo Braamcamp, 'O Almirantado da Índia: Data da sua criação', *Archivo Histórico Portuguez* 1 (1903), 25–32.

Freire, Anselmo Braamcamp, 'O livro das tenças del Rei', *Archivo Histórico Portuguez* 2, 3 (1904), 31–60.

Freire, Anselmo Braamcamp, 'O livro das tenças d'el Rei', *Arquivo Histórico Portuguez* 2, 6 (1904), 201–27.

Freire, Anselmo Braamcamp, *Brasões da Sala de Sintra*, rev. edn Luís Bivar Guerra, 3 vols. (Lisbon: Imprensa Nacional, 1973).

Gama, D. Maria Telles da, *Le Comte-Amiral D. Vasco da Gama* (Paris: A. Roger and F. Chernoviz, 1902).

Gayo, Felgueiras, *Nobiliário de Famílias de Portugal*, Vol. XIV (Braga, 1939).

Genot-Bismuth, Jacqueline, 'Le Mythe de l'Orient dans l'eschatologie des juifs de l'Espagne à l'époque des conversions forcées et de l'expulsion', *Annales ESC* 45, 4 (1990), 819–38.

Godinho, Vitorino Magalhães, *Os Descobrimentos e a Economia Mundial*, 4 vols., 2nd edn (Lisbon: Editorial Presença, 1981–3).

Godinho, Vitorino Magalhães, *Mito e mercadoria, utopia e prática de navegar, séculos XIII–XVIII* (Lisbon: Difel, 1990).

Goitein, Salomon D., 'From the Mediterranean to India: Documents on the trade to India, South Arabia and East Africa from the eleventh and twelfth centuries', *Speculum* 29, 2, 1 (1954), 181–97.

Goitein, S.D., 'Two eyewitness reports on an expedition of the King of Kîsh (Qais) against Aden', *Bulletin of the School of Oriental and African Studies* 16, 2 (1954), 247–57.

Greenlee, W.B., *The voyage of Pedro Álvares Cabral to Brazil and India* (London: The Hakluyt Society, 1937).

Guerreiro, Inácio, and Rodrigues, Vítor Luís Gaspar, 'O "grupo de Cochim" e a oposição a Afonso de Albuquerque', *Studia* 51 (1992), 119–44.

Guillot, Claude, 'Libre entreprise contre économie dirigée: Guerres civiles à Banten, 1580–1609', *Archipel* 43 (1992), 57–72.

Gumilev, L.N., *Searches for an Imaginary Kingdom: The Legend of the Kingdom of Prester John*, tr. R.E.F. Smith (Cambridge: Cambridge University Press, 1987).

Hallam, Elizabeth M., *Capetian France, 987–1328* (London and New York: Longman, 1980).

Hamdani, Abbas, 'Columbus and the Recovery of Jerusalem', *Journal of the American Oriental Society* 99, 1 (1979).

Harrisse, Henry, *Los restos de Don Cristoval Colón* (Seville: F. Alvarez y Cª, 1878).

Heers, Jacques, *Christophe Colomb* (Paris: Hachette, 1981).

Heifetz, Hank, and Narayana Rao, Velcheru, *For the Lord of the Animals – Poems from the Telugu: The Kâḷahastîśvara Śatakamu of Dhûrjati* (Berkeley: University of California Press, 1987).

Hespanha, António Manuel, *As vésperas do Leviathan: Instituições e poder político, Portugal – século XVII* (Coimbra: Almedina, 1994).

Hook, Sidney, *The Hero in History: A Study in Limitation and Possibility* (Boston: Beacon Press, 1955).

Hourani, George F., *Arab Seafaring in the Indian Ocean in Ancient and early Medieval Times* (Princeton: Princeton University Press, 1951).

Huebner, Steven, 'Africaine, L' ("The African Maid")', in Stanley Sadie (ed.), *The New Grove Dictionary of Opera*, Vol. I (London: Macmillan, 1992), pp. 31–3.

Jayne, K.G., *Vasco da Gama and His Successors, 1460–1580* (London: Methuen and Co., 1910).

João, Maria Isabel, 'A festa cívica: O tricentenário de Camões nos Açores (10 de junho de 1880)', *Revista de História Económica e Social* 20 (1987), 87–111.

Jouanna, Arlette, *Le devoir de revolte: La noblesse française et la gestation de l'Etat Moderne, 1559–1661* (Paris: Librairie Arthème Fayard, 1989).

Kadir, Djelal, *Columbus and the Ends of the Earth: Europe's Prophetic*

Rhetoric as Conquering Ideology (Berkeley: University of California Press, 1992).

Kantorowicz, Ernst H., *The King's Two Bodies: A Study in Mediaeval Political Theology* (Princeton: Princeton University Press, 1958).

Keil, Luís, *As Assinaturas de Vasco da Gama: Uma falsa assinatura do navegador português, Críticas, comentários e documentos* (Lisbon, 1934).

Khoury, Ibrahim, *As-Sufaliyya, 'The Poem of Sofala', by Ahmad ibn Mâğid* (Coimbra: Junta de Investigações Científicas do Ultramar, 1983).

Kieniewicz, Jan, 'Pepper gardens and market in precolonial Malabar', *Moyen Orient et Océan Indien* 3 (1986), 1–36.

Krishna Ayyar, K.V. *The Zamorins of Calicut (From the earliest times down to A.D. 1806)*, (Calicut: Norman Printing Bureau, 1938).

Kroell, Anne, 'Le voyage de Lazarus Nürnberger en Inde (1517–1518)', *Bulletin des Etudes Portugaises et Brésiliennes* 41 (1980), 59–87.

Lefebure, Docteur Francis, *Expériences initiatiques*, 3 vols. (Paris: Omnium Littéraire, 1954).

Leite, Duarte, *Descobrimentos portugueses*, 2 vols., ed. V.M. Godinho (Lisbon: Cosmos, 1958–60).

Lesure, Michel, 'Un document ottoman de 1525 sur l'Inde portugaise et les pays de la Mer Rouge', *Mare Luso-Indicum* 3 (1976), 137–60.

Levi, Giovanni, 'Les usages de la biographie', *Annales ESC* 44, 6 (1989), 1325–36.

Lieberman, Samuel (ed.), *Roman Drama*, (New York: Bantam Books, 1964).

Lima, João Paulo de Abreu, 'Vasco da Gama e os Frescos das "Casas Pintadas" da Cidade de Évora', *Panorama* (Revista Portuguesa de Arte e Turismo), 4th Series, 31 (1969), 51–63.

Lipiner, Elias, *Gaspar da Gama: Um converso na Frota de Cabral* (Rio de Janeiro: Nova Fronteira, 1987).

Lobato, Alexandre, *Da época e dos feitos de António de Saldanha* (Lisbon: CEHU, 1964).

Lobato, Alexandre, 'Dois novos fragmentos do regimento de Cabral para a viagem da Índia em 1500', *Studia* 25 (1968), 31–50.

Lombard, Denys, *Le Sultanat d'Atjéh au temps d'Iskandar Muda (1607–1636)* (Paris: EFEO, 1967).

Macedo, Jorge Borges de, *Um caso de luta pelo poder e a sua interpretação n'"Os Lusíadas'* (Lisbon: Academia Portuguesa de História, 1976).

Macedo, Jorge Borges de, 'Para o estudo da mentalidade portuguesa do século XVI – uma ideologia de cortesão: As sentenças de D. Francisco de Portugal', *ICALP – Revista* 7–8 (1987), 73–106.

Machado, José Pedro and Campos, Viriato, *Vasco da Gama e a sua viagem de descobrimento* (Lisbon: Câmara Municipal de Lisboa, 1969).

Magalhães, Joaquim Antero Romero, *Para o estudo do Algarve económico durante o século XVI* (Lisbon: Edições Cosmos, 1970).

Major, Richard Henry, *The Life of Prince Henry of Portugal, Surnamed the Navigator, and its results: comprising the discovery, within one century, of half the world* (London: A. Asher, 1868).

Martín, José Luís, *Origenes de la Orden Militar de Santiago (1170–1195)* (Barcelona: Consejo Superior de Investigaciones Científicas, 1974).

Martins, António Coimbra, 'Sobre as Décadas que Diogo do Couto deixou inéditas', *Arquivos do Centro Cultural Português* 3 (1971), 272–355.

Martins, António Coimbra, 'Camões et Couto', in *Les cultures ibériques en devenir: Essais publiés en hommage à la mémoire de Marcel Bataillon* (Paris, 1979), pp. 691–705.

Martins, António Coimbra, 'Diogo do Couto et la famille Da Gama: Un traité inédit', *Revue des Littératures Comparées* (1979), 279–92.

Martins, António Coimbra, 'Correia, Castanheda e as Diferenças da Índia', *Revista da Universidade de Coimbra* 30 (1984), 1–86.

Martins, António Coimbra, *Em torno de Diogo do Couto* (Coimbra: Biblioteca Geral da Universidade de Coimbra, 1985).

Martins, Joaquim Pedro Oliveira, *The golden age of Prince Henry the Navigator*, tr. J.J. Abraham and W.E. Reynolds (London: Chapman and Hall, 1914).

Matos, Luís de, 'António Maldonado de Hontiveras e a "questão das Molucas"', in A. Teixeira da Mota (ed.), *A viagem de Fernão de Magalhães e a questão das Molucas* (Lisbon: IICT, 1975), pp. 548–77.

Mattoso, José, *Identificação de um país: Ensaio sobre as origens de Portugal, 1096–1325*, 2 vols. (Lisbon: Editorial Estampa, 1985).

Mayer, Hans Eberhard, *The Crusades*, tr. John Gillingham (New York and Oxford: Oxford University Press, 1972).

Mazzaoui, M.M., 'Global policies of Sultan Selim', in Donald P. Little (ed.), *Essays on Islamic Civilization presented to Niyazi Berkes* (Leiden, 1976), pp. 224–43.

Meilink-Roelofsz, M.A.P., *Asian trade and European influence in the Indonesian archipelago from 1500 to about 1630* (The Hague: Martinus Nijhoff, 1962).

Mendonça, Manuela, *D. João II: Um percurso humano e político nas origens da modernidade em Portugal* (Lisbon: Editorial Estampa, 1991).

Meyerbeer, Giacomo and Scribe, Eugène, *L'Africaine: Opéra en cinq actes de G. Meyerbeer* (Paris: A. Lacroix, Verbockhoven et Cie, 1865).

Momigliano, Arnaldo, *Le radici classiche della storiografia moderna* (Florence: Sansoni, 1992).

Monteiro, José de Souza, 'Vasco da Gama – A Psychologia d'um Heroe', *Revista Portugueza Colonial e Marítima* (special commemorative number, 20 May 1898), 3–12.

Moreno, Humberto Baquero, 'A presença dos corregedores nos municípios e os conflitos de competências (1332–1459)', *Revista de História* (Oporto) 9 (1989), 77–88.

Moreno, Humberto Baquero, 'La lutte de la noblesse portugaise contre la royauté à la fin du Moyen Age', *Arquivos do Centro Cultural Português* 26 (1989), 49–66.

Moreno, Humberto Baquero, 'Vasco da Gama, alcaide das sacas de Olivença', *in Encontros: Revista hispano-portuguesa de investigadores en Ciencias Humanas y Sociales* 1 (1989).

Moreno, Humberto Baquero, *Exilados, Marginais e Contestatários da sociedade portuguesa medieval: Estudos de história* (Lisbon: Editorial Presença, 1990).

Moricca, Humbertus (ed.), *Medea – Oedipus – Agememnon – Hercules (Oetaeus)* (Turin: G.B. Paravia and Co. 1925).

Moser, Gerald M., 'What did the Old Man of Restelo Mean?', *Luso-Brazilian Review* 17 (1980), 139–51.

Mundadan, A. Mathias, *A History of Christianity in India*, vol. I (*From the beginning up to the middle of the sixteenth century*) (Bangalore, 1984).

Nandy, Ashis, *et al.*, *The Blinded Eye: 500 Years of Christopher Columbus* (Goa: The Other India Press, 1993).

Nowell, Charles E., *The Great Discoveries and the First Colonial Empires* (Ithaca: Cornell University Press, 1954).

Obeyesekere, Gananath, *The Apotheosis of Captain Cook: European Mythmaking in the Pacific* (Princeton: Princeton University Press, 1992).

O'Callaghan, Joseph F., *A History of Medieval Spain* (Ithaca and London: Cornell University Press, 1975).

Oyarzun, J., *Expediciones españolas al estrecho de Magallanes y Tierra de Fuego* (Madrid, 1976).

Özbaran, Salih, 'The Ottoman Turks and the Portuguese in the Persian Gulf, 1534–1581', *Journal of Asian History* 6, 1 (1972), 45–87.

Özbaran, Salih, 'A Turkish Report on the Red Sea and the Portuguese in the Indian Ocean (1525)', *Arabian Studies* 4 (1978), 81–8.

Özbaran, Salih, *The Ottoman Response to European Expansion: Studies on Ottoman–Portuguese Relations in the Indian Ocean and Ottoman Administration in the Arab Lands during the Sixteenth Century* (Istanbul: The Isis Press, 1994).

Panikkar, K.M., *Asia and Western Dominance* (1st edn, London, 1959; reprint Kuala Lumpur: The Other Press, 1993).

Parker, Geoffrey, *The Military Revolution: Military Innovation and the Rise of the West, 1500–1800* (Cambridge: Cambridge University Press, 1987).

Parry, J.H., *The Discovery of the Sea* (Berkeley: University of California Press, 1981).

Pavoni, Rosanna, *Christophe Colomb: Images d'un visage inconnu*, tr. Federica Bevilacqua-Mariat (Paris: Vilo, 1990).

Pearson, Michael N., *Merchants and Rulers in Gujarat: The response to the Portuguese in the sixteenth century*, (Berkeley–Los Angeles: University of California Press, 1976).

Pearson, M.N., *The Portuguese in India* (Cambridge: Cambridge University Press, 1987).

Pereira, Isaías da Rosa, 'Documentos inéditos sobre Gonçalo Gil Barbosa, Pêro Vaz de Caminha, Martinho Neto e Afonso Furtado, escrivães da despesa e receita do feitor Aires Correia (1500)', in *Congresso Internacional, Bartolomeu Dias e a sua época: Actas*, Vol. II (Oporto: Universidade do Porto, 1989), pp. 505–13.

Pereira, Isaías da Rosa, *Matrícula de Ordens da diocese de Évora (1480–1483): Qual dos dois Vasco da Gama foi à Índia em 1497?* (Lisbon: Academia Portuguesa da História, 1990).

Pereira, Moacir Soares, 'Capitães, Naus e Caravelas da Armada de Cabral', *Revista da Universidade de Coimbra* 27 (1979), 31–134.

Phelan, John L., *The millenial kingdom of the Franciscans in the New World* (Berkeley: University of California Press, 1970).

Phillips, Carla Rahn, and William D. Phillips jr, *The Worlds of Christopher Columbus* (Cambridge: Cambridge University Press, 1992).

Phillips, William D. jr, 'Christopher Columbus in Portugal: Years of Preparation', *Terrae Incognitae* 24 (1992), 31–42.

Pinto, J. Estêvão, and Reis, Maria Alice, *Vasco da Gama* (Lisbon: Comissão Executiva das Comemorações do V Centenário do Nascimento de Vasco da Gama, 1969).

Pissurlencar, P.S.S., *Agentes da diplomacia portuguesa na Índia* (Bastorá–Goa, 1952).

Pouwels, Randall L., *Horn and Crescent: Cultural Change and Traditional Islam on the East African Coast, 800–1900* (Cambridge: Cambridge University Press, 1987).

Quint, David, 'Voices of Resistance: The Epic Curse and Camões's Adamastor', *Representations* 27 (1989), 118–41.

Radulet, Carmen M., 'Girolamo Sernigi e a importância económica do Oriente', *Revista da Universidade de Coimbra* 32 (1985), 67–77.

Radulet, Carmen M., 'As viagens de Diogo Cão: Um problema ainda em aberto', *Revista da Universidade de Coimbra* 34 (1988), 105–19.

Radulet, Carmen M., 'As viagens de descobrimento de Diogo Cão: Nova proposta de interpretação', *Mare Liberum* 1 (1990), 175–204.

Radulet, Carmen M., and Saldanha, António Vasconcelos de (eds.), *O Regimento do Almirantado da Índia* (Lisbon: Inapa, 1989).

Ramos, Demetrio, '¿Quien decidiò a Fernando el Católico a aceptar el proyecto descubridor de Colón?', *Mare Liberum* 3 (1991), 184–91.

Rau, Virgínia, *A exploração e o comércio do sal de Setubal* (Lisbon: Instituto para a Alta Cultura, 1951).

Rebelo, Luís de Sousa, 'Millénarisme et historiographie dans les chroniques de Fernão Lopes', *Arquivos do Centro Cultural Português* 26 (1989), 97–120.

Reeves, Marjorie, *The Influence of Prophecy in the Later Middle Ages: A Study in Joachimism* (Oxford: Oxford University Press, 1969).

Rego, António da Silva, 'O início do segundo governo do vice-rei da Índia D. Francisco da Gama, 1622–1623', *Memórias da Academia das Ciências de Lisboa, Classe de Letras* 19 (1978), 323–45.

Roberts, John Howell, 'The Genesis of Meyerbeer's "L'Africaine"', PhD dissertation in Music (Berkeley, University of California, 1977).

Rodrigues, Vítor Luís Gaspar, 'A organização militar do Estado Português da Índia (1500–1580)', unpublished thesis (Lisbon, Instituto de Investigação Científica Tropical, 1990).

Round, Nicholas Grenville, *The Greatest Man Uncrowned: A study of the fall of don Álvaro de Luna* (London: Tamesis Books, 1986).

Saldanha, António Vasconcelos de, 'O Almirante de Portugal: Estatuto quatrocentista e quinhetista de um cargo medieval', *Revista da Universidade de Coimbra* 34 (1988), 137–56.

Sampaio, L.M. Vaz de, *Subsídios para uma biografia de Pedro Álvares Cabral* (Coimbra, 1971).

Sanceau, Elaine, *Henry the Navigator* (New York: W.W. Norton, 1947).

Santos, Carlos, and Santos, Fernando, *Vasco da Gama e a Índia* (Odivelas: Europress, 1992).

Saraiva, José Hermano (ed.), *Ditos Portugueses Dignos de Memória: História íntima do Século XVI* (Lisbon, n.d.).

Schurhammer, Georg, *Gesammelte Studien*, ed. László Szilas, 4 vols. in 5 parts (Lisbon: Centro de Estudos Históricos Ultramarinos, 1962–65).

Schurhammer, Georg, 'Orientalische Briefe aus der Zeit des Hl. Franz Xaver (1500–1552)', *Euntes Docente* 21 (1968), 255–301.

Schwartz, Stuart B., *Sovereignty and Society in Colonial Brazil: The High Court of Bahia and Its Judges, 1609–1751* (Berkeley: University of California Press, 1973).

Sérgio, António, *Breve interpretação da História de Portugal* (Lisbon: Sá da Costa, 1972).

Serjeant, R.B., 'Materials for South Arabian History: Notes on new MSS from Hadramawt', *Bulletin of the School of Oriental and African Studies* 13, 2 (1950), 281–307.

Serjeant, R.B., 'Yemeni Merchants and Trade in Yemen, 13th–16th Centuries', in Denys Lombard and Jean Aubin (eds.), *Marchands et hommes d'affaires asiatiques dans l'Océan Indien et la Mer de Chine (13e–20e siècles)* (Paris: EHESS, 1988), pp. 61–82.

Serjeant, R.B., 'Fifteenth Century "Interlopers" on the Coast of Rasûlid

Yemen', *Res Orientales VI: Itinéraires d'Orient, Hommages à Claude Cahen* (1994), 83–94.

Serrão, Joaquim Veríssimo, *História de Portugal*, Vol. II: *Formação do Estado Moderno (1415–1495)* (Lisbon: Editorial Verbo, 1978).

Servières, Georges, 'Les transformations et tribulations de *L'Africaine*', *Rivista Musicale Italiana* 34 (1927), 80–99.

Shephard, Robert, 'Court Factions in Early Modern England', *The Journal of Modern History* 64, 4 (1992), 721–45.

Sherburne, Sir Edward, trans., *The Tragedies of L. Annaeus Seneca, the Philosopher* (London: S. Smith and B. Walford, 1702).

Shokoohy, Mehrdad, 'Architecture of the Sultanate of Ma^c^bar in Madura, and other Muslim Monuments in South India', *Journal of the Royal Asiatic Society of Great Britain and Ireland*, 3rd Series, 1, 1 (1991), 31–92.

Siddiqui, Iqtidar Hussain, *Perso-Arabic Sources of Information on the Life and Conditions in the Sultanate of Delhi* (New Delhi: Munshiram Manoharlal, 1992).

Smith, Ronald Bishop, *Diogo Lopes de Sequeira* (Lisbon: The author, 1975).

Smith, Ronald Bishop, 'Diogo Lopes de Sequeira, the Governor of India, in the Collection of São Bento da Saude in the Torre do Tombo', *Mare Luso-Indicum* 4 (1980), 129–41.

Smith, Ronald Bishop, *Diogo Lopes de Sequeira: Elements on his office of Almotacé-mor* (Lisbon: Silvas, 1993).

Stanley, Henry E.J., *The Three Voyages of Vasco da Gama and His Viceroyalty* (London: The Hakluyt Society, 1869).

Subrahmanyam, Sanjay, *Improvising Empire: Portuguese Trade and Settlement in the Bay of Bengal, 1500–1700* (Delhi: Oxford University Press, 1990).

Subrahmanyam, Sanjay, *The Political Economy of Commerce: Southern India, 1500–1650* (Cambridge: Cambridge University Press, 1990).

Subrahmanyam, Sanjay, *The Portuguese Empire in Asia, 1500–1700: A Political and Economic History* (London: Longman, 1993).

Subrahmanyam, Sanjay, 'Of *Imârat* and *Tijârat*: Asian merchants and state power in the western Indian Ocean, 1400–1750', *Comparative Studies in Society and History* 37, 4 (1995), 750–80.

Subrahmanyam, Sanjay, '"The Life and Actions of Mathias de Albuquerque" (1547–1609): A Portuguese Source for Deccan History', *Portuguese Studies* 11 (1995), 62–77.

Taviani, P.E., *Cristoforo Colombo: La Genesi della Grande Scopertà*, 2 vols. (Novara, 1974).

Tavim, José Alberto Rodrigues da Silva, 'Os Judeus e a Expansão Portuguesa na Índia durante o século XVI: O exemplo de Isaac do Cairo: Espião, "língua" e "judeu de Cochim de Cima"', *Arquivos do Centro Cultural Calouste Gulbenkian* 33 (1994), 137–260.

Teixeira da Mota, Avelino, *A viagem de António de Saldanha e a rota de Vasco da Gama no Atlântico Sul* (Lisbon, 1971).
Teixeira da Mota, A., 'Duarte Pacheco Pereira: Capitão e Governador de S. Jorge da Mina', *Mare Liberum* 1 (1990), 1–27.
Thomaz, Luís Filipe F.R., 'Expansão portuguesa e expansão européia: Reflexões em torno da génese dos descobrimentos', *Studia* 47 (1989), 371–413.
Thomaz, Luís Filipe F.R., 'O projecto imperial joanino: Tentativa de interpretação global da política ultramarina de D. João II', in *Congresso Internacional, Bartolomeu Dias e a sua época – Actas*, Vol. I (Oporto, 1989).
Thomaz, Luís Filipe F.R., 'L'idée impériale manueline', in Jean Aubin (ed.), *La découverte, le Portugal et l'Europe* (Paris: Centre Culturel Portugais, 1990), pp. 35–103.
Thomaz, Luís Filipe F.R., 'Factions, Interests and Messianism: The Politics of Portuguese Expansion in the East, 1500–1521', *The Indian Economic and Social History Review* 28, 1 (1991), 97–109.
Thomaz, Luís Filipe F.R., 'A "Carta que mandaram os Padres da Índia, da China e da Magna China" – um relato siriaco da chegada dos Portugueses ao Malabar', *Revista da Universidade de Coimbra* 36 (1991), 119–81.
Thomaz, Luís Filipe F.R., 'Do Cabo Espichel a Macau: Vicissitudes do corso português', in Artur Teodoro de Matos and Luís Filipe F. Reis Thomaz, eds., *As Relações entre a Índia Portuguesa, a Ásia do Sueste e o Extremo Oriente: Actas do VI Seminário Internacional de História Indo-Portuguesa* (Macau/Lisbon: The Editors, 1993), pp. 537–68.
Thomaz, Luís Filipe F.R., 'Diogo Pereira, *O Malabar*', *Mare Liberum* 5 (1993), 49–64.
Thomaz, Luís Filipe F.R., *De Ceuta a Timor* (Lisbon: Difel, 1994).
Triaud, Jean-Louis, 'L'expansion de l'Islam en Afrique', in Jean-Claude Garcin, *et al.*, *Etats, Sociétés et Cultures du Monde Musulman Médiéval, Xe–XVe siècles* Vol. I (Paris: Presses Universitaires de France, 1995), pp. 397–429.
Valensi, Lucette, *Fables de la Mémoire: La glorieuse bataille des trois rois* (Paris: Editions du Seuil, 1992).
Vatsa, Rajendra Singh, *Vâsko da Gâmâ: Bhârat ke liye samudrî mârg ḍhûnḍhne vâlâ pahlâ yûropîya* (New Delhi: NCERT, 1993).
Veluthat, Kesavan, *The Political Structure of Early Medieval South India* (New Delhi: Orient Longman, 1993).
Ventura, Margarida Garcez, *O Messias de Lisboa: Um Estudo de Mitologia Política (1383–1415)* (Lisbon: Edições Cosmos, 1992).
Wake, C.H.H., 'The changing pattern of Europe's pepper and spice imports, ca. 1400–1700', *Journal of European Economic History* 8, 2 (1979), 361–403.
Wake, C.H.H., 'The volume of European spice imports at the beginning

and end of the fifteenth century', *Journal of European Economic History* 15, 3 (1986), 621–35.

Watts, Pauline Moffitt, 'Prophecy and Discovery: On the Spiritual Origins of Christopher Columbus's "Enterprise of the Indies"', *American Historical Review* 90, 1 (1985), 73–102.

Weinstein, Donald, *Ambassador from Venice: Pietro Pasqualigo in Lisbon, 1501* (Minneapolis: University of Minnesota Press, 1960).

Weinstein, Donald, *Savonarola and Florence: Prophecy and Patriotism in the Renaissance* (Princeton: Princeton University Press, 1970).

West, Delno C., and Sandra Zimdars-Swartz, *Joachim of Fiore: A study in spiritual perception and history* (Bloomington, 1983).

Wink, André, *Al-Hind: The Making of the Indo-Islamic World*, Vol. I (*Early medieval India and the expansion of Islam, seventh to eleventh centuries)* (Delhi: Oxford University Press, 1990).

INDEX

Names in italics generally refer to ships. Proper names have been listed as mentioned in the 'Note on transliteration'. Relationships in brackets after persons' names refer to their ties with Vasco da Gama, e.g. (son), (sister), etc.

Printed in the United States
By Bookmasters